Advanced Textbooks in Control and Signal Processing

Series Editors

Professor Michael J. Grimble, Professor of Industrial Systems and Director
Professor Michael A. Johnson, Professor Emeritus of Control Systems and Deputy Director
Industrial Control Centre, Department of Electronic and Electrical Engineering,
University of Strathclyde, Graham Hills Building, 50 George Street, Glasgow G1 1QE, UK

Other titles published in this series:

Genetic Algorithms
K.F. Man, K.S. Tang and S. Kwong

Neural Networks for Modelling and Control of Dynamic Systems
M. Nørgaard, O. Ravn, L.K. Hansen and N.K. Poulsen

Modelling and Control of Robot Manipulators (2nd Edition)
L. Sciavicco and B. Siciliano

Fault Detection and Diagnosis in Industrial Systems
L.H. Chiang, E.L. Russell and R.D. Braatz

Soft Computing
L. Fortuna, G. Rizzotto, M. Lavorgna, G. Nunnari, M.G. Xibilia and R. Caponetto

Statistical Signal Processing
T. Chonavel

Discrete-time Stochastic Processes (2nd Edition)
T. Söderström

Parallel Computing for Real-time Signal Processing and Control
M.O. Tokhi, M.A. Hossain and M.H. Shaheed

Multivariable Control Systems
P. Albertos and A. Sala

Control Systems with Input and Output Constraints
A.H. Glattfelder and W. Schaufelberger

Analysis and Control of Non-linear Process Systems
K. Hangos, J. Bokor and G. Szederkényi

Model Predictive Control (2nd Edition)
E.F. Camacho and C. Bordons

Principles of Adaptive Filters and Self-learning Systems
A. Zaknich

Digital Self-tuning Controllers
V. Bobál, J. Böhm, J. Fessl and J. Macháček

Control of Robot Manipulators in Joint Space
R. Kelly, V. Santibáñez and A. Loría

Active Noise and Vibration Control
M.O. Tokhi
Publication due November 2005

D.-W. Gu, P. Hr. Petkov and M. M. Konstantinov

Robust Control Design with MATLAB®

With 288 Figures

Springer

Da-Wei Gu, PhD DIC, CEng
Engineering Department, University of Leicester, University Road, Leicester, LE1 7RH, UK

Petko Hristov Petkov, PhD,
Department of Automatics, Technical University of Sofia, 1756 Sofia, Bulgaria

Mihail Mihaylov Konstantinov, PhD
University of Architecture, Civil Engineering and Geodesy,
1 Hristo Smirnenski Blvd., 1046 Sofia, Bulgaria

British Library Cataloguing in Publication Data
Gu, D, -W.
Robust control design with MATLAB.- (Advanced textbooks in control and signal processing)
1. MATLAB (Computer file) 2. Robust control 3. Control Theory
I. Title II. Petkov, P. Hr (Petko Hr.), 1948-
III. Konstantinov, M. M. (Mihail M.), 1948-
629.8'312
ISBN-10: 1852339837

Library of Congress Control Number: 2005925110

Advanced Textbooks in Control and Signal Processing series ISSN 1439-2232
ISBN-10 1-85233-983-7
ISBN-13: 978-1-85233-983-8
Springer Science+Business Media
springer.com

Gu, Petkov and Konstantinov: Robust Control Design with MATLAB

First Indian Reprint 2006
Second Indian Reprint 2009

ISBN 978-81-8128-512-6

This edition is published by Springer (India) Private Limited,
A part of Springer Science+Business Media, Registered Office: 906-907,
Akash Deep Building, Barakhamba Road, New Delhi – 110 001, India.

Printed in India by Rajkamal Electric Press, Delhi.

To our families

Series Editors' Foreword

The topics of control engineering and signal processing continue to flourish and develop. In common with general scientific investigation, new ideas, concepts and interpretations emerge quite spontaneously and these are then discussed, used, discarded or subsumed into the prevailing subject paradigm. Sometimes these innovative concepts coalesce into a new sub-discipline within the broad subject tapestry of control and signal processing. This preliminary battle between old and new usually takes place at conferences, through the Internet and in the journals of the discipline. After a little more maturity has been acquired by the new concepts then archival publication as a scientific or engineering monograph may occur.

A new concept in control and signal processing is known to have arrived when sufficient material has evolved for the topic to be taught as a specialised tutorial workshop or as a course to undergraduate, graduate or industrial engineers. *Advanced Textbooks in Control and Signal Processing* are designed as a vehicle for the systematic presentation of course material for both popular and innovative topics in the discipline. It is hoped that prospective authors will welcome the opportunity to publish a structured and systematic presentation of some of the newer emerging control and signal processing technologies in the textbook series.

It is always interesting to look back at how a particular field of control systems theory developed. The impetus for change and realization that a new era in a subject is dawning always seems to be associated with short, sharp papers that make the academic community think again about the prevalent theoretical paradigm. In the case of the evolution of robust control theory, the conference papers of Zames (circa. 1980) on robustness and the very short paper of Doyle on the robustness of linear quadratic Gaussian control systems seem to stand as landmarks intimating that control theory was going to change direction again. And the change did come; all through the 1980s came a steady stream of papers re-writing control theory, introducing system uncertainty, H_∞ robust control and μ-synthesis as part of a new control paradigm.

Change, however did not come easily to the industrial applications community because the new theories and methods were highly mathematical. In the early stages even the classical feedback diagram which so often opened control engineering courses was replaced by a less intuitively obvious diagram. Also it was difficult to see the benefits to be gained from the new development. Throughout the 1990s the robust control theory and methods consolidated and the first major textbooks and software toolboxes began to appear. Experience with some widely disseminated benchmark problems such as control design for distillation columns, the control design for hard-disk drives, and the inverted-pendulum control problem helped the industrial community see how to apply the new method and the control benefits that accrued.

This advanced course textbook on robust control system design using MATLAB® by Da-Wei Gu, Petko Petkov and Mihail Konstantinov has arrived at a very opportune time. More than twenty years of academic activity in the robust control field forms the bedrock on which this course book and its set of insightful applications examples are developed. Part I of the volume presents the theory – a systematic presentation of: systems notation, uncertainty modelling, robust design specification, the H_∞ design method, H_∞ loop shaping, μ-analysis and synthesis and finally the algorithms for providing the low-order controllers that will be implemented. This is a valuable and concise presentation of all the necessary theoretical concepts prior to their application which is covered in Part II.

Inspired by the adage "practice makes perfect", Part II of the volume comprises six fully worked-out extended examples. To learn how to apply the complex method of H_∞ design and μ-synthesis there can be no surer route than to work through a set of carefully scripted examples. In this volume, the examples range from the academic mass-damper-spring system through to the industrially relevant control of a distillation column and a flexible manipulator system. The benchmark example of the ubiquitous hard-disk drive control system is also among the examples described. The MATLAB® tools of the Robust Control Toolbox, the Control System Toolbox and Simulink® are used in these application examples. The CD-ROM contains all the necessary files and instructions together with a pdf containing colour reproductions of many of the figures in the book.

In summary, after academic development of twenty years or so, the robust control paradigm is now fully fledged and forms a vital component of advanced control engineering courses. This new volume in our series of advanced control and signal processing course textbooks on applying the methods of H_∞ and μ-synthesis control design will be welcomed by postgraduate students, lecturers and industrial control engineers alike.

M.J. Grimble and M.A. Johnson
Glasgow, Scotland, U.K.
February 2005

Preface

Robustness has been an important issue in control-systems design ever since 1769 when James Watt developed his flyball governor. A successfully designed control system should be always able to maintain stability and performance level in spite of uncertainties in system dynamics and/or in the working environment to a certain degree. Design requirements such as gain margin and phase margin in using classical frequency-domain techniques are solely for the purpose of robustness. The robustness issue was not that prominently considered during the period of 1960s and 1970s when system models could be much more accurately described and design methods were mainly mathematical optimisations in the time domain. Due to its importance, however, the research on robust design has been going on all the time. A breakthrough came in the late 1970s and early 1980s with the pioneering work by Zames [170] and Zames and Francis [171] on the theory, now known as the $\mathcal{H}_\infty$ optimal control theory. The $\mathcal{H}_\infty$ optimisation approach and the μ-synthesis/analysis method are well developed and elegant. They provide systematic design procedures of robust controllers for linear systems, though the extension into nonlinear cases is being actively researched.

Many books have since been published on $\mathcal{H}_\infty$ and related theories and methods [26, 38, 65, 137, 142, 145, 174, 175]. The algorithms to implement the design methods are readily available in software packages such as MATLAB® and Slicot [119]. However, from our experience in teaching and research projects, we have felt that a reasonable percentage of people, students as well as practising engineers, still have difficulties in applying the $\mathcal{H}_\infty$ and related theory and in using MATLAB® routines. The mathematics behind the theory is quite involved. It is not straightforward to formulate a practical design problem, which is usually nonlinear, into the $\mathcal{H}_\infty$ or μ design framework and then apply MATLAB® routines. This hinders the application of such a powerful theory. It also motivated us to prepare this book.

This book is for people who want to learn how to deal with robust control-system design problems but may not want to research the relevant theoretic developments. Methods and solution formulae are introduced in the first part of the book, but kept to a minimum. The majority of the book is devoted to several practical design case studies (Part II). These design examples, ranging from teaching laboratory experiments such as a mass-damper-spring system to complex systems such as a supersonic rocket autopilot and a flexible-link manipulator, are discussed with detailed presentations. The design exercises are all conducted using the new *Robust Control Toolbox v3.0* and are in a hands-on, tutorial manner. Studying these examples with the attached MATLAB® and Simulink® programs (170 plus M- and MDL-files) used in all designs will help the readers learn how to deal with nonlinearities involved in the system, how to parameterise dynamic uncertainties and how to use MATLAB® routines in the analysis and design, *etc.*. It is also hoped that by going through these exercises the readers will understand the essence of robust control system design and develop their own skills to design real, industrial, robust control systems.

The readership of this book is postgraduates and control engineers, though senior undergraduates may use it for their final-year projects. The material included in the book has been adopted in recent years for MSc and PhD engineering students at Leicester University and at the Technical University of Sofia. The design examples are independent of each other. They have been used extensively in the laboratory projects on the course *Robust and Optimal Control Systems* taught in a masters programme in the Technical University of Sofia.

The authors are indebted to several people and institutions who helped them in the preparation of the book. We are particularly grateful to The MathWorks, Inc. for their continuous support, to Professor Sigurd Skogestad of Norwegian University of Science and Technology who kindly provided the nonlinear model of the Distillation Column and to Associate Professor Georgi Lehov from Technical University of Russe, Bulgaria, who developed the uncertainty model of the Flexible-Link Manipulator.

Using the CD ROM

The attached CD ROM contains six folders with M- and MDL-files intended for design, analysis and simulation of the six design examples, plus a pdf file with colour hypertext version of the book. In order to use the M- and MDL-files the reader should have at his (her) disposition of MATLAB® v7.0.2 with Robust Control Toolbox v 3.0, Control System Toolbox v6.1 and Simulink® v6.1. Further information on the use of the files can be found in the file Readme.m on the disc.

Contents

Part I Basic Methods and Theory

Part I

Basic Methods and Theory

1

Introduction

Robustness is of crucial importance in control-system design because real engineering systems are vulnerable to external disturbance and measurement noise and there are always differences between mathematical models used for design and the actual system. Typically, a control engineer is required to design a controller that will stabilise a plant, if it is not stable originally, and satisfy certain performance levels in the presence of disturbance signals, noise interference, unmodelled plant dynamics and plant-parameter variations. These design objectives are best realised *via* the feedback control mechanism, although it introduces in the issues of high cost (the use of sensors), system complexity (implementation and safety) and more concerns on stability (thus internal stability and stabilising controllers).

Though always being appreciated, the need and importance of robustness in control-systems design has been particularly brought into the limelight during the last two decades. In classical single-input single-output control, robustness is achieved by ensuring good gain and phase margins. Designing for good stability margins usually also results in good, well-damped time responses, *i.e.* good performance. When multivariable design techniques were first developed in the 1960s, the emphasis was placed on achieving good performance, and not on robustness. These multivariable techniques were based on linear quadratic performance criteria and Gaussian disturbances, and proved to be successful in many aerospace applications where accurate mathematical models can be obtained, and descriptions for external disturbances/noise based on white noise are considered appropriate. However, application of such methods, commonly referred to as the linear quadratic Gaussian (LQG) methods, to other industrial problems made apparent the poor robustness properties exhibited by LQG controllers. This led to a substantial research effort to develop a theory that could explicitly address the robustness issue in feedback design. The pioneering work in the development of the forthcoming theory, now known as the $\mathcal{H}_\infty$ optimal control theory, was conducted in the early 1980s by Zames [170] and Zames and Francis [171]. In the $\mathcal{H}_\infty$ approach, the designer from the outset specifies a model of system uncertainty, such as additive perturbation and/or output disturbance (details in Chapter 2), that is most suited to the problem at hand. A constrained optimisation is then performed to maximise the robust stability of the closed-loop system to the type of uncertainty chosen, the constraint being the internal stability of the feedback system. In most cases, it would be sufficient to seek a feasible controller such that the closed-loop system achieves certain robust stability. Performance objectives can also be included in the optimisation cost function. Elegant solution formulae have been developed, which are based on the solutions of certain algebraic Riccati equations, and are readily available in software packages such as Slicot [119] and MATLAB®.

Despite the mature theory ([26, 38, 175]) and availability of software packages, commercial or licensed freeware, many people have experienced difficulties in solving industrial control-systems design problems with these $\mathcal{H}_\infty$ and related methods, due to the complex mathematics of the advanced approaches and numerous presentations of formulae as well as adequate translations of industrial design into relevant configurations. This book aims at bridging the gap between the theory and applications. By sharing the experiences in industrial case studies with minimum exposure to the theory and formulae, the authors hope readers will obtain an insight into robust industrial control-system designs using major $\mathcal{H}_\infty$ optimisation and related methods.

In this chapter, the basic concepts and representations of systems and signals will be discussed.

1.1 Control-system Representations

A control system or plant or process is an interconnection of components to perform certain tasks and to yield a desired response, *i.e.* to generate desired signal (the output), when it is driven by manipulating signal (the input). A control system is a causal, dynamic system, *i.e.* the output depends not only the present input but also the input at the previous time.

In general, there are two categories of control systems, the open-loop systems and closed-loop systems. An open-loop system uses a controller or control actuator to obtain the design response. In an open-loop system, the output has no effect on the input. In contrast to an open-loop system, a closed-loop control system uses sensors to measure the actual output to adjust the input in order to achieve desired output. The measure of the output is called the feedback signal, and a closed-loop system is also called a feedback system. It will be shown in this book that only feedback configurations are able to achieve the robustness of a control system.

Due to the increasing complexity of physical systems under control and rising demands on system properties, most industrial control systems are no longer single-input and single-output (SISO) but multi-input and multi-output (MIMO) systems with a high interrelationship (coupling) between these channels. The number of (state) variables in a system could be very large as well. These systems are called multivariable systems.

In order to analyse and design a control system, it is advantageous if a mathematical representation of such a relationship (a model) is available. The system dynamics is usually governed by a set of differential equations in either open-loop or closed-loop systems. In the case of linear, time-invariant systems, which is the case this book considers, these differential equations are linear ordinary differential equations. By introducing appropriate state variables and simple manipulations, a linear, time-invariant, continuous-time control system can be described by the following model,

$$\begin{aligned} \dot{x}(t) &= Ax(t) + Bu(t) \\ y(t) &= Cx(t) + Du(t) \end{aligned} \tag{1.1}$$

where $x(t) \in R^n$ is the state vector, $u(t) \in R^m$ the input (control) vector, and $y(t) \in R^p$ the output (measurement) vector.

With the assumption of zero initial condition of the state variables and using Laplace transform, a transfer function matrix corresponding to the system in (1.1) can be derived as

$$G(s) := C(sI_n - A)^{-1}B + D \tag{1.2}$$

and can be further denoted in a short form by

$$G(s) =: \left[\begin{array}{c|c} A & B \\ \hline C & D \end{array}\right] \tag{1.3}$$

It should be noted that the $\mathcal{H}_\infty$ optimisation approach is a frequency-domain method, though it utilises the time-domain description such as (1.1) to explore the advantages in numerical computation and to deal with multivariable systems. The system given in (1.1) is assumed in this book to be minimal, *i.e.* completely controllable and completely observable, unless described otherwise.

In the case of discrete-time systems, similarly the model is given by

$$\begin{aligned} x(k+1) &= Ax(k) + Bu(k) \\ y(k) &= Cx(k) + Du(k) \end{aligned} \tag{1.4}$$

or

$$\begin{aligned} x_{k+1} &= Ax_k + Bu_k \\ y_k &= Cx_k + Du_k \end{aligned}$$

with a corresponding transfer function matrix as

$$\begin{aligned} G(s) &:= C(zI_n - A)^{-1}B + D \\ &=: \left[\begin{array}{c|c} A & B \\ \hline C & D \end{array}\right] \end{aligned} \tag{1.5}$$

1.2 System Stabilities

An essential issue in control-systems design is the *stability*. An unstable system is of no practical value. This is because any control system is vulnerable to disturbances and noises in a real work environment, and the effect due to these signals would adversely affect the expected, normal system output in an unstable system. Feedback control techniques may reduce the influence generated by uncertainties and achieve desirable performance. However, an inadequate feedback controller may lead to an unstable closed-loop system though the original open-loop system is stable. In this section, control-system stabilities and stabilising controllers for a given control system will be discussed.

When a dynamic system is just described by its input/output relationship such as a transfer function (matrix), the system is stable if it generates bounded outputs for any bounded inputs. This is called the bounded-input-bounded-output (BIBO) stability. For a linear, time-invariant system modelled by a transfer function matrix ($G(s)$ in (1.2)), the BIBO stability is guaranteed if and only if all the poles of $G(s)$ are in the open-left-half complex plane, *i.e.* with negative real parts.

When a system is governed by a state-space model such as (1.1), a stability concept called *asymptotic stability* can be defined. A system is asymptotically stable if, for an identically zero input, the system state will converge to zero from any initial states. For a linear, time-invariant system described by a model of (1.1), it is asymptotically stable if and only if all the eigenvalues of the state matrix A are in the open-left-half complex plane, *i.e.* with positive real parts.

In general, the asymptotic stability of a system implies that the system is also BIBO stable, but not *vice versa*. However, for a system in (1.1), if $[A, B, C, D]$ is of minimal realisation, the BIBO stability of the system implies that the system is asymptotically stable.

The above stabilities are defined for open-loop systems as well as closed-loop systems. For a closed-loop system (interconnected, feedback system), it is more interesting and intuitive to look at the asymptotic stability from another point of view and this is called the *internal stability* [20]. An interconnected system is internally stable if the subsystems of all input-output pairs are asymptotically stable (or the corresponding transfer function matrices are BIBO stable when the state space models are minimal, which is assumed in this chapter). Internal stability is equivalent to asymptotical stability in an interconnected, feedback system but may reveal explicitly the relationship between the original, open-loop system and the controller that influences the stability of the whole system. For the system given in Figure 1.1, there are two inputs r and d (the disturbance at the output) and two outputs y and u (the output of the controller K).

The transfer functions from the inputs to the outputs, respectively, are

$$T_{yr} = GK(I + GK)^{-1}$$

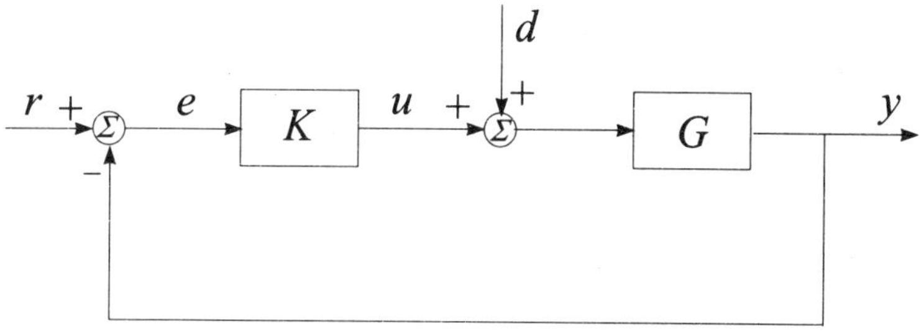

Fig. 1.1. An interconnected system of G and K

$$
\begin{aligned}
T_{yd} &= G(I+KG)^{-1} \\
T_{ur} &= K(I+GK)^{-1} \\
T_{ud} &= -KG(I+KG)^{-1}
\end{aligned}
\tag{1.6}
$$

Hence, the system is internally stable if and only if all the transfer functions in (1.6) are BIBO stable, or the transfer function matrix M from $\begin{bmatrix} r \\ d \end{bmatrix}$ to $\begin{bmatrix} y \\ u \end{bmatrix}$ is BIBO stable, where

$$
M := \begin{bmatrix} GK(I+GK)^{-1} & G(I+KG)^{-1} \\ K(I+GK)^{-1} & -KG(I+KG)^{-1} \end{bmatrix}
\tag{1.7}
$$

The stability of (1.7) is equivalent to the stability of

$$
\hat{M} := \begin{bmatrix} I - GK(I+GK)^{-1} & G(I+KG)^{-1} \\ K(I+GK)^{-1} & I - KG(I+KG)^{-1} \end{bmatrix}
\tag{1.8}
$$

By simple matrix manipulations, we have

$$
\begin{aligned}
\hat{M} &= \begin{bmatrix} (I+GK)^{-1} & G(I+KG)^{-1} \\ K(I+GK)^{-1} & (I+KG)^{-1} \end{bmatrix} \\
&= \begin{bmatrix} I & -G \\ -K & I \end{bmatrix}^{-1}
\end{aligned}
\tag{1.9}
$$

Hence, the feedback system in Figure 1.1 is internally stable if (1.9) is stable.

It can be shown [20] that if there is no unstable pole/zero cancellation between G and K, then any one of the four transfer functions being BIBO stable would be enough to guarantee that the whole system is internally stable.

1.3 Coprime Factorisation and Stabilising Controllers

Consider a system given in the form of (1.2) with $[A, B, C, D]$ assumed to be minimal. Matrices $(\tilde{M}(s),\ \tilde{N}(s)) \in \mathcal{H}_\infty$ $((M(s),\ N(s)) \in \mathcal{H}_\infty)$, where $\mathcal{H}_\infty$

denotes the space of functions with no poles in the closed right-half complex plane, constitute a left (right) coprime factorisation of $G(s)$ if and only if

(i) $\tilde{M}$ (M) is square, and $\det(\tilde{M})(\det(M)) \neq 0$.
(ii) the plant model is given by

$$G = \tilde{M}^{-1}\tilde{N}(= NM^{-1}) \tag{1.10}$$

(iii) There exists $(\tilde{V},\ \tilde{U})((V, U)) \in \mathcal{H}_\infty$ such that

$$\begin{aligned} \tilde{M}\tilde{V}\ +\ \tilde{N}\tilde{U}\ &=\ I \\ (UN + VM =\ &I) \end{aligned} \tag{1.11}$$

Transfer functions (or rational, fractional) matrices are *coprime* if they share no common zeros in the right-half complex plane, including at the infinity. The two equations in (iii) above are called Bezout identities ([97]) and are necessary and sufficient conditions for $(\tilde{M}, \tilde{N})$ $((M,\ N))$ being left coprime (right coprime), respectively. The left and right coprime factorisations of $G(s)$ can be grouped together to form a Bezout double identity as the following

$$\begin{bmatrix} V & U \\ -\tilde{N} & \tilde{M} \end{bmatrix} \begin{bmatrix} M & -\tilde{U} \\ N & \tilde{V} \end{bmatrix} = I \tag{1.12}$$

For $G(s)$ of minimal realisation (1.2) (actually G is required to be stabilisable and detectable only), the formulae for the coprime factors can be readily derived ([98]) as in the following theorem.

Theorem 1.1. *Let constant matrices F and H be such that $A + BF$ and $A + HC$ are both stable. Then the transfer function matrices $\tilde{M}$ and $\tilde{N}$ (M and N) defined in the following constitute a left (right) coprime factorisation of $G(s)$,*

$$\begin{bmatrix} \tilde{N}(s) & \tilde{M}(s) \end{bmatrix} = \left[\begin{array}{c|cc} A + HC & B + HD & -H \\ \hline C & D & I \end{array}\right] \tag{1.13}$$

$$\begin{bmatrix} N(s) \\ M(s) \end{bmatrix} = \left[\begin{array}{c|c} A + BF & B \\ \hline C + DF & D \\ F & I \end{array}\right] \tag{1.14}$$

Furthermore, the following $\tilde{U}(s)$, $\tilde{V}(s)$, $U(s)$ and $V(s)$ satisfy the Bezout double identity (1.12),

$$\begin{bmatrix} \tilde{U}(s) & \tilde{V}(s) \end{bmatrix} = \left[\begin{array}{c|cc} A + HC & H & B + HD \\ \hline F & 0 & I \end{array}\right] \tag{1.15}$$

$$\begin{bmatrix} U(s) \\ V(s) \end{bmatrix} = \left[\begin{array}{c|c} A + BF & H \\ \hline F & 0 \\ C + DF & I \end{array}\right] \tag{1.16}$$

■

It can be easily shown that the pairs $(\tilde{U},\ \tilde{V})$ and $(U,\ V)$ are stable and coprime. Using (1.9), it is straightforward to show the following lemma.

Lemma 1.2.

$$K := \tilde{V}^{-1}\tilde{U} = UV^{-1} \tag{1.17}$$

is a stabilising controller, i.e. the closed-loop system in Figure 1.6 is internally stable. ■

Further, the set of all stabilising controllers for $G = \tilde{M}^{-1}\tilde{N} = NM^{-1}$ can be obtained in the following *Youla Parameterisation Theorem* ([98, 167, 168]).

Theorem 1.3. *The set of all stabilising controllers for G is*

$$\{(\tilde{V} + Q\tilde{N})^{-1}(\tilde{U} + Q\tilde{M}) : \quad Q \in \mathcal{H}_\infty\} \tag{1.18}$$

The set can also be expressed as

$$\{(U + MQ)(V + NQ)^{-1} : \quad Q \in \mathcal{H}_\infty\} \tag{1.19}$$

■

1.4 Signals and System Norms

In this section the basic concepts concerning signals and systems will be reviewed in brief. A control system interacts with its environment through command signals, disturbance signals and noise signals, *etc.* Tracking error signals and actuator driving signals are also important in control systems design. For the purpose of analysis and design, appropriate measures, the norms, must be defined for describing the "size" of these signals. From the signal norms, we can then define induced norms to measure the "gain" of the operator that represents the control system.

1.4.1 Vector Norms and Signal Norms

Let the linear space X be $\mathcal{F}^m$, where $\mathcal{F} = \mathcal{R}$ for the field of real numbers, or $\mathcal{F} = \mathcal{C}$ for complex numbers. For $x = [x_1, x_2, ..., x_m]^T \in X$, the p-norm of the vector x is defined by

1-norm $\|x\|_1 := \sum_{i=1}^{m} |x_i|$, for $p = 1$
p-norm $\|x\|_p := \left(\sum_{i=1}^{m} |x_i|^p\right)^{1/p}$, for $1 < p < \infty$
∞-norm $\|x\|_\infty := \max_{1 \le i \le m} |x_i|$, for $p = \infty$

When $p = 2$, $\|x\|_2$ is the familiar Euclidean norm.

When X is a linear space of continuous or piecewise continuous time scalar-valued signals $x(t)$, $t \in \mathcal{R}$, the p-norm of a signal $x(t)$ is defined by

1-norm $\|x\|_1 := \int_{-\infty}^{\infty} |x(t)|\mathrm{d}t$, for $p = 1$

p-norm $\|x\|_p := \left(\int_{-\infty}^{\infty} |x(t)|^p \mathrm{d}t\right)^{1/p}$, for $1 < p < \infty$

∞-norm $\|x\|_\infty := \sup_{t\in\mathcal{R}} |x(t)|$, for $p = \infty$

The normed spaces, consisting of signals with finite norm as defined correspondingly, are called $L^1(\mathcal{R})$, $L^p(\mathcal{R})$ and $L^\infty(\mathcal{R})$, respectively. From a signal point of view, the 1-norm, $\|x\|_1$ of the signal $x(t)$ is the integral of its absolute value. The square of the 2-norm, $\|x\|_2^2$, is often called the *energy* of the signal $x(t)$ since that is what it is when $x(t)$ is the current through a 1 Ω resistor. The ∞-norm, $\|x\|_\infty$, is the amplitude or peak value of the signal, and the signal is bounded in magnitude if $x(t) \in L^\infty(\mathcal{R})$.

When X is a linear space of continuous or piecewise continuous *vector-valued* functions of the form $x(t) = [x_1(t), x_2(t), \cdots, x_m(t)]^T$, $t \in \mathcal{R}$, we may have

$$L_m^p(\mathcal{R}) := \{x(t) : \|x\|_p = \left(\int_{-\infty}^{\infty} \sum_{i=1}^m |x(t)|^p \mathrm{d}t\right)^{1/p} < \infty, \quad \text{for } 1 \le p < \infty\}$$

$$L_m^\infty(\mathcal{R}) := \{x(t) : \|x\|_\infty = \sup_{t\in\mathcal{R}} \|x(t)\|_\infty < \infty\}$$

Some signals are useful for control systems analysis and design, for example, the sinusoidal signal, $x(t) = A\sin(\omega t + \phi)$, $t \in \mathcal{R}$. It is unfortunately not a 2-norm signal because of the infinite energy contained. However, the average power of $x(t)$

$$\lim_{T\to\infty} \frac{1}{2T} \int_{-T}^{T} x^2(t)\mathrm{d}t$$

exists. The signal $x(t)$ will be called a *power signal* if the above limit exists. The square root of the limit is the well-known r.m.s. (*root-mean-square*) value of $x(t)$. It should be noticed that the average power does not introduce a norm, since a nonzero signal may have zero average power.

1.4.2 System Norms

System norms are actually the input-output gains of the system. Suppose that $\mathcal{G}$ is a linear and bounded system that maps the input signal $u(t)$ into the output signal $y(t)$, where $u \in (U, \|\cdot\|_U)$, $y \in (Y, \|\cdot\|_Y)$. U and Y are the signal spaces, endowed with the norms $\|\cdot\|_U$ and $\|\cdot\|_Y$, respectively. Then the norm, maximum system gain, of $\mathcal{G}$ is defined as

$$\|\mathcal{G}\| := \sup_{u\neq 0} \frac{\|\mathcal{G}u\|_Y}{\|u\|_U} \tag{1.20}$$

or

$$\|\mathcal{G}\| = \sup_{\|u\|_U=1} \|\mathcal{G}u\|_Y = \sup_{\|u\|_U\le 1} \|\mathcal{G}u\|_Y$$

Obviously, we have

$$\|\mathcal{G}u\|_Y \leq \|\mathcal{G}\| \cdot \|u\|_U$$

If $\mathcal{G}_1$ and $\mathcal{G}_2$ are two linear, bounded and compatible systems, then

$$\|\mathcal{G}_1\mathcal{G}_2\| \leq \|\mathcal{G}_1\| \cdot \|\mathcal{G}_2\|$$

$\|\mathcal{G}\|$ is called the *induced norm* of $\mathcal{G}$ with regard to the signal norms $\|\cdot\|_U$ and $\|\cdot\|_Y$. In this book, we are particularly interested in the so-called ∞-norm of a system. For a linear, time-invariant, stable system $\mathcal{G}$: $L^2_m(\mathcal{R}) \rightarrow L^2_p(\mathcal{R})$, the ∞-norm, or the induced 2-norm, of $\mathcal{G}$ is given by

$$\|\mathcal{G}\|_\infty = \sup_{\omega \in \mathcal{R}} \|G(j\omega)\|_2 \tag{1.21}$$

where $\|G(j\omega)\|_2$ is the spectral norm of the $p \times m$ matrix $G(j\omega)$ and $G(s)$ is the transfer function matrix of $\mathcal{G}$. Hence, the ∞-norm of a system describes the maximum energy gain of the system and is decided by the peak value of the largest singular value of the frequency response matrix over the whole frequency axis. This norm is called the $\mathcal{H}_\infty$-norm, since we denote by $\mathcal{H}_\infty$ the linear space of all stable linear systems.

2

Modelling of Uncertain Systems

As discussed in Chapter 1, it is well understood that uncertainties are unavoidable in a real control system. The uncertainty can be classified into two categories: disturbance signals and dynamic perturbations. The former includes input and output disturbance (such as a gust on an aircraft), sensor noise and actuator noise, *etc.* The latter represents the discrepancy between the mathematical model and the actual dynamics of the system in operation. A mathematical model of any real system is always just an approximation of the true, physical reality of the system dynamics. Typical sources of the discrepancy include unmodelled (usually high-frequency) dynamics, neglected nonlinearities in the modelling, effects of deliberate reduced-order models, and system-parameter variations due to environmental changes and torn-and-worn factors. These modelling errors may adversely affect the stability and performance of a control system. In this chapter, we will discuss in detail how dynamic perturbations are usually described so that they can be well considered in system robustness analysis and design.

2.1 Unstructured Uncertainties

Many dynamic perturbations that may occur in different parts of a system can, however, be lumped into one single perturbation block Δ, for instance, some unmodelled, high-frequency dynamics. This uncertainty representation is referred to as "unstructured" uncertainty. In the case of linear, time-invariant systems, the block Δ may be represented by an unknown transfer function matrix. The unstructured dynamics uncertainty in a control system can be described in different ways, such as is listed in the following, where $G_{\mathrm{p}}(s)$ denotes the actual, perturbed system dynamics and $G_{\mathrm{o}}(s)$ a nominal model description of the physical system.

1. Additive perturbation:

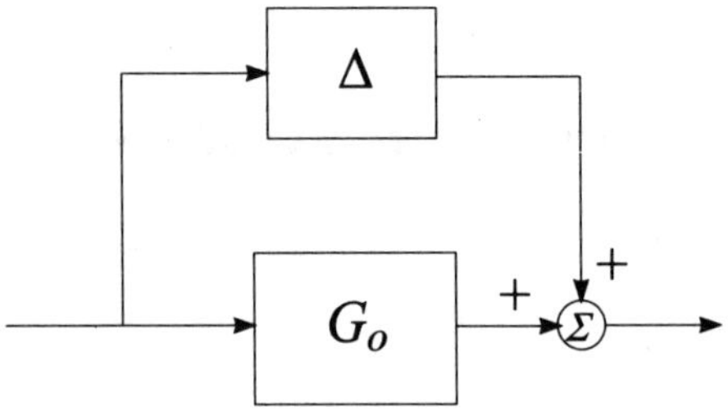

Fig. 2.1. Additive perturbation configuration

$$G_p(s) = G_o(s) + \Delta(s) \tag{2.1}$$

2. Inverse additive perturbation:

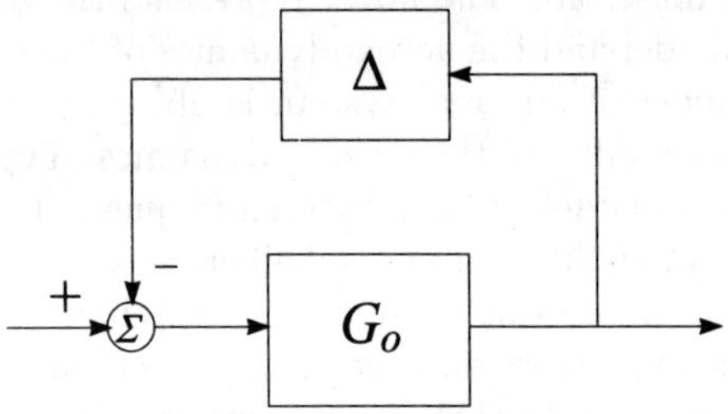

Fig. 2.2. Inverse additive perturbation configuration

$$(G_p(s))^{-1} = (G_o(s))^{-1} + \Delta(s) \tag{2.2}$$

3. Input multiplicative perturbation:

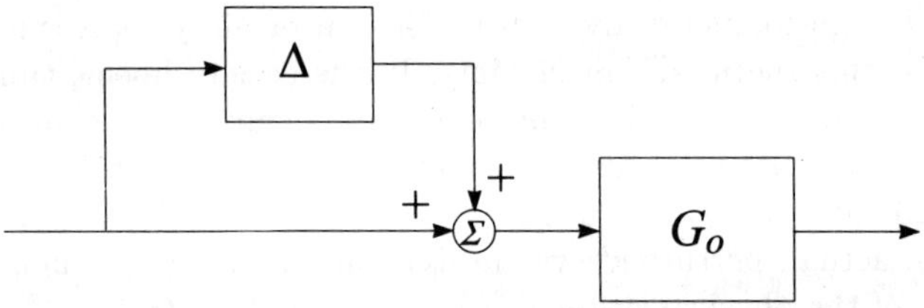

Fig. 2.3. Input multiplicative perturbation configuration

$$G_p(s) = G_o(s)[I + \Delta(s)] \tag{2.3}$$

4. Output multiplicative perturbation:

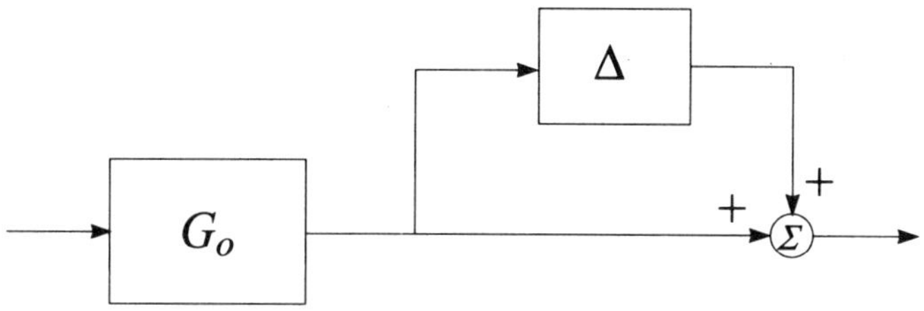

Fig. 2.4. Output multiplicative perturbation configuration

$$G_p(s) = [I + \Delta(s)]G_o(s) \tag{2.4}$$

5. Inverse input multiplicative perturbation:

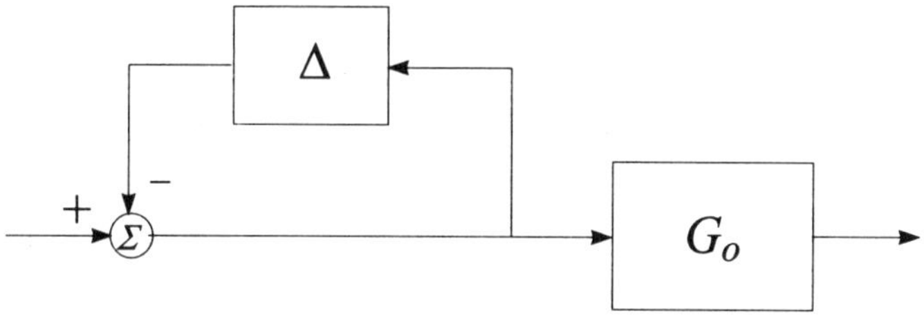

Fig. 2.5. Inverse input multiplicative perturbation configuration

$$(G_p(s))^{-1} = [I + \Delta(s)](G_o(s))^{-1} \tag{2.5}$$

6. Inverse output multiplicative perturbation:

$$(G_p(s))^{-1} = (G_o(s))^{-1}[I + \Delta(s)] \tag{2.6}$$

7. Left coprime factor perturbations:

$$G_p(s) = (\tilde{M} + \Delta_{\tilde{M}})^{-1}(\tilde{N} + \Delta_{\tilde{N}}) \tag{2.7}$$

8. Right coprime factor perturbations:

$$G_p(s) = (N + \Delta_N)(M + \Delta_M)^{-1} \tag{2.8}$$

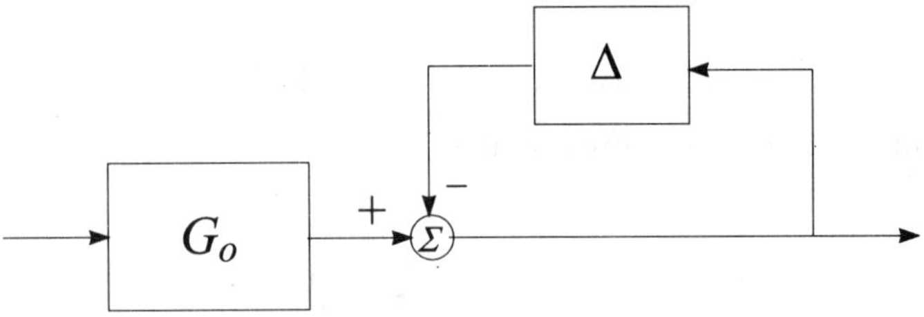

Fig. 2.6. Inverse output multiplicative perturbation configuration

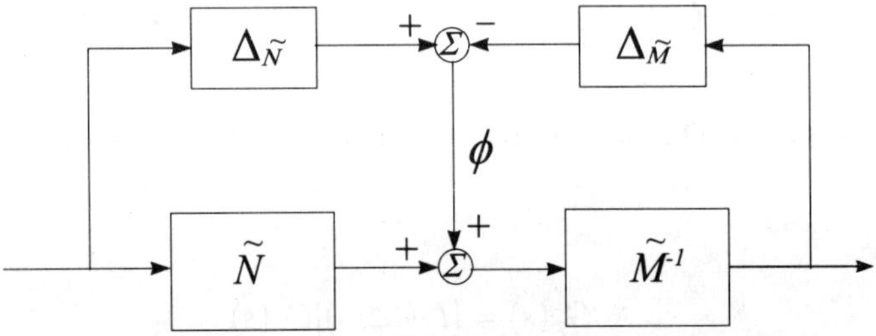

Fig. 2.7. Left coprime factor perturbations configuration

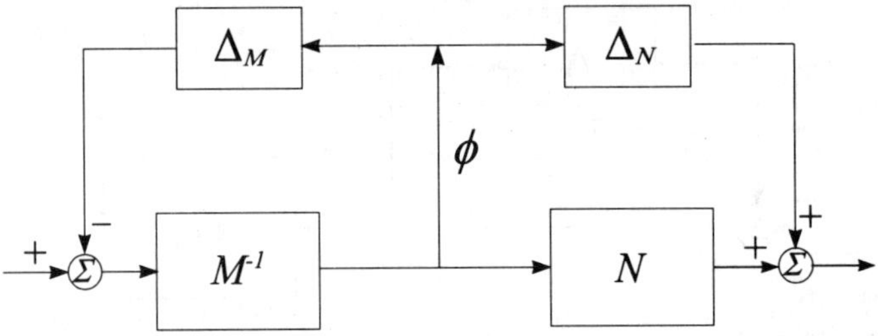

Fig. 2.8. Right coprime factor perturbations configuration

The additive uncertainty representations give an account of absolute error between the actual dynamics and the nominal model, while the multiplicative representations show relative errors.

In the last two representations, $(\tilde{M}, \tilde{N})/(M, N)$ are left/right coprime factorizations of the nominal system model $G_o(s)$, respectively; and $(\Delta_{\tilde{M}}, \Delta_{\tilde{N}})$ $/(\Delta_M, \Delta_N)$ are the perturbations on the corresponding factors [101].

The block Δ (or, $(\Delta_{\tilde{M}}, \Delta_{\tilde{N}})$ $/(\Delta_M, \Delta_N)$ in the coprime factor perturbations cases) is uncertain, but usually is norm-bounded. It may be bounded by a known transfer function, say $\overline{\sigma}[\Delta(j\omega)] \leq \delta(j\omega)$, for all frequencies ω, where δ is a known scalar function and $\overline{\sigma}[\cdot]$ denotes the largest singular value of a matrix. The uncertainty can thus be represented by a unit, norm-bounded block Δ cascaded with a scalar transfer function $\delta(s)$.

It should be noted that a successful robust control-system design would depend on, to certain extent, an appropriate description of the perturbation considered, though theoretically most representations are interchangeable.

Example 2.1

The dynamics of many control systems may include a "slow" part and a "fast" part, for instance in a dc motor. The actual dynamics of a scalar plant may be

$$G_{\rm p}(s) = g_{\rm gain} G_{\rm slow}(s) G_{\rm fast}(s)$$

where, $g_{\rm gain}$ is constant, and

$$G_{\rm slow}(s) = \frac{1}{1+sT}\ ; \quad G_{\rm fast}(s) = \frac{1}{1+\alpha sT}\ ; \qquad \alpha \underline{\leq\leq} 1.$$

In the design, it may be reasonable to concentrate on the slow response part while treating the fast response dynamics as a perturbation. Let $\Delta_{\rm a}$ and $\Delta_{\rm m}$ denote the additive and multiplicative perturbations, respectively. It can be easily worked out that

$$\begin{aligned}\Delta_{\rm a}(s) &= G_{\rm p} - g_{\rm gain} G_{\rm slow} = g_{\rm gain} G_{\rm slow}(G_{\rm fast} - 1) \\ &= g_{\rm gain} \frac{-\alpha sT}{(1+sT)(1+\alpha sT)} \\ \Delta_{\rm m}(s) &= \frac{G_{\rm p} - g_{\rm gain} G_{\rm slow}}{g_{\rm gain} G_{\rm slow}} = G_{\rm fast} - 1 = \frac{-\alpha sT}{1+\alpha sT}\end{aligned}$$

The magnitude Bode plots of $\Delta_{\rm a}$ and $\Delta_{\rm m}$ can be seen in Figure 2.9, where $g_{\rm gain}$ is assumed to be 1. The difference between the two perturbation representations is obvious: though the magnitude of the absolute error may be small, the relative error can be large in the high-frequency range in comparison to that of the nominal plant. Also $T = 10$ and $\alpha = 0.001$

2.2 Parametric Uncertainty

The unstructured uncertainty representations discussed in Section 2.1 are useful in describing unmodelled or neglected system dynamics. These complex uncertainties usually occur in the high-frequency range and may include unmodelled lags (time delay), parasitic coupling, hysteresis and other nonlinearities. However, dynamic perturbations in many industrial control systems may also be caused by inaccurate description of component characteristics, torn-and-worn effects on plant components, or shifting of operating points, etc. Such perturbations may be represented by variations of certain system parameters over some possible value ranges (complex or real). They affect the low-frequency range performance and are called "parametric uncertainties".

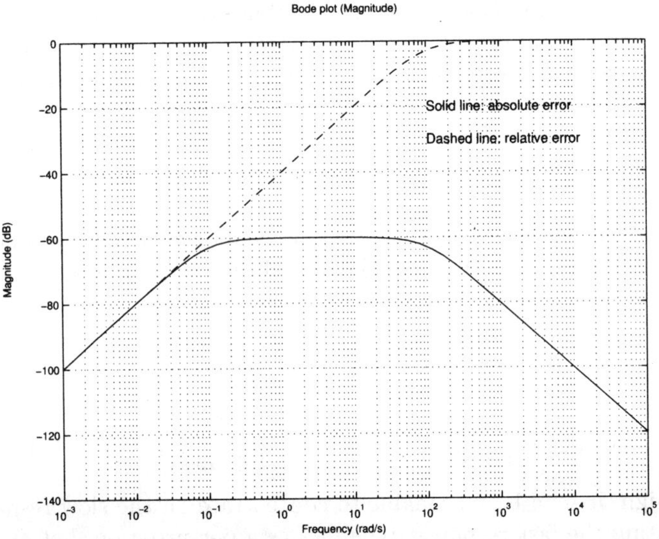

Fig. 2.9. Absolute and relative errors in Example 2.1

Example 2.2

A mass-spring-damper system can be described by the following second-order, ordinary differential equation

$$m\frac{d^2x(t)}{dt^2} + c\frac{dx(t)}{dt} + kx(t) = f(t)$$

where, m is the mass, c the damping constant, k the spring stiffness, $x(t)$ the displacement and $f(t)$ the external force. For imprecisely known parameter values, the dynamic behaviour of such a system is actually described by

$$(m_o + \delta_m)\frac{d^2x(t)}{dt^2} + (c_o + \delta_c)\frac{dx(t)}{dt} + (k_o + \delta_k)x(t) = f(t)$$

where, m_o, c_o and k_o denote the nominal parameter values and δ_m, δ_c and δ_k possible variations over certain ranges.

By defining the state variables x_1 and x_2 as the displacement variable and its first-order derivative (velocity), the 2nd-order differential equation (2.2) may be rewritten into a standard state-space form

$$\begin{aligned}\dot{x_1} &= x_2 \\ \dot{x_2} &= \frac{1}{m_o + \delta_m}\left[-(k_o + \delta_k)x_1 - (c_o + \delta_c)x_2 + f\right] \\ y &= x_1\end{aligned}$$

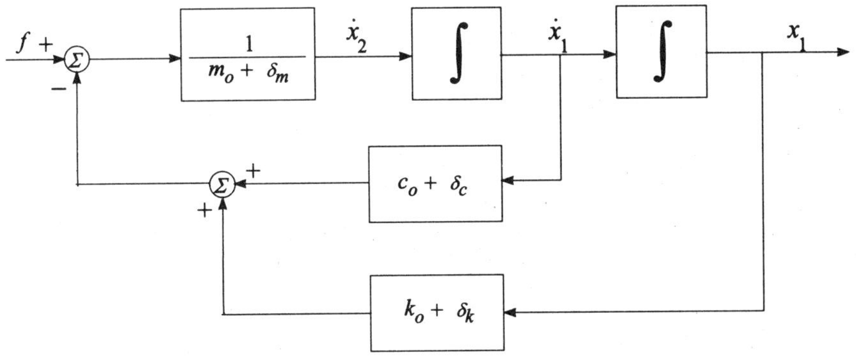

Fig. 2.10. Analogue block diagram of Example 2.2

Further, the system can be represented by an analogue block diagram as in Figure 2.10.

Notice that $\frac{1}{m_o+\delta_m}$ can be rearranged as a feedback in terms of $\frac{1}{m_o}$ and δ_m. Figure 2.10 can be redrawn as in Figure 2.11, by pulling out all the uncertain variations.

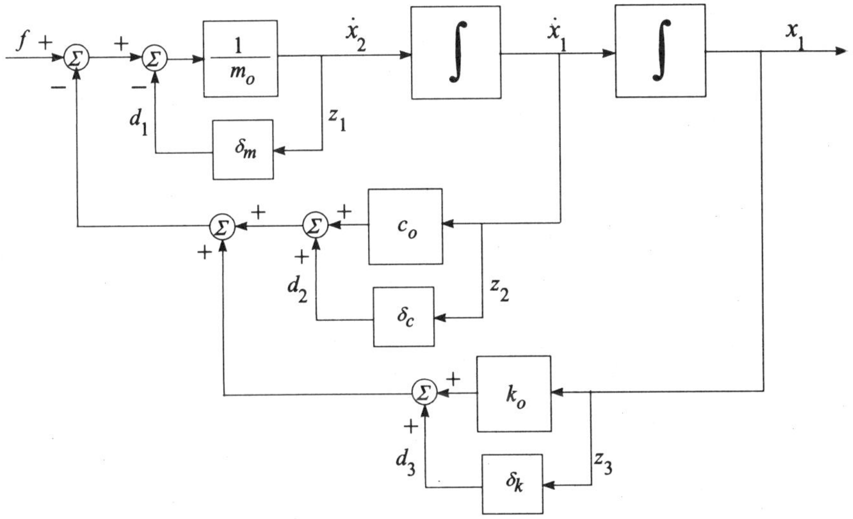

Fig. 2.11. Structured uncertainties block diagram of Example 2.2

Let z_1, z_2 and z_3 be $\dot{x}_2$, x_2 and x_1, respectively, considered as another, fictitious output vector; and, d_1, d_2 and d_3 be the signals coming out from the perturbation blocks δ_m, δ_c and δ_k, as shown in Figure 2.11. The perturbed system can be arranged in the following state-space model and represented as in Figure 2.12.

$$\begin{bmatrix} \dot{x}_1 \\ \dot{x}_2 \end{bmatrix} = \begin{bmatrix} 0 & 1 \\ -\frac{k_o}{m_o} & -\frac{c_o}{m_o} \end{bmatrix} \begin{bmatrix} x_1 \\ x_2 \end{bmatrix} + \begin{bmatrix} 0 & 0 & 0 \\ -1 & -1 & -1 \end{bmatrix} \begin{bmatrix} d_1 \\ d_2 \\ d_3 \end{bmatrix} + \begin{bmatrix} 0 \\ \frac{1}{m_o} \end{bmatrix} f$$
$$\begin{bmatrix} z_1 \\ z_2 \\ z_3 \end{bmatrix} = \begin{bmatrix} -\frac{k_o}{m_o} & -\frac{c_o}{m_o} \\ 0 & 1 \\ 1 & 0 \end{bmatrix} \begin{bmatrix} x_1 \\ x_2 \end{bmatrix} + \begin{bmatrix} -1 & -1 & -1 \\ 0 & 0 & 0 \\ 0 & 0 & 0 \end{bmatrix} \begin{bmatrix} d_1 \\ d_2 \\ d_3 \end{bmatrix} + \begin{bmatrix} \frac{1}{m_o} \\ 0 \\ 0 \end{bmatrix} f$$
$$y = \begin{bmatrix} 1 & 0 \end{bmatrix} \begin{bmatrix} x_1 \\ x_2 \end{bmatrix} \tag{2.9}$$

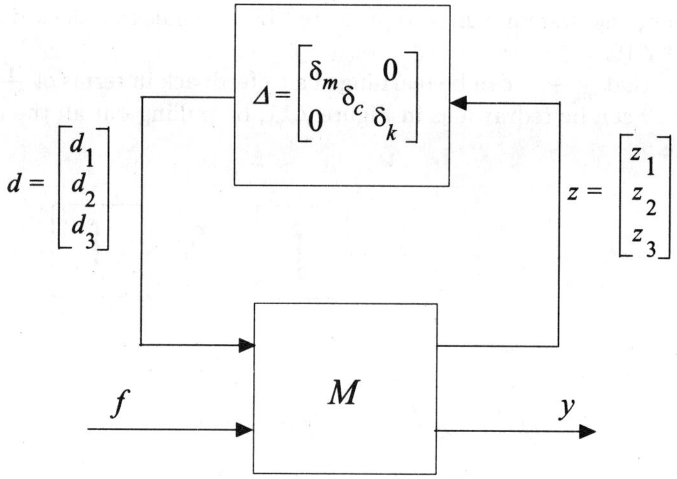

Fig. 2.12. Standard configuration of Example 2.2

The state-space model of (2.9) describes the augmented, interconnection system M of Figure 2.12. The perturbation block Δ in Figure 2.12 corresponds to parameter variations and is called "parametric uncertainty". The uncertain block Δ is not a full matrix but a diagonal one. It has certain structure, hence the terminology of "structured uncertainty". More general cases will be discussed shortly in Section 2.4.

2.3 Linear Fractional Transformations

The block diagram in Figure 2.12 can be generalised to be a standard configuration to represent how the uncertainty affects the input/output relationship of the control system under study. This kind of representation first appeared in the circuit analysis back in the 1950s ([128, 129]). It was later adopted in the robust control study ([132]) for uncertainty modelling. The general framework is depicted in Figure 2.13.

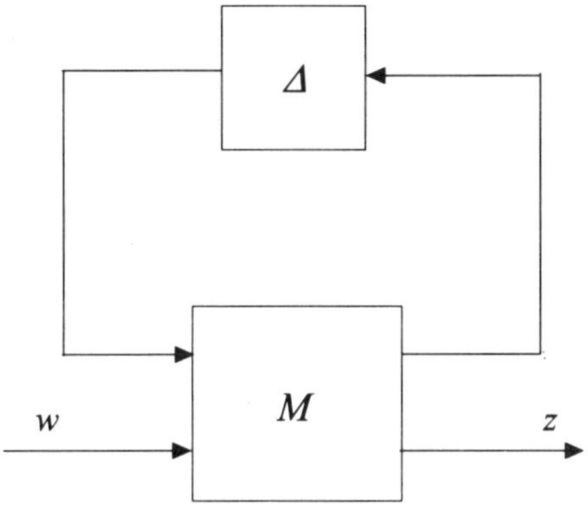

Fig. 2.13. Standard M-Δ configuration

The interconnection transfer function matrix M in Figure 2.13 is partitioned as

$$M = \begin{bmatrix} M_{11} & M_{12} \\ M_{21} & M_{22} \end{bmatrix}$$

where the dimensions of M_{11} conform with those of Δ. By routine manipulations, it can be derived that

$$z = \left[M_{22} + M_{21}\Delta(I - M_{11}\Delta)^{-1}M_{12}\right] w$$

if $(I - M_{11}\Delta)$ is invertible. When the inverse exists, we may define

$$F(M, \Delta) = M_{22} + M_{21}\Delta(I - M_{11}\Delta)^{-1}M_{12}$$

$F(M, \Delta)$ is called a *linear fractional transformation(LFT)* of M and Δ. Because the "upper"loop of M is closed by the block Δ, this kind of linear fractional transformation is also called an *upper linear fractional transformation(ULFT)*, and denoted with a subscript u, *i.e.* $F_u(M, \Delta)$, to show the way of connection. Similarly, there are also *lower linear fractional transformations(LLFT)* that are usually used to indicate the incorporation of a controller K into a system. Such a lower LFT can be depicted as in Figure 2.14 and defined by

$$F_l(M, K) = M_{11} + M_{12}K(I - M_{22}K)^{-1}M_{21}$$

With the introduction of linear fractional transformations, the unstructured uncertainty representations discussed in Section 2.1 may be uniformly described by Figure 2.13, with appropriately defined interconnection matrices Ms as listed below.

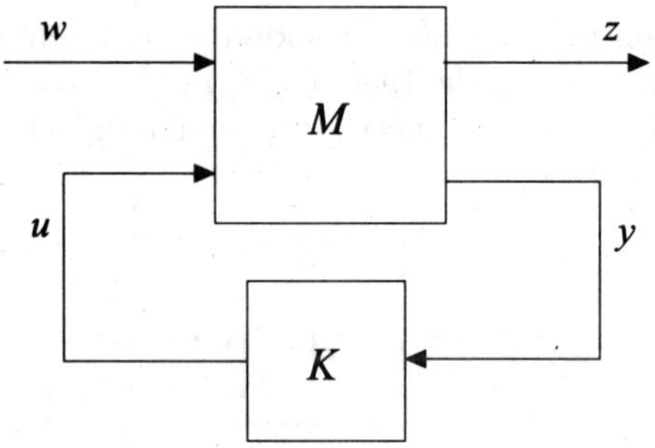

Fig. 2.14. Lower LFT configuration

1. Additive perturbation:

$$M = \begin{bmatrix} 0 & I \\ I & G_o \end{bmatrix} \tag{2.10}$$

2. Inverse additive perturbation:

$$M = \begin{bmatrix} -G_o & G_o \\ -G_o & G_o \end{bmatrix} \tag{2.11}$$

3. Input multiplicative perturbation:

$$M = \begin{bmatrix} 0 & I \\ G_o & G_o \end{bmatrix} \tag{2.12}$$

4. Output multiplicative perturbation:

$$M = \begin{bmatrix} 0 & G_o \\ I & G_o \end{bmatrix} \tag{2.13}$$

5. Inverse input multiplicative perturbation:

$$M = \begin{bmatrix} -I & I \\ -G_o & G_o \end{bmatrix} \tag{2.14}$$

6. Inverse output multiplicative perturbation:

$$M = \begin{bmatrix} -I & G_o \\ -I & G_o \end{bmatrix} \tag{2.15}$$

7. Left coprime factor perturbations:

$$M = \begin{bmatrix} \begin{bmatrix} -\tilde{M}_G^{-1} \\ 0 \end{bmatrix} & \begin{bmatrix} -G_o \\ I \end{bmatrix} \\ \tilde{M}_G^{-1} & G_o \end{bmatrix} \tag{2.16}$$

where $G_o = \tilde{M}_G^{-1}\tilde{N}_G$, a left coprime factorisation of the nominal plant; and, the perturbed plant is $G_p = (\tilde{M}_G + \Delta_{\tilde{M}})^{-1}(\tilde{N}_G + \Delta_{\tilde{N}})$.

8. Right coprime factor perturbations:

$$M = \begin{bmatrix} [-M_G^{-1}\ 0] & M_G^{-1} \\ [-G_o\ I] & G_o \end{bmatrix} \tag{2.17}$$

where $G_o = N_G {M_G}^{-1}$, a right coprime factorisation of the nominal plant; and, the perturbed plant is $G_p = (N_G + \Delta_N)(M_G + \Delta_M)^{-1}$.

In the above, it is assumed that $[I - M_{11}\Delta]$ is invertible. The perturbed system is thus

$$G_p(s) = F_u(M, \Delta)$$

In the coprime factor perturbation representations, (2.16) and (2.17), $\Delta = \begin{bmatrix} \Delta_{\tilde{M}} & \Delta_{\tilde{N}} \end{bmatrix}$ and $\Delta = \begin{bmatrix} \Delta_M \\ \Delta_N \end{bmatrix}$, respectively. The block Δ in (2.10)–(2.17) is supposed to be a "full" matrix, *i.e.* it has no specific *structure*.

2.4 Structured Uncertainties

In many robust design problems, it is more likely that the uncertainty scenario is a mixed case of those described in Sections 2.1 and 2.2. The uncertainties under consideration would include unstructured uncertainties, such as unmodelled dynamics, as well as parameter variations. All these uncertain parts still can be taken out from the dynamics and the whole system can be rearranged in a standard configuration of (upper) linear fractional transformation $F(M, \Delta)$. The uncertain block Δ would then be in the following general form

$$\Delta = \text{diag}\,[\delta_1 I_{r_1}, \cdots, \delta_s I_{r_s}, \Delta_1, \cdots, \Delta_f] : \delta_i \in \mathcal{C}, \Delta_j \in \mathcal{C}^{m_j \times m_j} \tag{2.18}$$

where $\sum_{i=1}^{s} r_i + \sum_{j=1}^{f} m_j = n$ with n is the dimension of the block Δ. We may define the set of such Δ as $\mathbf{\Delta}$. The total block Δ thus has two types of uncertain blocks: s repeated *scalar* blocks and f *full* blocks. The parameters δ_i of the repeated scalar blocks can be real numbers only, if further information of the uncertainties is available. However, in the case of real numbers, the analysis and design would be even harder. The full blocks in (2.18) need not be square, but by restricting them as such makes the notation much simpler.

When a perturbed system is described by an LFT with the uncertain block of (2.18), the Δ considered has a certain structure. It is thus called "structured uncertainty". Apparently, using a lumped, full block to model the uncertainty in such cases, for instance in Example 2.2, would lead to pessimistic analysis of the system behaviour and produce conservative designs.

3

Robust Design Specifications

A control system is *robust* if it remains stable and achieves certain performance criteria in the presence of possible uncertainties as discussed in Chapter 2. The robust design is to find a controller, for a given system, such that the closed-loop system is robust. The $\mathcal{H}_\infty$ optimisation approach and its related approaches, being developed in the last two decades and still an active research area, have been shown to be effective and efficient robust design methods for linear, time-invariant control systems. We will first introduce in this chapter the *Small-Gain Theorem*, which plays an important role in the $\mathcal{H}_\infty$ optimisation methods, and then discuss the stabilisation and performance requirements in robust designs using the $\mathcal{H}_\infty$ optimisation and related ideas.

3.1 Small-gain Theorem and Robust Stabilisation

The Small-Gain Theorem is of central importance in the derivation of many stability tests. In general, it provides only a sufficient condition for stability and is therefore potentially conservative. The Small-Gain Theorem is applicable to general operators. What will be included here is, however, a version that is suitable for the $\mathcal{H}_\infty$ optimisation designs, and in this case, it is a sufficient *and* necessary result.

Consider the feedback configuration in Figure 3.1, where $G_1(s)$ and $G_2(s)$ are the transfer function matrices of corresponding linear, time-invariant systems. We then have the following theorem.

Theorem 3.1. *[21] If $G_1(s)$ and $G_2(s)$ are stable, i.e. $G_1 \in \mathcal{H}_\infty$, $G_2 \in \mathcal{H}_\infty$, then the closed-loop system is internally stable if and only if*

$$\|G_1G_2\|_\infty < 1 \quad \text{and} \quad \|G_2G_1\|_\infty < 1$$

■

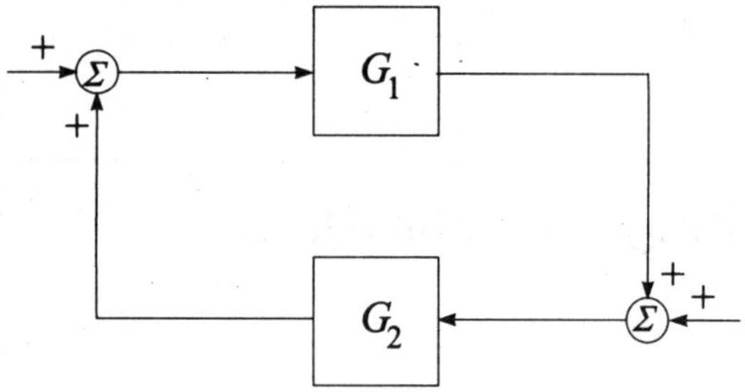

Fig. 3.1. A feedback configuration

A closed-loop system of the plant G and controller K is robustly stable if it remains stable for all possible, under certain definition, perturbations on the plant. This implies, of course, that K is a stabilising controller for the nominal plant G, since we always assume that the perturbation set includes zero (no perturbation). Let us consider the case of additive perturbation as depicted in Figure 3.2, where $\Delta(s)$ is the perturbation, a "full" matrix unknown but stable.

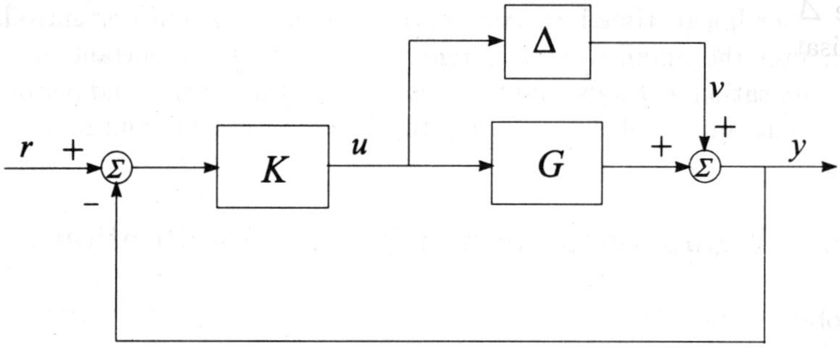

Fig. 3.2. Additive perturbation configuration

It is easy to work out that the transfer function from the signal v to u is $T_{uv} = -K(I + GK)^{-1}$. As mentioned earlier, the controller K should stabilise the nominal plant G. Hence, from the Small-Gain theorem, we have the following theorem.

Theorem 3.2. *[14, 109] For stable $\Delta(s)$, the closed-loop system is robustly stable if $K(s)$ stabilises the nominal plant and the following holds*

$$\|\Delta K(I + GK)^{-1}\|_\infty < 1$$

and

$$\|K(I + GK)^{-1}\Delta\|_\infty < 1$$

or, in a strengthened form,

$$\|K(I+GK)^{-1}\|_\infty < \frac{1}{\|\Delta\|_\infty} \tag{3.1}$$

The second condition becomes necessary, when the unknown Δ may have all phases. ■

If required to find a controller to robustly stabilise the largest possible set of perturbations, in the sense of ∞-norm, it is then clear that we need to solve the following minimisation problem

$$\min_{K\,stabilising} \|K(I+GK)^{-1}\|_\infty \tag{3.2}$$

In many cases, we may have *a priori* knowledge of the perturbation, say,

$$\overline{\sigma}(\Delta(j\omega)) \leq \overline{\sigma}(W_2(j\omega)) \quad \text{for all} \quad \omega \in \mathcal{R}$$

Then, we may rewrite the perturbation block as

$$\Delta(s) = \widetilde{\Delta}(s)W_2(s)$$

where $\widetilde{\Delta}(s)$ is the unit norm perturbation set. Correspondingly, the robust stabilisation condition becomes

$$\|W_2K(I+GK)^{-1}\|_\infty < 1$$

and the optimisation problem

$$\min_{K\,stabilising} \|W_2K(I+GK)^{-1}\|_\infty \tag{3.3}$$

Robust stabilisation conditions can be derived similarly for other perturbation representations discussed in Chapter 2 and are listed below (G_o is replaced by G for the sake of simplicity).

1. Inverse additive perturbation:

$$\|G(I+KG)^{-1}\|_\infty < \frac{1}{\|\Delta\|_\infty} \tag{3.4}$$

2. Input multiplicative perturbation:

$$\|KG(I+KG)^{-1}\|_\infty < \frac{1}{\|\Delta\|_\infty} \tag{3.5}$$

3. Output multiplicative perturbation:

$$\|GK(I+GK)^{-1}\|_\infty < \frac{1}{\|\Delta\|_\infty} \tag{3.6}$$

4. Inverse input multiplicative perturbation:

$$\|(I+KG)^{-1}\|_\infty < \frac{1}{\|\Delta\|_\infty} \tag{3.7}$$

5. Inverse output multiplicative perturbation:

$$\|(I+GK)^{-1}\|_\infty < \frac{1}{\|\Delta\|_\infty} \tag{3.8}$$

The cases of perturbation on coprime factors will be discussed in Chapter 5.

Remark:

In the above discussion, the stability of the perturbation block has been assumed. Actually, the conclusions are also true if the perturbed systems have the same number of closed right-half plane poles as the nominal system does (see [96]). If even this is not satisfied, then we will have to use the coprime factor perturbation models, as will be discussed in Chapter 5.

Robust stabilisation is an important issue not just as a design requirement. As a matter of fact, the $\mathcal{H}_\infty$ design and related approaches first formulate the stability as well as performance design specifications as a robust stabilisation problem and then solve the robust stabilisation problem to find a controller.

3.2 Performance Consideration

Figure 3.3 depicts a typical closed-loop system configuration, where G is the plant and K the controller to be designed. r, y, u, e, d, n are, respectively, the reference input, output, control signal, error signal, disturbance and measurement noise. With a little abuse of notations, we do not distinguish the notations of signals in the time or frequency domains. The following relationships are immediately available.

$$\begin{aligned}
y &= (I+GK)^{-1}GKr \;+\; (I+GK)^{-1}d \;-\; (I+GK)^{-1}GKn \\
u &= K(I+GK)^{-1}r \;-\; K(I+GK)^{-1}d \;-\; K(I+GK)^{-1}n \\
e &= (I+GK)^{-1}r \;-\; (I+GK)^{-1}d \;-\; (I+GK)^{-1}n
\end{aligned}$$

Assume that the signals r, d, n are energy bounded and have been normalised, *i.e.* lying in the unit ball of $\mathcal{L}_2$ space. We, however, do not know what exactly these signals are. It is required that the usual performance specifications, such as tracking, disturbance attenuation and noise rejection, should be as good as possible for any r, d or n whose energy does not exceed 1. From the discussions in Chapter 1 on signal and system norms, it is clear that we should

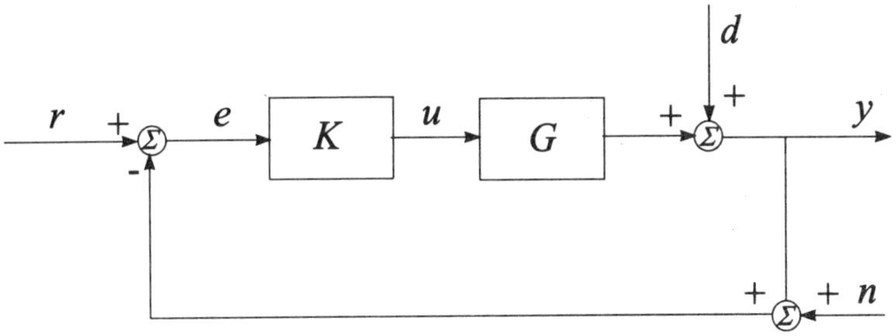

Fig. 3.3. A closed-loop configuration of G and K

minimise the ∞-norm, the gain, of corresponding transfer function matrices. Hence, the design problem is that over the set of all stabilising controller Ks, (*i.e.* those Ks make the closed-loop system internally stable), find the optimal one that minimises

- for good tracking,
  $\|(I + GK)^{-1}\|_\infty$
- for good disturbance attenuation,
  $\|(I + GK)^{-1}\|_\infty$
- for good noise rejection,
  $\| - (I + GK)^{-1}GK\|_\infty$
- for less control energy,
  $\|K(I + GK)^{-1}\|_\infty$

It is conventional to denote $\mathcal{S} := (I+GK)^{-1}$, the *sensitivity* function, and $\mathcal{T} := (I + GK)^{-1}GK$, the *complementary sensitivity* function.

In general, weighting functions would be used in the above minimisation to meet the design specifications. For instance, instead of minimising the sensitivity function alone, we would aim at solving

$$\min_{K\,stabilising} \|W_1 \mathcal{S} W_d\|_\infty$$

where W_1 is chosen to tailor the tracking requirement and is usually a high-gain low-pass filter type, W_d can be regarded as a generator that characterises all relevant disturbances in the case considered. Usually, the weighting functions are stable and of minimum phase.

3.3 Structured Singular Values

Systems with uncertain dynamics can all be put in the standard $M - \Delta$ configuration of Figure 2.13. The robust stabilisation conditions derived in Section 3.1 are sufficient and necessary conditions for unstructured uncertainties, *i.e.* Δ is a full block and will have all phases. In the case of structured uncertainty

(Section 2.4), these robust stabilization results could be very conservative. To deal with structured uncertainties, we need to introduce the so-called *structured singular values* (SSV).

In fact, these robust stabilisation results, such as that in Theorem 3.2, can be equivalently written as [14, 109]

$$\det\left[I - M(j\omega)\Delta(j\omega)\right] \neq 0, \ \forall \omega \in \mathcal{R}, \ \forall \Delta \tag{3.9}$$

where the nominal (closed-loop) system $M(s)$ is assumed to be stable as usual.

This condition for robust stability is sufficient and necessary even for structured uncertainty Δ. Roughly speaking, in order to have the closed-loop system robustly stable, all the uncertainties of a known structure of (2.18) should be small enough not to violate the condition (*i.e.* not make $I - M(j\omega)\Delta(j\omega)$ singular at any frequency ω). On the other hand, for a given M with a fixed controller K and a known structure of the uncertainties, the smallest "size" of the uncertainty that makes $I - M(j\omega)\Delta(j\omega)$ singular at some frequency ω describes how robustly stable the controller K is in dealing with such structured uncertainties. This measurement is the so-called structured singular values (SSV) introduced below.

First, as in Section 2.4, we define the structure of uncertainty of (2.18) and repeat here for convenience,

$$\mathbf{\Delta} = \{\mathrm{diag}[\delta_1 I_{r_1}, \cdots, \delta_s I_{r_s}, \Delta_1, \cdots, \Delta_f] : \delta_i \in \mathcal{C}, \Delta_j \in \mathcal{C}^{m_j \times m_j}\} \tag{3.10}$$

where $\sum_{i=1}^{s} r_i + \sum_{j=1}^{f} m_j = n$ with n is the dimension of the block Δ. We also assume the set of $\mathbf{\Delta}$ is bounded. And, we may thus define a normalised set of structured uncertainty by

$$\mathbf{B\Delta} := \{\Delta : \ \overline{\sigma}(\Delta) \leq 1, \ \Delta \in \mathbf{\Delta}\} \tag{3.11}$$

Definition 3.3. *For $M \in \mathcal{C}^{n \times n}$, the structured singular value $\mu_\Delta(M)$ of M with respect to $\mathbf{\Delta}$ is the number defined such that $\mu_\Delta^{-1}(M)$ is equal to the smallest $\overline{\sigma}(\Delta)$ needed to make $(I - M\Delta)$ singular (rank deficiency). That is*

$$\mu_\Delta^{-1}(M) := \min_{\Delta \in \mathbf{\Delta}} \{\overline{\sigma}(\Delta) : \ det(I - M\Delta) = 0\} \tag{3.12}$$

If there is no $\Delta \in \mathbf{\Delta}$ such that $det(I - M\Delta) = 0$, then $\mu_\Delta(M) := 0$. ■

When M is an interconnected transfer matrix as in Figure 2.13, the structured singular value, with respect to $\mathbf{\Delta}$, is defined by

$$\mu_\Delta(M(s)) := \sup_{\omega \in \mathcal{R}} \mu_\Delta(M(j\omega)) \tag{3.13}$$

Correspondingly, the uncertainty set may be defined as

$$\mathcal{M}(\mathbf{\Delta}) := \{\Delta(\cdot) \in \mathcal{RH}_\infty : \ \Delta(j\omega) \in \mathbf{\Delta} \ for\ all\ \omega \in \mathcal{R}\} \tag{3.14}$$

When the uncertainty structure is fixed, we may omit the subscript Δ of $\mu_\Delta(M)$ for brevity.

The reciprocal of the structured singular value denotes a frequency-dependent stability margin ([18, 132]). The robust stability result with regard to structured uncertainty is now given in the following theorem.

Theorem 3.4. *[23] Let the nominal feedback system $(M(s))$ be stable and Let $\beta > 0$ be an uncertainty bound, i.e. $\|\Delta\|_\infty \leq \beta, \forall\ \Delta(\cdot) \in \mathcal{M}(\mathbf{\Delta})$. The perturbed system of Figure 2.3 is robustly stable, with respect to $\mathbf{\Delta}$, if and only if $\mu_\Delta(M(s)) < \frac{1}{\beta}$.* ■

It is obvious that if the uncertainty lies in the unit ball $\mathbf{B\Delta}$, the robust stability condition is then $\mu_\Delta(M(s)) < 1$.

$\mu_\Delta(M(s))$ is frequency dependent and is calculated at "each" frequency over a reasonable range in practical applications.

In the literature, $\mu_\Delta(M(s))$ is sometimes redefined as $\|M\|_\mu$ for an interconnected transfer function matrix $M(s)$. This notation is convenient; however, it should be clear that it is not a norm. It does not satisfy the three basic properties of a norm. Also, it depends on $M(s)$ as well as the uncertainty structure of Δ.

The structured singular value plays an important role in robust design. As became clear in the early parts of this chapter, the $\mathcal{H}_\infty$ optimisation approach can deal with robust stabilisation problems with regard to unstructured uncertainties and can achieve nominal performance requirements. In the case of structured uncertainty, Theorem 3.4 gives a sufficient and necessary condition for robust stabilisation. Furthermore, the robust performance design can also be transformed into a robust stabilisation problem with regard to structured uncertainties, as will be described in Chapter 6. However, the computation of the structured singular value is not an easy task. It is still an open topic and requires further research.

A few properties of the structured singular value and its computation are listed below. Interested readers are referred to [23, 32, 33, 120, 121, 122, 175] for details. Coded routines for the μ computation are available in software packages MATLAB® [9] and Slicot [119].

Let $\rho(\cdot)$ denote the spectral radius of a square matrix. By direct manipulations, we have the following lemma.

Lemma 3.5. *For a square, constant matrix M,*

$$\mu(M) = \max_{\Delta \in \mathbf{B\Delta}} \rho(M\Delta)$$

■

The following properties of μ can also be easily derived.

- $\mu(\alpha M) = |\alpha| \cdot \mu(M), \quad \forall \alpha \in \mathcal{C}$
- $det(I - M\Delta) \neq 0, \ \forall \Delta \in \mathbf{B\Delta} \Longleftrightarrow \mu(M) < 1$

- if $\mathbf{\Delta} = \{\delta I_n : \delta \in \mathcal{C}\}$ $(s = 1, f = 0; r_1 = n) \longrightarrow \mu(M) = \rho(M)$
- if $\mathbf{\Delta} = \mathcal{C}^{n\times n}$ $(s = 0, f = 1; m_1 = n) \Longrightarrow \mu(M) = \overline{\sigma}(M)$

In the above, s, f and n are dimensions of the uncertainty set $\mathbf{\Delta}$ as defined in (3.10).

Further, we have the following lemma.

Lemma 3.6.

$$\rho(M) \leq \mu(M) \leq \overline{\sigma}(M)$$

■

Lemma 3.6 gives upper and lower bounds for $\mu(M)$. They are, however, not particularly useful for the computation of $\mu(M)$, because the gap between $\rho(M)$ and $\overline{\sigma}(M)$ could be arbitrarily large. More accurate bounds are needed and these can be obtained by using some transformations on M that may change the values of $\rho(M)$ and $\overline{\sigma}(M)$ but not change $\mu(M)$. The following two constant matrix sets are introduced for this purpose.

Define

$$\mathbf{U} = \{U \in \mathbf{\Delta} : \ UU^* = I_n\}$$

and

$$\mathbf{D} = \{D = \text{diag}\left[D_1, \cdots, D_s, d_1 I_{m_1}, \cdots, d_f I_{m_f}\right] : \\ D_i \in \mathcal{C}^{r_i \times r_i}, D_i = D_i^* > 0, d_j > 0\}$$

The matrix sets $\mathbf{U}$ and $\mathbf{D}$ match the structure of $\mathbf{\Delta}$. $\mathbf{U}$ is of a (block-)diagonal structure of unitary matrices and for any $D \in \mathbf{D}$ and $\Delta \in \mathbf{\Delta}$, D (D^{-1}) commutes with Δ. Furthermore, for any $\Delta \in \mathbf{\Delta}$, $U \in \mathbf{U}$, and $D \in \mathbf{D}$, we have

- $U^* \in \mathbf{U}$, $U\Delta \in \mathbf{\Delta}$, $\Delta U \in \mathbf{\Delta}$, and $\overline{\sigma}(U\Delta) = \overline{\sigma}(\Delta U) = \overline{\sigma}(\Delta)$
- $D\Delta D^{-1} = \Delta$, $D\Delta D^{-1} \in \mathbf{\Delta}$, and $\overline{\sigma}(D\Delta D^{-1}) = \overline{\sigma}(\Delta)$

More importantly, we have

$$\rho(MU) \leq \mu(MU) = \mu(M) = \mu(DMD^{-1}) \leq \overline{\sigma}(DMD^{-1}) \qquad (3.15)$$

In the above, $\mu(MU) = \mu(M)$ is derived from $\det(I - M\Delta) = \det(I - MUU^*\Delta)$ and $U^*\Delta \in \mathbf{\Delta}$, $\overline{\sigma}(U^*\Delta) = \overline{\sigma}(\Delta)$. Also, $\mu(M) = \mu(DMD^{-1})$ can be seen from $\det(I - DMD^{-1}\Delta) = \det(I - DM\Delta D^{-1}) = \det(I - M\Delta)$.

The relations in (3.15) directly lead to the following theorem.

Theorem 3.7.

$$\max_{U\in\mathbf{U}} \rho(MU) \leq \mu(M) \leq \inf_{D\in\mathbf{D}} \overline{\sigma}(DMD^{-1})$$

■

Theorem 3.7 provides tighter upper and lower bounds on $\mu(M)$. In [23], it was shown that the above lower bound is actually an equality,

$$\max_{U \in \mathbf{U}} \rho(MU) = \mu(M)$$

Unfortunately, this optimisation problem is not convex. $\rho(MU)$ may have multiple local maxima. Direct computation of $\max_{U \in \mathbf{U}} \rho(MU)$ may not find a global maximum. On the other hand, the upper bound of $\mu(M)$ in Theorem 3.7 is easier to find, since $\overline{\sigma}(DMD^{-1})$ is convex in lnD ([25, 139]). However, this upper bound is not always equal to $\mu(M)$. For the cases of $2s + f \leq 3$, it can be shown that

$$\mu(M) = \inf_{D \in \mathbf{D}} \overline{\sigma}(DMD^{-1})$$

The problem of calculating $\mu(M)$ is therefore reduced to an optimal diagonal scaling problem. Most algorithms proposed so far for the structured singular values compute this upper bound.

4

$\mathcal{H}_\infty$ Design

A control system is *robust* if it remains stable and achieves certain performance criteria in the presence of possible uncertainties as discussed in Chapter 2. The robust design is to find a controller, for a given system, such that the closed-loop system is robust. The $\mathcal{H}_\infty$ optimisation approach, being developed in the last two decades and still an active research area, has been shown to be an effective and efficient robust design method for linear, time-invariant control systems. In the previous chapter, various robust stability considerations and nominal performance requirements were formulated as a minimisation problem of the infinitive norm of a closed-loop transfer function matrix. Hence, in this chapter, we shall discuss how to formulate a robust design problem into such a minimisation problem and how to find the solutions. The $\mathcal{H}_\infty$ optimisation approach solves, in general, the robust stabilisation problems and nominal performance designs.

4.1 Mixed Sensitivity $\mathcal{H}_\infty$ Optimisation

Every practising control engineer knows very well that it will never be appropriate in any industrial design to use just a single cost function, such as those formulated in Chapter 3. A reasonable design would use a combination of these functions. For instance, it makes sense to require a good tracking as well as to limit the control signal energy, as depicted in Figure 4.1. We may then want to solve the following *mixed sensitivity* (or, so-called *S over KS*) problem,

$$\min_{K\,stabilising} \left\| \begin{bmatrix} (I+GK)^{-1} \\ K(I+GK)^{-1} \end{bmatrix} \right\|_\infty \tag{4.1}$$

This cost function can also be interpreted as the design objectives of nominal performance, good tracking or disturbance attenuation, and robust stabilisation, with regard to additive perturbation.

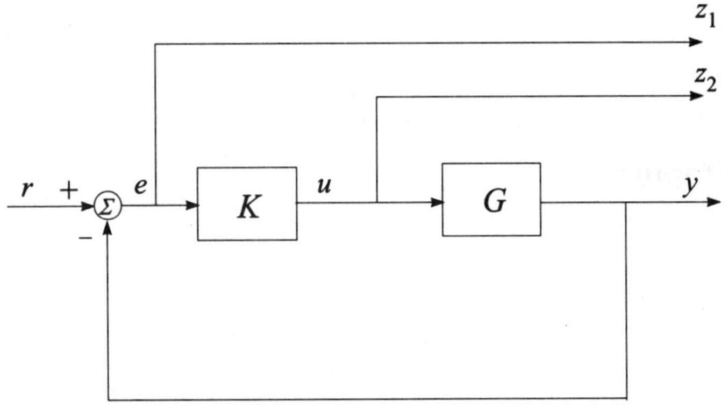

Fig. 4.1. A mixed sensitivity consideration

In order to adopt a unified solution procedure, the above cost function (4.1) can be recast into a standard configuration as in Figure 4.2. This can be obtained by using the LFT technique introduced in Chapter 2 and by specifying/grouping signals into sets of external inputs, outputs, input to the controller and output from the controller, which of course is the control signal. Note that in Figure 4.2 all the external inputs are denoted by w, z denotes the output signals to be minimised/penalised that includes both performance and robustness measures, y is the vector of measurements available to the controller K and u the vector of control signals. $P(s)$ is called the *generalised* plant or *interconnected* system. The objective is to find a stabilising controller K to minimise the output z, in the sense of energy, over all w with energy less than or equal to 1. Thus, it is equivalent to minimising the $\mathcal{H}_\infty$-norm of the transfer function from w to z.

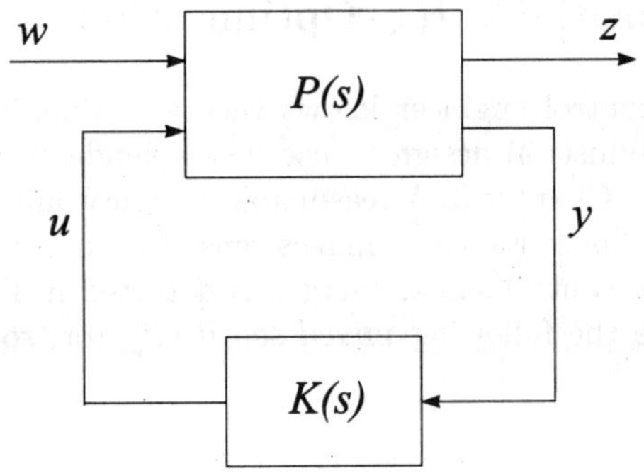

Fig. 4.2. The standard $\mathcal{H}_\infty$ configuration

Partitioning the interconnected system P as:

$$P(s) = \begin{bmatrix} P_{11}(s) & P_{12}(s) \\ P_{21}(s) & P_{22}(s) \end{bmatrix}$$

it can be obtained directly

$$\begin{aligned} z &= [P_{11} + P_{12}K(I - P_{22}K)^{-1}P_{21}]w \\ &=: \mathcal{F}_l(P, K)w \end{aligned}$$

where $\mathcal{F}_l(P, K)$ is the *lower linear fractional transformation* of P and K. The design objective now becomes

$$\min_{K\,stabilising} \|\mathcal{F}_l(P, K)\|_\infty \tag{4.2}$$

and is referred to as the $\mathcal{H}_\infty$ *optimisation problem.*

Referring to the problem in (4.1), it is easy to derive its standard form by defining $w = r$, $z = \begin{bmatrix} z_1 \\ z_2 \end{bmatrix} = \begin{bmatrix} e \\ u \end{bmatrix}$, $y = e$ and $u = u$. Consequently, the interconnected system

$$P = \begin{bmatrix} I & -G \\ 0 & I \\ I & -G \end{bmatrix} \tag{4.3}$$

where we may set

$$\begin{aligned} P_{11} &= \begin{bmatrix} I \\ 0 \end{bmatrix} & P_{12} &= \begin{bmatrix} -G \\ I \end{bmatrix} \\ P_{21} &= I & P_{22} &= -G \end{aligned}$$

Other mixed cases of cost transfer function matrices such as *S over T*, *S over T over KS*, *etc.*, can be dealt with similarly to formulate into the standard configuration. In practical designs, it is often necessary to include (closed-loop) weights with these cost functions. For instance, instead of (4.1), we may have to consider with $z_1 = W_1 e$ and $z_2 = W_2 u$,

$$\min_{K\,stabilising} \left\| \begin{matrix} W_1(I + GK)^{-1} \\ W_2K(I + GK)^{-1} \end{matrix} \right\|_\infty \tag{4.4}$$

These weighting functions can be easily absorbed into the interconnected system $P(s)$, as in this case,

$$P = \begin{bmatrix} W_1 & -W_1G \\ 0 & W_2 \\ I & -G \end{bmatrix} \tag{4.5}$$

4.2 2-Degree-Of-Freedom $\mathcal{H}_\infty$ Design

Among control-system design specifications, reference-signal tracking is often a requirement. The output of a designed control system is required to follow a preselected signal, or in a more general sense, the system is forced to track, for instance, the step response of a specified model (infinite-time model following task). A *2-degree-of-freedom(2DOF)* control scheme suits naturally this situation. The idea of a 2DOF scheme is to use a feedback controller (K_2) to achieve the internal and robust stability, disturbance rejection, *etc.*, and to design another controller (K_1) on the feedforward path to meet the tracking requirement, which minimises the difference between the output of the overall system and that of the reference model. 2DOF control schemes were successfully applied in some practical designs ([67, 87, 114]).

Figure 4.3 shows one structure of these 2DOF control schemes. In this configuration, in addition to the internal stability requirement, two signals e and u are to be minimised. The signal e shows the difference between the system output and the reference model output. u is the control signal and also is related to robust stability in the additive perturbation case. In Figure 4.3, two weighting functions are included to reflect the trade-off between and/or characteristics of these two penalised signals.

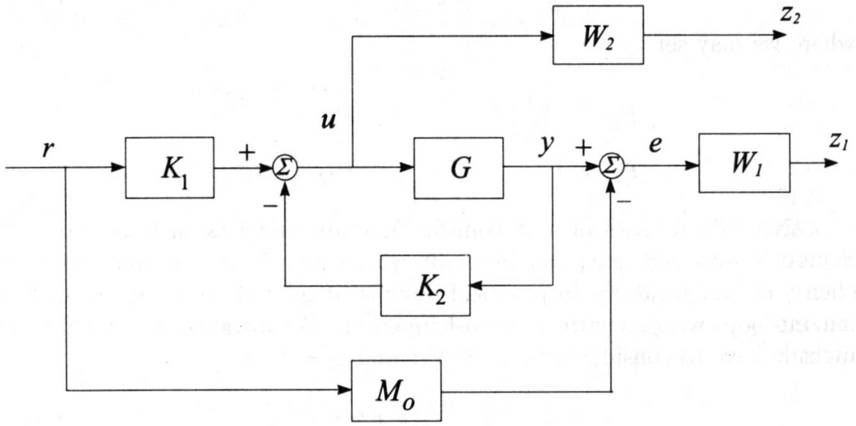

Fig. 4.3. A 2DOF design configuration

The configuration of Figure 4.3 can be rearranged as the standard configuration of Figure 4.2 by defining $w = r$, $z = \begin{bmatrix} z_1 \\ z_2 \end{bmatrix} = \begin{bmatrix} W_1 e \\ W_2 u \end{bmatrix}$, $y = \begin{bmatrix} r \\ y \end{bmatrix}$ and $u = u$. Note that in this configuration, $K = [K_1 \quad -K_2]$. Consequently, the interconnected system

$$P = \begin{bmatrix} -W_1 M_o & W_1 G \\ 0 & W_2 \\ I & 0 \\ 0 & G \end{bmatrix}$$

where we may set

$$P_{11} = \begin{bmatrix} -W_1 M_o \\ 0 \end{bmatrix} \qquad P_{12} = \begin{bmatrix} W_1 G \\ W_2 \end{bmatrix}$$

$$P_{21} = \begin{bmatrix} I \\ 0 \end{bmatrix} \qquad P_{22} = \begin{bmatrix} 0 \\ G \end{bmatrix}$$

4.3 $\mathcal{H}_\infty$ Suboptimal Solutions

The solution to the optimisation problem (4.2) is not unique except in the scalar case ([175, 59]). Generally speaking, there are no analytic formulae for the solutions. In practical design, it is usually sufficient to find a stabilising controller K such that the $\mathcal{H}_\infty$-norm of the closed-loop transfer function is less than a given positive number, *i.e.*,

$$\|\mathcal{F}_l(P, K)\|_\infty < \gamma \tag{4.6}$$

where $\gamma > \gamma_o := \min_{K\,stabilising} \|\mathcal{F}_l(P, K)\|_\infty$. This is called the $\mathcal{H}_\infty$ *suboptimal problem*. When certain conditions are satisfied, there are formulae to construct a set of controllers that solve the problem (4.6). The solution set is characterised by a free parameter $Q(s)$, which is stable and of ∞-norm less than γ.

It is imaginable that if we successively reduce the value of γ, starting from a relatively large number to ensure the existence of a suboptimal solution, we may obtain an optimal solution. It should, however, be pointed out here that when γ is approaching its minimum value γ_o the problem would become more and more ill-conditioned numerically. Hence, the "solution" thus obtained might be very unreliable.

4.3.1 Solution Formulae for Normalised Systems

Let the state-space description of the generalised (interconnected) system P in Figure 4.2 be given by

$$\begin{aligned} \dot{x}(t) &= Ax(t) + B_1 w(t) + B_2 u(t) \\ z(t) &= C_1 x(t) + D_{11} w(t) + D_{12} u(t) \\ y(t) &= C_2 x(t) + D_{21} w(t) \end{aligned}$$

where $x(t) \in R^n$ is the state vector, $w(t) \in R^{m_1}$ the exogenous input vector, $u(t) \in R^{m_2}$ the control input vector, $z(t) \in R^{p_1}$ the error (output) vector, and

$y(t) \in R^{p_2}$ the measurement vector, with $p_1 \geq m_2$ and $p_2 \leq m_1$. $P(s)$ may be further denoted as

$$
\begin{aligned}
P(s) &= \begin{bmatrix} P_{11}(s) & P_{12}(s) \\ P_{21}(s) & P_{22}(s) \end{bmatrix} \\
&= \left[\begin{array}{c|cc} A & B_1 & B_2 \\ \hline C_1 & D_{11} & D_{12} \\ C_2 & D_{21} & 0 \end{array}\right] \\
&=: \left[\begin{array}{c|c} A & B \\ \hline C & D \end{array}\right]
\end{aligned}
\tag{4.7}
$$

Note that in the above definition it is assumed that there is no direct link between the control input and the measurement output, *i.e.* $D_{22} = 0$. This assumption is reasonable because most industrial control systems are strictly proper and the corresponding $P(s)$ would have a zero D_{22} in a sensible design configuration. The case of a nonzero direct term between $u(t)$ and $y(t)$ will be, however, considered in the next subsection for the sake of completeness.

The $\mathcal{H}_\infty$ solution formulae use solutions of two *algebraic Riccati equations(ARE)*. An algebraic Riccati equation

$$E^T X + XE - XWX + Q = 0$$

where $W = W^T$ and $Q = Q^T$, uniquely corresponds to a Hamiltonian matrix $\begin{bmatrix} E & -W \\ -Q & -E^T \end{bmatrix}$. The stabilising solution X, if it exists, is a symmetric matrix that solves the ARE and is such that $E - WX$ is a stable matrix. The stabilising solution is denoted as

$$X := \mathbf{Ric} \begin{bmatrix} E & -W \\ -Q & -E^T \end{bmatrix}$$

Define

$$R_n := {D_{1*}}^T D_{1*} - \begin{bmatrix} \gamma^2 I_{m_1} & 0 \\ 0 & 0 \end{bmatrix}$$

and

$$\tilde{R}_n := D_{*1} D_{*1}^T - \begin{bmatrix} \gamma^2 I_{p_1} & 0 \\ 0 & 0 \end{bmatrix}$$

where

$$D_{1*} = \begin{bmatrix} D_{11} & D_{12} \end{bmatrix} \quad \text{and} \quad D_{*1} = \begin{bmatrix} D_{11} \\ D_{21} \end{bmatrix}$$

Assume that R_n and $\tilde{R}_n$ are nonsingular. We define two Hamiltonian matrices $\mathbf{H}$ and $\mathbf{J}$ as

$$\mathbf{H} := \begin{bmatrix} A & 0 \\ -C_1^T C_1 & -A^T \end{bmatrix} - \begin{bmatrix} B \\ -C_1^T D_{1*} \end{bmatrix} R_n^{-1} \begin{bmatrix} D_{1*}^T C_1 & B^T \end{bmatrix}$$

$$\mathbf{J} := \begin{bmatrix} A^T & 0 \\ -B_1 B_1^T & -A \end{bmatrix} - \begin{bmatrix} C^T \\ -B_1 D_{*1}^T \end{bmatrix} \tilde{R}_n^{-1} \begin{bmatrix} D_{*1} B_1^T & C \end{bmatrix}$$

Let

$$X := \textbf{Ric}\ (\mathbf{H})$$
$$Y := \textbf{Ric}\ (\mathbf{J})$$

Based on X and Y, a state feedback matrix F and an observer gain matrix L can be constructed, which will be used in the solution formulae,

$$F := -R_n^{-1}(D_{1*}{}^T C_1 + B^T X) =: \begin{bmatrix} F_1 \\ F_2 \end{bmatrix} =: \begin{bmatrix} F_{11} \\ F_{12} \\ F_2 \end{bmatrix}$$
$$L := -(B_1 D_{*1}{}^T + Y C^T)\tilde{R}_n^{-1} =: \begin{bmatrix} L_1 & L_2 \end{bmatrix} =: \begin{bmatrix} L_{11} & L_{12} & L_2 \end{bmatrix}$$

where F_1, F_2, F_{11} and F_{12} are of $m_1, m_2, m_1 - p_2$ and p_2 rows, respectively, and L_1, L_2, L_{11} and L_{12} of $p_1, p_2, p_1 - m_2$ and m_2 columns, respectively.

Glover and Doyle [48] derived necessary and sufficient conditions for the existence of an $\mathcal{H}_\infty$ suboptimal solution and further parameterised all such controllers. The results are obtained under the following assumptions.

A1 $(A,\ B_2)$ is stabilisable and $(C_2,\ A)$ detectable;

A2 $D_{12} = \begin{bmatrix} 0 \\ I_{m_2} \end{bmatrix}$ and $D_{21} = \begin{bmatrix} 0 & I_{p_2} \end{bmatrix}$;

A3 $\begin{bmatrix} A - j\omega I & B_2 \\ C_1 & D_{12} \end{bmatrix}$ has full column rank for all ω;

A4 $\begin{bmatrix} A - j\omega I & B_1 \\ C_2 & D_{21} \end{bmatrix}$ has full row rank for all ω.

Together with appropriate partition of $D_{11} = \begin{bmatrix} D_{1111} & D_{1112} \\ D_{1121} & D_{1122} \end{bmatrix}$, where D_{1122} has m_2 rows and p_2 columns, the solution formulae are given in the following theorem.

Theorem 4.1. *[175] Suppose $P(s)$ satisfies the assumptions* **A1 – A4**.
(a) There exists an internally stabilising controller $K(s)$ such that $\|\mathcal{F}_l(P,\ K)\|_\infty < \gamma$ if and only if
(i)

$$\gamma > max(\overline{\sigma}[D_{1111}, D_{1112}], \overline{\sigma}[D_{1111}{}^T, D_{1121}{}^T])$$

and
(ii) there exist stabilising solutions $X \geq 0$ and $Y \geq 0$ satisfying the two AREs

corresponding to the Hamiltonian matrices **H** *and* **J**, *respectively, and such that*

$$\rho(XY) < \gamma^2$$

where $\rho(\cdot)$ *denotes the spectral radius.*

(b) Given that the conditions of part (a) are satisfied, then all rational, internally stabilising controllers, $K(s)$, *satisfying* $\|\mathcal{F}_l(P,\ K)\|_\infty < \gamma$ *are given by*

$$K(s) = \mathcal{F}_l(M, \Phi)$$

for any rational $\Phi(s) \in \mathcal{H}_\infty$ *such that* $\|\Phi(s)\|_\infty < \gamma$, *where* $M(s)$ *has the realisation*

$$M(s) = \left[\begin{array}{c|cc} \hat{A} & \hat{B}_1 & \hat{B}_2 \\ \hline \hat{C}_1 & \hat{D}_{11} & \hat{D}_{12} \\ \hat{C}_2 & \hat{D}_{21} & 0 \end{array}\right]$$

and

$$\hat{D}_{11} = -D_{1121}{D_{1111}}^T(\gamma^2 I - D_{1111}{D_{1111}}^T)^{-1}D_{1112} - D_{1122}\,,$$

$\hat{D}_{12} \in R^{m_2 \times m_2}$ *and* $\hat{D}_{21} \in R^{p_2 \times p_2}$ *are any matrices (e.g. Cholesky factors) satisfying*

$$\hat{D}_{12}\hat{D}_{12}^T = I - D_{1121}(\gamma^2 I - {D_{1111}}^T D_{1111})^{-1}{D_{1121}}^T$$

$$\hat{D}_{21}^T\hat{D}_{21} = I - {D_{1112}}^T(\gamma^2 I - D_{1111}{D_{1111}}^T)^{-1}D_{1112}$$

and

$$\hat{B}_2 = Z(B_2 + L_{12})\hat{D}_{12}$$

$$\hat{C}_2 = -\hat{D}_{21}(C_2 + F_{12})$$

$$\begin{aligned}\hat{B}_1 &= -ZL_2 + \hat{B}_2\hat{D}_{12}^{-1}\hat{D}_{11} \\ &= -ZL_2 + Z(B_2 + L_{12})\hat{D}_{11}\end{aligned}$$

$$\begin{aligned}\hat{C}_1 &= F_2 + \hat{D}_{11}\hat{D}_{21}^{-1}\hat{C}_2 \\ &= F_2 - \hat{D}_{11}(C_2 + F_{12})\end{aligned}$$

$$\begin{aligned}\hat{A} &= A + BF + \hat{B}_1\hat{D}_{21}^{-1}\hat{C}_2 \\ &= A + BF - \hat{B}_1(C_2 + F_{12})\end{aligned}$$

where

$$Z = (I - \gamma^{-2} YX)^{-1}$$

■

When $\Phi(s) = 0$ is chosen, the corresponding suboptimal controller is called the *central* controller that is widely used in the $\mathcal{H}_\infty$ optimal design and has the state-space form

$$K_o(s) = \left[\begin{array}{c|c} \hat{A} & \hat{B}_1 \\ \hline \hat{C}_1 & \hat{D}_{11} \end{array}\right]$$

In the assumptions made earlier, **A2** assumes that the matrices D_{12} and D_{21} are in normalised forms and the system $P(s)$ is thus a so-called normalised system. When these two matrices are of full rank but not necessarily in the normalised forms will be discussed later.

4.3.2 Solution to S-over-KS Design

The mixed sensitivity optimisation S-over-KS is a common design problem in practice. The problem is formulated as in (4.1) or (4.4), and the interconnected system is in (4.3) or (4.5). In this case, the second algebraic Riccati equation is of a special form, *i.e.* the constant term is zero. We list the solution formulae for this special design problem here.

For simplicity, we consider the suboptimal design of (4.1). That is, for a given γ, we want to find a stabilising controller $K(s)$ such that,

$$\left\| \begin{matrix} (I + GK)^{-1} \\ K(I + GK)^{-1} \end{matrix} \right\|_\infty < \gamma$$

The interconnected system in the standard configuration is

$$P = \begin{bmatrix} I_p & -G \\ 0 & I_m \\ I_p & -G \end{bmatrix} = \left[\begin{array}{c|cc} A & 0 & -B \\ \hline \begin{bmatrix} C \\ 0 \end{bmatrix} & \begin{bmatrix} I_p \\ 0 \end{bmatrix} & \begin{bmatrix} 0 \\ I_m \end{bmatrix} \\ C & I_p & 0 \end{array}\right]$$

where we assume that $G(s)$ has m inputs and p outputs, and

$$G(s) = \left[\begin{array}{c|c} A & B \\ \hline C & 0 \end{array}\right]$$

It is clear that γ must be larger than 1 from the assumption $(a)/(i)$ of Theorem 4.1, because the 2-norm of the constant matrix D_{11} of the above $P(s)$ is 1. In this case, the two algebraic Riccati equations are

$$A^TX + XA - XBB^TX + (1-\gamma^{-2})^{-1}C^TC = 0$$
$$AY + YA^T - YC^TCY = 0$$

And, the formula of the central controller is

$$K_o = \left[\begin{array}{c|c} A - BB^TX - (1-\gamma^{-2})^{-1}ZYC^TC & ZYC^T \\ \hline B^TX & 0 \end{array}\right]$$

where $Z = (I - \gamma^{-2}YX)^{-1}$.
The formulae for all suboptimal controllers are

$$K(s) = \mathcal{F}_l(M, \Phi)$$

with $M(s)$ having the state-space realisation

$$M(s) = \left[\begin{array}{c|cc} A - BB^TX - (1-\gamma^{-2})^{-1}ZYC^TC & ZYC^T & -ZB \\ \hline B^TX & 0 & I_m \\ -(1-\gamma^{-2})^{-1}C & I_p & 0 \end{array}\right]$$

and $\Phi(s) \in \mathcal{H}_\infty$, such that $\|\Phi\|_\infty < \gamma$.

4.3.3 The Case of $D_{22} \neq 0$

When there is a direct link between the control input and the measurement output, the matrix D_{22} will not disappear in (4.7). The controller formulae for the case $D_{22} \neq 0$ are discussed here.

As a matter of fact, the D_{22} term can be easily separated from the rest of the system as depicted in Figure 4.4. A controller $K(s)$ for the system with zero D_{22} will be synthesised first, and then the controller $\tilde{K}(s)$ for the original system can be recovered from $K(s)$ and D_{22} by

$$\tilde{K}(s) = K(s)(I + D_{22}K(s))^{-1}$$

The state-space model of $\tilde{K}(s)$ can be derived as

$$\tilde{K}(s) = \left[\begin{array}{c|c} A_K - B_KD_{22}(I + D_KD_{22})^{-1}C_K & B_K(I + D_{22}D_K)^{-1} \\ \hline (I + D_KD_{22})^{-1}C_K & D_K(I + D_{22}D_K)^{-1} \end{array}\right]$$

where we assume that

$$K(s) = \left[\begin{array}{c|c} A_K & B_K \\ \hline C_K & D_K \end{array}\right]$$

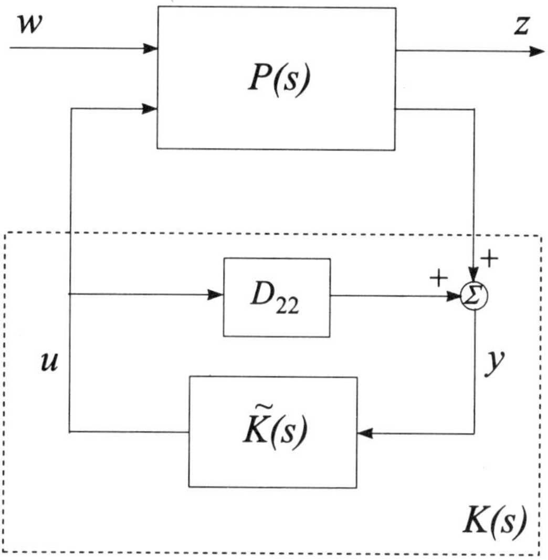

Fig. 4.4. The case of nonzero D_{22}

4.3.4 Normalisation Transformations

In general, the system data given would not be in the normalisation form as discussed in Section 4.3.1, though D_{12} will be of full column rank and D_{21} of full row rank, respectively, for any realistic control systems. In order to apply the results in Theorem 4.1, certain transformations must be used first. Using the *singular value decomposition(SVD)* or otherwise, we may find orthonormal matrices U_{12}, V_{12}, U_{21} and V_{21}, such that

$$U_{12}D_{12}V_{12}^T = \begin{bmatrix} 0 \\ \Sigma_{12} \end{bmatrix} \tag{4.8}$$

$$U_{21}D_{21}V_{21}^T = \begin{bmatrix} 0 \ \Sigma_{21} \end{bmatrix} \tag{4.9}$$

where $\Sigma_{12} : m_2 \times m_2$ and $\Sigma_{21} : p_2 \times p_2$ are nonsingular. Furthermore, we have

$$U_{12}D_{12}V_{12}^T\Sigma_{12}^{-1} = \begin{bmatrix} 0 \\ I \end{bmatrix} \tag{4.10}$$

$$\Sigma_{21}^{-1}U_{21}D_{21}V_{21}^T = \begin{bmatrix} 0 \ I \end{bmatrix} \tag{4.11}$$

The right-hand sides of the above equations are now in the normalised form.

When $p_1 > m_2$ and $p_2 < m_1$, the matrices U_{12} and V_{21} can be partitioned as

$$U_{12} = \begin{bmatrix} U_{121} \\ U_{122} \end{bmatrix} \tag{4.12}$$

$$V_{21} = \begin{bmatrix} V_{211} \\ V_{212} \end{bmatrix} \tag{4.13}$$

with $U_{121} : (p_1 - m_2) \times p_1, U_{122} : m_2 \times p_1, V_{211} : (m_1 - p_2) \times m_1$ and $V_{212} : p_2 \times m_1$.

The normalisation of $P(s)$ into $\overline{P}(s)$ is based on the above transformations and shown in Figure 4.5.

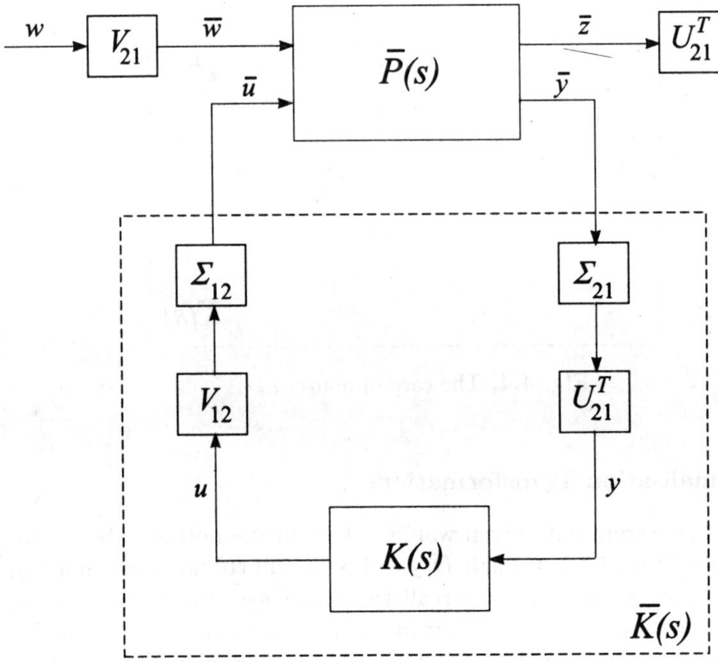

Fig. 4.5. Normalisation configuration

Given $P(s)$, the state-space form of $\overline{P}(s)$ is obtained as follows

$$\overline{B}_1 = B_1 V_{21}^T$$
$$\overline{B}_2 = B_2 V_{12}^T \Sigma_{12}^{-1}$$
$$\overline{C}_1 = U_{12} C_1$$
$$\overline{C}_2 = \Sigma_{21}^{-1} U_{21} C_2$$
$$\overline{D}_{11} = U_{12} D_{11} V_{21}^T = \begin{bmatrix} U_{121} D_{11} V_{211}^T & U_{121} D_{11} V_{212}^T \\ U_{122} D_{11} V_{211}^T & U_{122} D_{11} V_{212}^T \end{bmatrix}$$
$$\overline{D}_{12} = \begin{bmatrix} 0 \\ I \end{bmatrix} = U_{12} D_{12} V_{12}^T \Sigma_{12}^{-1}$$
$$\overline{D}_{21} = \begin{bmatrix} 0 & I \end{bmatrix} = \Sigma_{21}^{-1} U_{21} D_{21} V_{21}^T$$

Since V_{21} and U_{12} are orthonormal, the infinitive norms of the transfer function matrices from w to z and from $\overline{w}$ to $\overline{z}$ are the same, *i.e.* $||T_{zw}||_\infty = ||T_{\overline{z}\overline{w}}||_\infty$, with obviously $K(s) = V_{12}^T \Sigma_{12}^{-1} \overline{K}(s) \Sigma_{21}^{-1} U_{21}$, where $\overline{K}(s)$ is a suboptimal solution with regard to $\overline{P}(s)$.

4.3.5 Direct Formulae for $\mathcal{H}_\infty$ Suboptimal Central Controller

Based on the transformations discussed in the last subsection and by further manipulations, direct solution formulae can be derived for the general, non-normalised system data (4.7) [57].

We first discuss the two algebraic Riccati equations that are essential in the computations of $\mathcal{H}_\infty$ suboptimal controllers. The conclusion is a little surprising: the two AREs remain the same with matrices in the normalised system data (matrices with bars) replaced by corresponding matrices in the general form (matrices without bars) of $P(s)$. This is obtained by the following routine matrix manipulations,

$$
\begin{aligned}
\overline{C}_1^T \overline{C}_1 &= C_1^T U_{12}^T U_{12} C_1 = C_1^T C_1 \\
\begin{bmatrix} \overline{B} \\ -\overline{C}_1^T \overline{D}_{1*} \end{bmatrix} &= \begin{bmatrix} B_1 V_{21}^T & B_2 V_{12}^T \Sigma_{12}^{-1} \\ -C_1^T U_{12}^T U_{12} D_{11} V_{21}^T & -C_1^T U_{12}^T U_{12} D_{12} V_{12}^T \Sigma_{12}^{-1} \end{bmatrix} \\
&= \begin{bmatrix} B \\ -C_1^T D_{1*} \end{bmatrix} \begin{bmatrix} V_{21}^T & 0 \\ 0 & V_{12}^T \Sigma_{12}^{-1} \end{bmatrix} \\
\overline{D}_{1*}{}^T \overline{D}_{1*} &= \begin{bmatrix} V_{21} D_{11}^T U_{12}^T \\ \Sigma_{12}^{-T} V_{12} D_{12}^T U_{12}^T \end{bmatrix} \begin{bmatrix} U_{12} D_{11} V_{21}^T & U_{12} D_{12} V_{12}^T \Sigma_{12}^{-1} \end{bmatrix} \\
&= \begin{bmatrix} V_{21} & 0 \\ 0 & \Sigma_{12}^{-T} V_{12} \end{bmatrix} D_{1*}{}^T D_{1*} \begin{bmatrix} V_{21}^T & 0 \\ 0 & V_{12}^T \Sigma_{12}^{-1} \end{bmatrix}
\end{aligned}
$$

where

$$
D_{1*} := \begin{bmatrix} D_{11} & D_{12} \end{bmatrix} \tag{4.14}
$$

Thus,

$$
R_n = \begin{bmatrix} V_{21} & 0 \\ 0 & \Sigma_{12}^{-T} V_{12} \end{bmatrix} R \begin{bmatrix} V_{21}^T & 0 \\ 0 & V_{12}^T \Sigma_{12}^{-1} \end{bmatrix}
$$

with

$$
R := D_{1*}{}^T D_{1*} - \begin{bmatrix} \gamma^2 I_{m_1} & 0 \\ 0 & 0 \end{bmatrix} \tag{4.15}
$$

Consequently, the Hamiltonian $\mathbf{H}$ defined for normalised system data is given by

$$
\mathbf{H} = \begin{bmatrix} A & 0 \\ -C_1^T C_1 & -A^T \end{bmatrix} - \begin{bmatrix} B \\ -C_1^T D_{1*} \end{bmatrix} R^{-1} \begin{bmatrix} D_{1*}^T C_1 & B^T \end{bmatrix} \tag{4.16}
$$

Similarly, the Hamiltonian $\mathbf{J}$ is given by

$$\mathbf{J} = \begin{bmatrix} A^T & 0 \\ -B_1B_1^T & -A \end{bmatrix} - \begin{bmatrix} C^T \\ -B_1D_{*1}^T \end{bmatrix} \tilde{R}^{-1} \begin{bmatrix} D_{*1}B_1^T \ C \end{bmatrix} \tag{4.17}$$

where

$$D_{*1} := \begin{bmatrix} D_{11} \\ D_{21} \end{bmatrix} \tag{4.18}$$

and

$$\tilde{R} := D_{*1}D_{*1}{}^T - \begin{bmatrix} \gamma^2 I_{p_1} & 0 \\ 0 & 0 \end{bmatrix} \tag{4.19}$$

with

$$\tilde{R}_n = \begin{bmatrix} U_{12} & 0 \\ 0 & \Sigma_{21}^{-1}U_{21} \end{bmatrix} \tilde{R} \begin{bmatrix} U_{12}^T & 0 \\ 0 & U_{21}^T\Sigma_{21}^{-T} \end{bmatrix}$$

Hence, we have shown that the two AREs are exactly the same and so we may, in the following, let X and Y be the stabilising solutions of the two AREs corresponding to the general system data.

Another pair of matrices used in Theorem 4.1 may also be defined from the solutions X and Y,

$$F := -R^{-1}(D_{1*}{}^T C_1 + B^T X) =: \begin{bmatrix} F_1 \\ F_2 \end{bmatrix} \tag{4.20}$$

$$L := -(B_1 D_{*1}{}^T + YC^T)\tilde{R}^{-1} =: \begin{bmatrix} L_1 \ L_2 \end{bmatrix} \tag{4.21}$$

with $F_1 : m_1 \times n, F_2 : m_2 \times n, L_1 : n \times p_1$ and $L_2 : n \times p_2$. It is easy to check that

$$\overline{F} = \begin{bmatrix} V_{21} & 0 \\ 0 & \Sigma_{12}V_{12} \end{bmatrix} F$$

and

$$\begin{aligned} \overline{F}_{11} &= V_{211}F_1 \\ \overline{F}_{12} &= V_{212}F_1 \\ \overline{F}_2 &= \Sigma_{12}V_{12}F_2 \end{aligned}$$

Similarly, we have

$$\overline{L} = L \begin{bmatrix} U_{12}^T & 0 \\ 0 & U_{21}^T\Sigma_{21} \end{bmatrix}$$

and

$$\begin{aligned}\overline{L}_{11} &= L_1 U_{121}^T\\ \overline{L}_{12} &= L_1 U_{122}^T\\ \overline{L}_2 &= L_2 U_{21}^T \Sigma_{21}\end{aligned}$$

Using the above results and Theorem 4.1, with further manipulations, we can get the following corollary. Note that the assumption **A2** is relaxed to the general case

A2 D_{12} is of full column rank and D_{21} of full row rank

Also, note that for the sake of simplicity, only the formulae for the central suboptimal controller is given. Formulae for all suboptimal controllers can be derived accordingly.

Corollary 4.2. *Suppose $P(s)$ satisfies the assumptions* **A1** – **A4**.
(a) There exists an internally stabilising controller $K(s)$ such that $||F_l(P,K)||_\infty < \gamma$ if and only if
(i)

$$\gamma > max\{\overline{\sigma}(U_{121}D_{11}), \overline{\sigma}(D_{11}V_{211}^T)\} \tag{4.22}$$

where U_{121} and V_{211} are as defined in (4.12) and (4.13), and
(ii) there exist solutions $X \geq 0$ and $Y \geq 0$ satisfying the two AREs corresponding to (4.16) and (4.17), respectively, and such that

$$\rho(XY) < \gamma^2 \tag{4.23}$$

where $\rho(\bullet)$ denotes the spectral radius.

(b) Given that the conditions of part (a) are satisfied, then the central $\mathcal{H}_\infty$ suboptimal controller for $P(s)$ is given by

$$K(s) = \left[\begin{array}{c|c} A_K & B_K \\ \hline C_K & D_K \end{array}\right]$$

where

$$\begin{aligned}D_K &= -\gamma^2 (U_{122}D_{12})^{-1} U_{122} (\gamma^2 I - D_{11}V_{211}^T V_{211} D_{11}^T U_{121}^T U_{121})^{-1}\\ &\qquad D_{11} V_{212}^T (D_{21} V_{212}^T)^{-1} && (4.24)\\ B_K &= -Z\,[L_2 - (B_2 + L_1 D_{12}) D_K] && (4.25)\\ C_K &= F_2 - D_K (C_2 + D_{21} F_1) && (4.26)\\ A_K &= A + BF - B_K (C_2 + D_{21} F_1) && (4.27)\end{aligned}$$

where

$$Z = (I - \gamma^{-2} Y X)^{-1} \tag{4.28}$$

and U_{12}, V_{21}, F and L are as defined in (4.8) and (4.12), (4.9) and (4.13), (4.20), and (4.21), respectively. ■

Remark:

The above corollary addresses the general case of $p_1 > m_2$ and $p_2 < m_1$. There are three special cases of the dimensions in which the solution formulae would be simpler.

Case 1: $p_1 = m_2$ and $p_2 < m_1$.
In this case, the orthogonal transformation on D_{12} is not needed. The condition $(a)/(i)$ in Corollary 4.2 is reduced to $\gamma > \overline{\sigma}(D_{11}V_{211}^T)$, and

$$D_K = -D_{12}^{-1}D_{11}V_{212}^T(D_{21}V_{212}^T)^{-1}$$

Case 2: $p_1 > m_2$ and $p_2 = m_1$.
In this case, the orthogonal transformation on D_{21} is not needed. The condition $(a)/(i)$ in Corollary 4.2 is reduced to $\gamma > \overline{\sigma}(U_{121}D_{11})$, and

$$D_K = -(U_{122}D_{12})^{-1}U_{122}D_{11}D_{21}^{-1}$$

Case 3: $p_1 = m_2$ and $p_2 = m_1$.
In this case, both orthogonal transformations are not needed. The condition $(a)/(i)$ in Corollary 4.2 is reduced to any positive γ, and

$$D_K = -D_{12}^{-1}D_{11}D_{21}^{-1}$$

Another special case is when $D_{11} = 0$, in which the central controller is simply given by

$$K(s) = \left[\begin{array}{c|c} A + BF + ZL_2(C_2 + D_{21}F_1) & -ZL_2 \\ \hline F_2 & 0 \end{array}\right]$$

4.4 Formulae for Discrete-time Cases

In this section, formulae for $\mathcal{H}_\infty$ solutions in discrete-time cases ([53, 60]) will be given. Consider a generalised, linear, discrete-time system, described by the equations

$$\begin{aligned} x_{k+1} &= Ax_k + B_1w_k + B_2u_k \\ z_k &= C_1x_k + D_{11}w_k + D_{12}u_k \\ y_k &= C_2x_k + D_{21}w_k + D_{22}u_k \end{aligned} \tag{4.29}$$

where $x_k \in \mathcal{R}^n$ is the state vector, $w_k \in \mathcal{R}^{m_1}$ is the exogenous input vector (the disturbance), $u_k \in \mathcal{R}^{m_2}$ is the control input vector, $z_k \in \mathcal{R}^{p_1}$ is the error vector, and $y_k \in \mathcal{R}^{p_2}$ is the measurement vector, with $p_1 \geq m_2$ and $p_2 \leq m_1$. The transfer function matrix of the system will be denoted by

$$
\begin{aligned}
P(z) &= \begin{bmatrix} P_{11}(z) & P_{12}(z) \\ P_{21}(z) & P_{22}(z) \end{bmatrix} \\
&= \left[\begin{array}{c|cc} A & B_1 & B_2 \\ \hline C_1 & D_{11} & D_{12} \\ C_2 & D_{21} & D_{22} \end{array}\right] \\
&=: \left[\begin{array}{c|c} A & B \\ \hline C & D \end{array}\right]
\end{aligned}
\tag{4.30}
$$

Similar to the continuous-time case, the $\mathcal{H}_\infty$ suboptimal discrete-time control problem is to find an internally stabilising controller $K(z)$ such that, for a prespecified positive value γ,

$$
\|F_\ell(P,K)\|_\infty < \gamma \tag{4.31}
$$

We need the following assumptions.

A1 (A, B_2) is stabilisable and (C_2, A) is detectable;

A2 $\begin{bmatrix} A - e^{j\Theta} I_n & B_2 \\ C_1 & D_{12} \end{bmatrix}$ has full column rank for all $\Theta \in [0, 2\pi)$;

A3 $\begin{bmatrix} A - e^{j\Theta} I_n & B_1 \\ C_2 & D_{21} \end{bmatrix}$ has full row rank for all $\Theta \in [0, 2\pi)$.

We shall also assume that a loop-shifting transformation that enables us to set $D_{22} = 0$ has been carried out. The general case ($D_{22} \neq 0$) will be dealt with at the end of this section.

Note that the method under consideration does not involve reduction of the matrices D_{12} and D_{21} to some special form, as is usually required in the design of continuous-time $\mathcal{H}_\infty$ controllers.

Let

$$
\overline{C} = \begin{bmatrix} C_1 \\ 0 \end{bmatrix}, \quad \overline{D} = \begin{bmatrix} D_{11} & D_{12} \\ I_{m_1} & 0 \end{bmatrix}
$$

and define

$$
J = \begin{bmatrix} I_{p_1} & 0 \\ 0 & -\gamma^2 I_{m_1} \end{bmatrix}, \quad \hat{J} = \begin{bmatrix} I_{m_1} & 0 \\ 0 & -\gamma^2 I_{m_2} \end{bmatrix}, \quad \tilde{J} = \begin{bmatrix} I_{m_1} & 0 \\ 0 & -\gamma^2 I_{p_1} \end{bmatrix}
$$

Let X_∞ be the solution to the discrete-time Riccati equation

$$
X_\infty = \overline{C}^T J \overline{C} + A^T X_\infty A - L^T R^{-1} L \tag{4.32}
$$

where

$$
R = \overline{D}^T J \overline{D} + B^T X_\infty B =: \begin{bmatrix} R_1 & R_2^T \\ R_2 & R_3 \end{bmatrix}
$$

$$L = \overline{D}^T J \overline{C} + B^T X_\infty A =: \begin{bmatrix} L_1 \\ L_2 \end{bmatrix}$$

Assume that there exists an $m_2 \times m_2$ matrix V_{12} such that

$$V_{12}^T V_{12} = R_3$$

and an $m_1 \times m_1$ matrix V_{21} such that

$$V_{21}^T V_{21} = -\gamma^{-2} \nabla, \quad \nabla = R_1 - R_2^T R_3^{-1} R_2 < 0$$

Define the matrices

$$\begin{bmatrix} A_t & \tilde{B}_t \\ C_t & \tilde{D}_t \end{bmatrix} =: \left[\begin{array}{c|c} A_t & \tilde{B}_{t_1} \tilde{B}_{t_2} \\ \hline C_{t_1} & \tilde{D}_{t_{11}} \tilde{D}_{t_{12}} \\ C_{t_2} & \tilde{D}_{t_{21}} \tilde{D}_{t_{22}} \end{array}\right] =$$

$$\left[\begin{array}{c|cc} A - B_1 \nabla^{-1} L_\nabla & B_1 V_{21}^{-1} & 0 \\ \hline V_{12} R_3^{-1} (L_2 - R_2 \nabla^{-1} L_\nabla) & V_{12} R_3^{-1} R_2 V_{21}^{-1} & I \\ C_2 - D_{21} \nabla^{-1} L_\nabla & D_{21} V_{21}^{-1} & 0 \end{array}\right]$$

where

$$L_\nabla = L_1 - R_2^T R_3^{-1} L_2$$

Let Z_∞ be the solution to the discrete-time Riccati equation

$$Z_\infty = \tilde{B}_t \hat{J} \tilde{B}_t^T + A_t Z_\infty A_t^T - M_t S_t^{-1} M_t^T \tag{4.33}$$

in which

$$S_t = \tilde{D}_t \hat{J} \tilde{D}_t^T + C_t Z_\infty C_t^T =: \begin{bmatrix} S_{t_1} & S_{t_2} \\ S_{t_2}^T & S_{t_3} \end{bmatrix}$$

$$M_t = \tilde{B}_t \hat{J} \tilde{D}_t^T + A_t Z_\infty C_t^T =: [M_{t_1} \, M_{t_2}]$$

Equations (4.32) and (4.33) are referred as the *X-Riccati equation* and *Z-Riccati equation*, respectively.

A stabilising controller that satisfies

$$\|F_\ell(P, K)\|_\infty < \gamma$$

exists, if and only if ([53])

1. there exists a solution to the Riccati equation (4.32) satisfying

$$X_\infty \geq 0$$

$$\nabla < 0$$

such that $A - BR^{-1}L$ is asymptotically stable;

2. there exists a solution to the Riccati equation (4.33) such that

$$Z_\infty \geq 0$$

$$S_{t_1} - S_{t_2} S_{t_3}^{-1} S_{t_2}^T < 0$$

with $A_t - M_t S_t^{-1} C_t$ asymptotically stable. ■

In this case, a controller that achieves the objective may be given by ([53])

$$\hat{x}_{k+1} = A_t \hat{x}_k + B_2 u_k + M_{t_2} S_{t_3}^{-1}(y_k - C_{t_2}\hat{x}_k)$$

$$V_{12} u_k = -C_{t_1}\hat{x}_k - S_{t_2} S_{t_3}^{-1}(y_k - C_{t_2}\hat{x}_k)$$

that yields

$$K_0 = \left[\begin{array}{c|c} A_t - B_2 V_{12}^{-1}(C_{t_1} - S_{t_2}S_{t_3}^{-1}C_{t_2}) - M_{t_2}S_{t_3}^{-1}C_{t_2} & -B_2V_{12}^{-1}S_{t_2}S_{t_3}^{-1} + M_{t_2}S_{t_3}^{-1} \\ \hline -V_{12}^{-1}(C_{t_1} - S_{t_2}S_{t_3}^{-1}C_{t_2}) & -V_{12}^{-1}S_{t_2}S_{t_3}^{-1} \end{array}\right] \tag{4.34}$$

This is the so-called *central controller* that is widely used in practice.

Consider now the general case of $D_{22} \neq 0$. Suppose

$$\hat{K} = \left[\begin{array}{c|c} \hat{A}_k & \hat{B}_k \\ \hline \hat{C}_k & \hat{D}_k \end{array}\right]$$

is a stabilising controller for D_{22} set to zero, and satisfies

$$\|F_\ell\left(P - \begin{bmatrix} 0 & 0 \\ 0 & D_{22} \end{bmatrix}, \hat{K}\right)\|_\infty < \gamma$$

Then

$$\begin{aligned} F_\ell(P, \hat{K}(I + D_{22}\hat{K})^{-1}) &= P_{11} + P_{12}\hat{K}(I + D_{22}\hat{K} - P_{22}\hat{K})^{-1}P_{21} \\ &= F_\ell\left(P - \begin{bmatrix} 0 & 0 \\ 0 & D_{22} \end{bmatrix}, \hat{K}\right) \end{aligned}$$

In this way, a controller $\hat{K}$ for

$$P - \begin{bmatrix} 0 & 0 \\ 0 & D_{22} \end{bmatrix}$$

yields a controller $K = \hat{K}(I + D_{22}\hat{K})^{-1}$ for P. It may be shown that

$$K = \left[\begin{array}{c|c} \hat{A}_k - \hat{B}_k D_{22}(I_{m_2} + \hat{D}_k D_{22})^{-1}\hat{C}_k & \hat{B}_k - \hat{B}_k D_{22}(I_{m_2} + \hat{D}_k D_{22})^{-1}\hat{D}_k \\ \hline (I_{m_2} + \hat{D}_k D_{22})^{-1}\hat{C}_k & (I_{m_2} + \hat{D}_k D_{22})^{-1}\hat{D}_k \end{array}\right]$$

In order to find K from $\hat{K}$ we must exclude the possibility of the feedback system becoming ill-posed, *i.e.* $\det(I + \hat{K}(\infty)D_{22}) = 0$.

5

$\mathcal{H}_\infty$ Loop-shaping Design Procedures

In the previous chapter the $\mathcal{H}_\infty$ optimisation approach was introduced that formulated robust stabilisation and nominal performance requirements as (sub)optimisation problems of the $\mathcal{H}_\infty$ norm of certain cost functions. Several formulations of cost function are applicable in the robust controller design, for instance, the weighted S/KS and S/T design methods. The optimisation of S/KS, where S is the sensitivity function and K the controller to be designed, could achieve the nominal performance in terms of tracking or output disturbance rejection and robustly stabilise the system against additive model perturbations. On the other hand, the mixed sensitivity optimisation of S/T, where T is the complementary sensitivity function, could achieve robust stability against multiplicative model perturbations in addition to the nominal performance. Both of them are useful robust controller design methods, but the model perturbation representations are limited by the condition on the number of right-half complex plane poles. Also, there may exist undesirable pole-zero cancellations between the nominal model and the $\mathcal{H}_\infty$ controllers ([138]). In this chapter, an alternative way to represent the model uncertainty is introduced. The uncertainty is described by the perturbations directly on the coprime factors of the nominal model ([153, 154]). The $\mathcal{H}_\infty$ robust stabilisation against such perturbations and the consequently developed design method, the $\mathcal{H}_\infty$ loop-shaping design procedure (LSDP) ([100, 101]), could relax the restrictions on the number of right-half plane poles and produce no pole-zero cancellations between the nominal model and controller designed. This method does not require an iterative procedure to obtain an optimal solution and thus raises the computational efficiency. Furthermore, the $\mathcal{H}_\infty$ LSDP inherits classical loop-shaping design ideas so that practising control engineers would feel more comfortable to use it.

5.1 Robust Stabilisation Against Normalised Coprime Factor Perturbations

Matrices $(\tilde{M},\ \tilde{N}) \in \mathcal{H}_\infty^+$, where $\mathcal{H}_\infty^+$ denotes the space of functions with no poles in the closed right-half complex plane, constitute a left coprime factorisation of a given plant model G if and only if

(i) $\tilde{M}$ is square, and $\det(\tilde{M}) \neq 0$.
(ii) the plant model is given by

$$G = \tilde{M}^{-1}\tilde{N} \tag{5.1}$$

(iii) There exists $(\tilde{V},\ \tilde{U}) \in \mathcal{H}_\infty^+$ such that

$$\tilde{M}\tilde{V} \ + \ \tilde{N}\tilde{U} \ = \ I \tag{5.2}$$

A left coprime factorisation of a plant model G as defined in (5.1) is *normalised* if and only if

$$\tilde{N}\tilde{N}^- \ + \ \tilde{M}\tilde{M}^- = \ I, \quad \forall s \tag{5.3}$$

where $\tilde{N}^-(s) = \tilde{N}^T(-s)$, *etc.*

For a minimal realisation of $G(s)$,

$$\begin{aligned} G(s) &= D + C(sI - A)^{-1}B \\ &=: \left[\begin{array}{c|c} A & B \\ \hline C & D \end{array}\right] \end{aligned} \tag{5.4}$$

A state-space construction for the normalised left coprime factorisation can be obtained in terms of the solution to the generalised filter algebraic Riccati equation

$$\begin{aligned} (A - BD^TR^{-1}C)Z + Z(A - BD^TR^{-1}C)^T - ZC^TR^{-1}CZ \\ + B(I - D^TR^{-1}D)B^T = 0 \end{aligned} \tag{5.5}$$

where $R \ := \ I + DD^T$ and $Z \geq 0$ is the unique stabilising solution. If $H = -(ZC^T + BD^T)R^{-1}$, then

$$[\tilde{N}\ \tilde{M}] := \left[\begin{array}{c|cc} A + HC & B + HD & H \\ \hline R^{-1/2}C & R^{-1/2}D & R^{-1/2} \end{array}\right] \tag{5.6}$$

is a normalised left coprime factorisation of G such that $G = \tilde{M}^{-1}\tilde{N}$.

The normalised right coprime factorisation can be defined similarly and a state-space representation can be similarly obtained in terms of the solution to the generalised control algebraic Riccati equation [101],

$$\begin{aligned} (A - BS^{-1}D^TC)^TX + X(A - BS^{-1}D^TC) - XBS^{-1}B^TX \\ + C^T(I - DS^{-1}D^T)C = 0 \end{aligned} \tag{5.7}$$

where $S := I + D^T D$ and $X \geq 0$ is the unique stabilising solution.
A perturbed plant transfer function can be described by

$$G_\Delta = (\tilde{M} + \Delta_{\tilde{M}})^{-1}(\tilde{N} + \Delta_{\tilde{N}})$$

where $(\Delta_{\tilde{M}}, \Delta_{\tilde{N}})$ are unknown but stable transfer functions that represent the uncertainty (perturbation) in the nominal plant model. The design objective of robust stabilisation is to stabilise not only the nominal model G, but the family of perturbed plants defined by

$$\mathcal{G}_\epsilon = \{(\tilde{M} + \Delta_{\tilde{M}})^{-1}(\tilde{N} + \Delta_{\tilde{N}}) : \|[\Delta_{\tilde{M}}, \Delta_{\tilde{N}}]\|_\infty < \epsilon\}$$

where $\epsilon > 0$ is the stability margin. Using a feedback controller K as shown schematically in Figure 5.1 and the Small-Gain Theorem, the feedback system $(\tilde{M}, \tilde{N}, K, \epsilon)$ is robustly stable if and only if (G, K) is internally stable and

$$\left\| \begin{bmatrix} K(I - GK)^{-1}\tilde{M}^{-1} \\ (I - GK)^{-1}\tilde{M}^{-1} \end{bmatrix} \right\|_\infty \leq \epsilon^{-1}$$

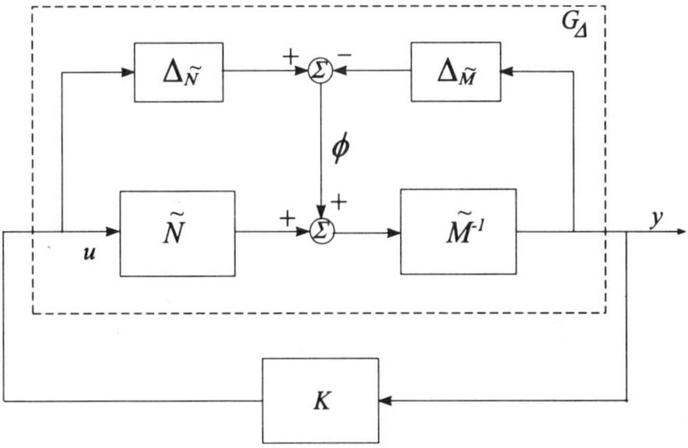

Fig. 5.1. Robust stabilisation with regard to coprime factor uncertainty

To maximise the robust stability of the closed-loop system given in Figure 5.1, one must minimise

$$\gamma := \left\| \begin{bmatrix} K \\ I \end{bmatrix} (I - GK)^{-1}\tilde{M}^{-1} \right\|_\infty$$

The lowest achievable value of γ for all stabilising controllers K is

$$\gamma_o = \inf_{K \ stabilizing} \left\| \begin{bmatrix} K \\ I \end{bmatrix} (I - GK)^{-1}\tilde{M}^{-1} \right\|_\infty \tag{5.8}$$

and is given in [101] by

$$\gamma_o = (1 - \|[\tilde{N} \;\; \tilde{M}]\|_H^2)^{-1/2} \tag{5.9}$$

where $\|\cdot\|_H$ denotes the Hankel norm. From [101]

$$\|[\tilde{N} \;\; \tilde{M}]\|_H^2 = \lambda_{max}(ZX(I+ZX)^{-1}) \tag{5.10}$$

where $\lambda_{\max}(\cdot)$ represents the maximum eigenvalue, hence from (5.9),

$$\gamma_o = (1 + \lambda_{\max}(ZX))^{1/2} \tag{5.11}$$

From [101], all controllers optimising γ are given by $K = UV^{-1}$, where U and V are stable and are right coprime factorisations of K, and where

$$\left\| \begin{bmatrix} -\tilde{N}^* \\ \tilde{M}^* \end{bmatrix} + \begin{bmatrix} U \\ V \end{bmatrix} \right\|_\infty = \|[\tilde{N} \;\; \tilde{M}]\|_H$$

This is a Hankel approximation problem and can be solved using an algorithm developed by Glover [46].

A controller that achieves a $\gamma > \gamma_0$ is given in [101] by

$$K := \left[\begin{array}{c|c} A + BF + \gamma^2(L^T)^{-1}ZC^T(C+DF) & \gamma^2(L^T)^{-1}ZC^T \\ \hline B^TX & -D^T \end{array}\right] \tag{5.12}$$

where $F = -S^{-1}(D^TC + B^TX)$ and $L = (1-\gamma^2)I + XZ$.

However, if $\gamma = \gamma_o$, then $L = XZ - \lambda_{\max}(XZ)I$, which is singular, and thus (5.12) cannot be implemented. This problem can be resolved using the descriptor system [134, 135]. A controller that achieves a $\gamma \geq \gamma_o$ can be given in the descriptor form by

$$K := \left[\begin{array}{c|c} -L^Ts + L^T(A+BF) + \gamma^2 ZC^T(C+DF) & \gamma^2 ZC^T \\ \hline B^TX & -D^T \end{array}\right] \tag{5.13}$$

5.2 Loop-shaping Design Procedures

In the classical control systems design with single-input-single-output (SISO) systems, it is a well-known and practically effective technique to use a compensator to alter the frequency response (the Bode diagram) of the open-loop transfer function so that the unity feedback system will achieve stability, good performance and certain robustness. Indeed, these performance requirements discussed in Chapter 3 can be converted into corresponding frequency requirements on the open-loop system. For instance, in order to achieve good tracking, it is required that $\|(I+GK)^{-1}\|_\infty$ should be small. That is $\overline{\sigma}((I+GK)^{-1}(j\omega)) \ll 1$, for ω over a low-frequency range, since the signals to be tracked are usually of low frequency. This in turn implies that

$\underline{\sigma}(GK(j\omega)) \gg 1$, for ω over that frequency range. Similar deductions can be applied to other design-performance specifications.

However, a direct extension of the above method into multivariable systems is difficult not only because multi-input and multi-output (MIMO) are involved but also due to the lack of phase information in MIMO cases that makes it impossible to predict the stability of the closed-loop system formed by the unity feedback. However, based on the robust stabilisation against perturbations on normalised coprime factorisations, a design method, known as the $\mathcal{H}_\infty$ loop-shaping design procedure (LSDP), has been developed [100]. The LSDP method augments the plant with appropriately chosen weights so that the frequency response of the open-loop system (the weighted plant) is reshaped in order to meet the closed-loop performance requirements. Then a robust controller is synthesised to meet the stability.

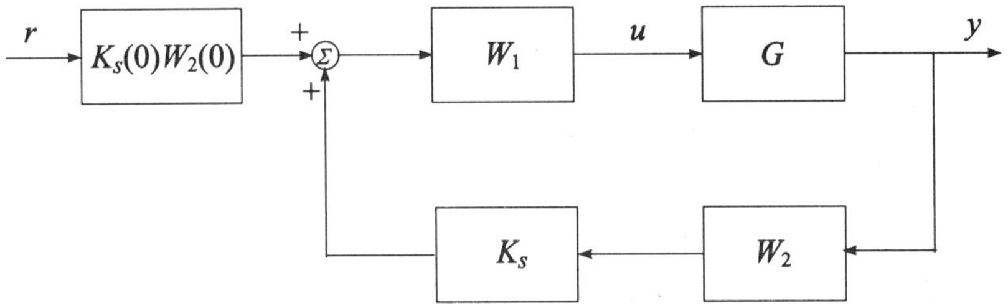

Fig. 5.2. One-degree-of-freedom LSDP configuration

This loop-shaping design procedure can be carried out in the following steps.

(i) Using a precompensator, W_1, and/or a postcompensator, W_2, as depicted in Figure 5.2, the singular values of the nominal system G are modified to give a desired loop shape. Usually, the least singular value of the weighted system should be made large over the low-frequency range to achieve good performance such as the tracking, the largest singular value is small over the high-frequency range to deal with unmodelled dynamics, and the bandwidth affects the system response speed, while the slope of the singular values near the bandwidth frequency should not be too steep. The nominal system and weighting functions W_1 and W_2 are combined to form the shaped system, G_s, where $G_s = W_2 G W_1$. It is assumed that W_1 and W_2 are such that G_s contains no hidden unstable modes.

(ii) A feedback controller, K_s, is synthesised that robustly stabilises the normalized left coprime factorisation of G_s, with a stability margin ϵ. It can be shown [100] that if ϵ is not less than 0.2, the frequency response of $K_s W_2 G W_1$ will be similar to that of $W_2 G W_1$. On the other hand, if the achievable ϵ is too large, this would probably indicate an overdesigned case in respect of the robustness, which means that the performance of the system may possibly be improved by using a larger γ in computing K_s.

(iii) The final feedback controller, K_{final}, is then constructed by combining the $\mathcal{H}_\infty$ controller K_s, with the weighting functions W_1 and W_2 such that

$$K_{final} = W_1 K_s W_2$$

For a tracking problem, the reference signal generally comes between K_s and W_1. In order to reduce the steady-state tracking error, a constant gain $K_s(0)W_2(0)$ is placed on the feedforward path, where

$$K_s(0)W_2(0) = \lim_{s \to 0} K_s(s)W_2(s)$$

The closed-loop transfer function between the reference r and the plant output y then becomes

$$Y(s) = [I - G(s)K_{final}(s)]^{-1} G(s)W_1(s)K_s(0)W_2(0)R(s)$$

The above design procedure can be developed further into a two-degree-of-freedom (2DOF) scheme as shown in Figure 5.3.

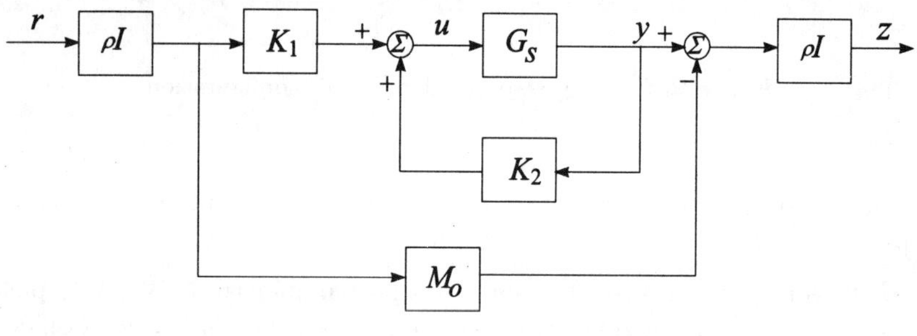

Fig. 5.3. Two-degree-of-freedom LSDP

The philosophy of the 2DOF scheme is to use the feedback controller $K_2(s)$ to meet the requirements of internal and robust stability, disturbance rejection, measurement noise attenuation, and sensitivity minimisation. The precompensator K_1 is applied to the reference signal, which optimises the response of the overall system to the command input such that the output of the system would be "near" to that of a chosen ideal system M_o. The feedforward compensator K_1 depends on design objectives and can be synthesised together with the feedback controller in a single step via the $\mathcal{H}_\infty$ LSDP [67]. In the 2DOF LSDP, the postcompensator W_2 is usually chosen as an identity matrix. The constant ρ is a scaling parameter used to emphasise the model-matching part in the design.

5.3 Formulae for the Discrete-time Case

In this section, the formulae for the (1-block) $\mathcal{H}_\infty$ LSDP controller in the discrete-time case will be introduced ([58]).

5.3.1 Normalised Coprime Factorisation of Discrete-time Plant

Let $G(z)$ be a minimal realisation, discrete-time model of a plant,

$$\begin{aligned} G(z) &= D + C(zI - A)^{-1}B \\ &:= \left[\begin{array}{c|c} A & B \\ \hline C & D \end{array}\right] \end{aligned}$$

with $A : n \times n$, $B : n \times m$, $C : p \times n$, and $D : p \times m$.

Matrices $(\tilde{M}(z),\ \tilde{N}(z)) \in \mathcal{H}_\infty^+$, where $\mathcal{H}_\infty^+$ denotes the space of functions with all poles in the open unit disc of the complex plane, constitute a left coprime factorisation of $G(z)$ if and only if

(i) $\tilde{M}$ is square, and $\det(\tilde{M}) \neq 0$.
(ii) the plant model is given by

$$G = \tilde{M}^{-1}\tilde{N} \tag{5.14}$$

(iii) There exists $(\tilde{V},\ \tilde{U}) \in \mathcal{H}_\infty^+$ such that

$$\tilde{M}\tilde{V} \ + \ \tilde{N}\tilde{U} \ = \ I_p \tag{5.15}$$

A left coprime factorisation of G as defined in (5.14) is *normalised* if and only if

$$\tilde{N}(z)\tilde{N}^T(\frac{1}{z}) \ + \ \tilde{M}(z)\tilde{M}^T(\frac{1}{z}) \ = \ I_p \tag{5.16}$$

The concept of right coprime factorisation and normalised right coprime factorisation can be introduced similarly. However, the work presented here will follow the (normalised) left coprime factorisation, although all results concerning the (normalised) right coprime factorisation can be derived similarly.

State-space constructions for the normalised coprime factorisations can be obtained in terms of the solutions to the following two *discrete algebraic Riccati equations(DARE)*,

$$\Phi^T P \Phi - P - \Phi^T P B Z_1 {Z_1}^T B^T P \Phi + C^T {R_1}^{-1} C = 0 \tag{5.17}$$

and

$$\Phi Q \Phi^T - Q - \Phi Q C^T Z_2^T Z_2 C Q \Phi^T + B {R_2}^{-1} B^T = 0 \tag{5.18}$$

where $R_1 = I_p + DD^T$, $R_2 = I_m + D^TD$, $\Phi = A - BR_2^{-1}D^TC$, $Z_1Z_1^T = (R_2 + B^TPB)^{-1}$, $Z_2^TZ_2 = (R_1 + CQC^T)^{-1}$. And, $P \geq 0$ and $Q \geq 0$ are the unique stabilising solutions, respectively.

Without loss of generality, we may assume that both Z_1 and Z_2 are square matrices, and $Z_1 = Z_1^T$, $Z_2 = Z_2^T$.

Further, by defining $H = -(AQC^T + BD^T)Z_2^TZ_2$ and $F = -Z_1Z_1^T(B^TPA + D^TC)$, then

$$[\tilde{N}\ \tilde{M}] := \left[\begin{array}{c|cc} A+HC & B+HD & H \\ \hline Z_2C & Z_2D & Z_2 \end{array}\right] \tag{5.19}$$

and

$$\begin{bmatrix} N \\ M \end{bmatrix} := \left[\begin{array}{c|c} A+BF & BZ_1 \\ \hline C+DF & DZ_1 \\ F & Z_1 \end{array}\right] \tag{5.20}$$

are the normalised left, and right, coprime factorisations of G, correspondingly.

5.3.2 Robust Controller Formulae

As in the continuous-time case, the discrete-time $\mathcal{H}_\infty$ loop-shaping design procedure is based on the construction of a robust stabilising controller against the perturbations on the coprime factors. The same form of cost function for robust stabilisation will be derived, and the optimal achievable γ is given by

$$\gamma_o = (1 + \lambda_{max}(QP))^{1/2} \tag{5.21}$$

where Q and P are the solutions to (5.17) and (5.18), respectively.

For a given $\gamma > \gamma_o$, a suboptimal $\mathcal{H}_\infty$ LSDP controller K in the discrete-time case can be recast as a standard $\mathcal{H}_\infty$ suboptimal control problem as was discussed in Section 4.4. The generalised (interconnected) system in this case is

$$\begin{aligned} P(z) &= \begin{bmatrix} P_{11}(z) & P_{12}(z) \\ P_{21}(z) & P_{22}(z) \end{bmatrix} \\ &= \left[\begin{array}{c|c} 0 & I_m \\ \tilde{M}^{-1} & G \\ \hline \tilde{M}^{-1} & G \end{array}\right] \\ &= \left[\begin{array}{c|cc} A & -HZ_2^{-1} & B \\ \hline 0 & 0 & I_m \\ C & Z_2^{-1} & D \\ C & Z_2^{-1} & D \end{array}\right] \end{aligned} \tag{5.22}$$

Following the solution procedure given in Section 4.4, one more DARE needs to be solved in order to compute the required controller. In general $\mathcal{H}_\infty$

suboptimal problems, two more algebraic Riccati equations are to be solved. Here, however, due to the structure of $P(z)$ in (5.22), it can be shown that the solution to one of the DARE is always zero. The third DARE is the following

$$A^T X_\infty A - X_\infty - \tilde{F}^T (R + \begin{bmatrix} -Z_2{}^{-1} H^T \\ R_2{}^{-1/2} B^T \end{bmatrix} X_\infty \\ \begin{bmatrix} -HZ_2{}^{-1} & BR_2{}^{-1/2} \end{bmatrix}) \tilde{F} + C^T C = 0 \qquad (5.23)$$

where

$$\tilde{F} = -(R + \begin{bmatrix} -Z_2{}^{-1} H^T \\ R_2{}^{-1/2} B^T \end{bmatrix} X_\infty \begin{bmatrix} -HZ_2{}^{-1} & BR_2{}^{-1/2} \end{bmatrix})^{-1} \\ (\begin{bmatrix} -Z_2{}^{-1} C \\ D^T R_1{}^{-1/2} C \end{bmatrix} + \begin{bmatrix} -Z_2{}^{-1} H^T \\ R_2{}^{-1/2} B^T \end{bmatrix} X_\infty A)$$

and

$$R = \begin{bmatrix} Z_2{}^{-2} - \gamma^2 I_p & Z_2{}^{-1} R_1{}^{-1/2} D \\ D^T R_1{}^{-1/2} Z_2{}^{-1} & I_m \end{bmatrix}$$

Further, by defining $\tilde{F} = \begin{bmatrix} F_1 \\ F_2 \end{bmatrix}$, where, $F_1 : p \times n$ and $F_2 : m \times n$, a suboptimal $\mathcal{H}_\infty$ *discrete-time LSDP(DLSDP)* controller K can be constructed as

$$K(z) = \left[\begin{array}{c|c} A_K & B_K \\ \hline C_K & D_K \end{array}\right]$$

where

$$\begin{aligned} A_K &= \hat{A}_K - \hat{B}_K D (I + \hat{D}_K D)^{-1} \hat{C}_K \\ B_K &= \hat{B}_K (I + D\hat{D}_K)^{-1} \\ C_K &= (I + \hat{D}_K D)^{-1} \hat{C}_K \\ D_K &= \hat{D}_K (I + D\hat{D}_K)^{-1} \end{aligned} \qquad (5.24)$$

with

$$\begin{aligned} \hat{D}_K &= -(R_2 + B^T X_\infty B)^{-1} (D^T - B^T X_\infty H) \\ \hat{B}_K &= -H + B\hat{D}_K \\ \hat{C}_K &= R_2^{-1/2} F_2 - \hat{D}_K (C + Z_2{}^{-1} F_1) \\ \hat{A}_K &= A + HC + B\hat{C}_K \end{aligned} \qquad (5.25)$$

5.3.3 The Strictly Proper Case

It may be appropriate to say that most (formulated) plants considered in the practical, discrete-time control-systems design are, in using the $\mathcal{H}_\infty$ optimisation approach in particular, strictly proper, *i.e.* $D = 0$. This is not only because most physical plants in industrial studies are strictly proper, as in the continuous-time case, but also because the $\mathcal{H}_\infty$ controllers designed tend to be proper due to the "flatness" of the optimality sought in the synthesis. Hence, when the (formulated) plant is just proper, it is possible to encounter the problem of an algebraic loop in the implementation of the resultant controller.

When the plant under consideration is strictly proper, all the computations and formulae described in Section 5.3.2 will be significantly simpler. The two DAREs (5.17) and (5.18) become

$$A^TPA - P - A^TPBZ_1{Z_1}^TB^TPA + C^TC = 0 \tag{5.26}$$

and

$$AQA^T - Q - AQC^TZ_2^TZ_2CQA^T + BB^T = 0 \tag{5.27}$$

where $Z_1{Z_1}^T = (I_m + B^TPB)^{-1}$, ${Z_2}^TZ_2 = (I_p + CQC^T)^{-1}$.

The third DARE (5.23) is now the following

$$\begin{aligned} A^TX_\infty A - X_\infty - \tilde{F}^T(R + \begin{bmatrix} -{Z_2}^{-1}H^T \\ B^T \end{bmatrix} X_\infty \\ [-H{Z_2}^{-1}\ B])\tilde{F} + C^TC = 0 \end{aligned} \tag{5.28}$$

where

$$\begin{aligned} \tilde{F} = -(R + \begin{bmatrix} -{Z_2}^{-1}H^T \\ B^T \end{bmatrix} X_\infty\, [-H{Z_2}^{-1}\ B])^{-1} \\ (\begin{bmatrix} -{Z_2}^{-1}C \\ 0 \end{bmatrix} + \begin{bmatrix} -{Z_2}^{-1}H^T \\ B^T \end{bmatrix} X_\infty A) \end{aligned}$$

and

$$\begin{aligned} R &= \begin{bmatrix} {Z_2}^{-2} - \gamma^2 I_p & 0 \\ 0 & I_m \end{bmatrix} \\ H &= -AQC^T{Z_2}^TZ_2 \end{aligned}$$

Further, by defining $\tilde{F} = \begin{bmatrix} F_1 \\ F_2 \end{bmatrix}$, where, $F_1 : p \times n$ and $F_2 : m \times n$, the suboptimal $\mathcal{H}_\infty$ DLSDP controller K in the case of a strictly proper G can be constructed as

$$K(z) = \left[\begin{array}{c|c} A_K & B_K \\ \hline C_K & D_K \end{array}\right]$$

where

$$\begin{aligned} D_K &= (I_m + B^TX_\infty B)^{-1}B^TX_\infty H \\ B_K &= -H + BD_K \\ C_K &= F_2 - D_K(C + {Z_2}^{-1}F_1) \\ A_K &= A + HC + BC_K \end{aligned} \tag{5.29}$$

5.3.4 On the Three DARE Solutions

As discussed above, the discrete-time $\mathcal{H}_\infty$ LSDP suboptimal controller formulae require the solutions to the three discrete-time algebraic Riccati equations, (5.17), (5.18) and (5.23), or (5.26), (5.27) and (5.28) in the strictly proper case. In this subsection, we will show that there is a relation between these three solutions, namely the solution X_∞ to the third DARE can be calculated directly from the first two solutions P and Q. This fact is important and useful, especially in the numerical implementation of the DLSDP routines.

We start with a general DARE, hence the notations are not related to those defined earlier in the chapter,

$$F^TXF - X - F^TXG_1(G_2 + G_1^TXG_1)^{-1}G_1^TXF + C^TC = 0 \qquad (5.30)$$

where $F,\ H,\ X \in \mathcal{R}^{n\times n}$, $G_1 \in \mathcal{R}^{n\times m}$, $G_2 \in \mathcal{R}^{m\times m}$, and $G_2 = G_2^T > 0$. We assume that $(F,\ G_1)$ is a stabilisable pair and that $(F,\ C)$ is a detectable pair. We also define $G = G_1G_2^{-1}G_1^T$.

It is well known ([124]) that solutions to DARE (5.30) are closely linked with a matrix pencil pair $(M,\ L)$, where

$$M = \begin{bmatrix} F & 0 \\ -H & I \end{bmatrix} \qquad (5.31)$$

$$L = \begin{bmatrix} I & G \\ 0 & F^T \end{bmatrix}$$

It also can be shown that if there exist $n \times n$ matrices S, U_1 and U_2, with U_1 invertible, such that

$$M\begin{bmatrix} U_1 \\ U_2 \end{bmatrix} = L\begin{bmatrix} U_1 \\ U_2 \end{bmatrix} S \qquad (5.32)$$

then, $X = U_2U_1^{-1}$ is a solution to (5.30). Further, the matrix $F - G_1(G_2 + G_1^TXG_1)^{-1}G_1^TXF$ shares the same spectrum as S. Hence, if S is stable, *i.e.* all the eigenvalues are within the open unit disc, $F - G_1(G_2 + G_1^TXG_1)^{-1}G_1^TXF$ is also stable. Such an X is non-negative definite and unique, and is called the stabilising solution to (5.30).

Under the above assumptions on (5.30), it was shown in [124] that none of the generalised eigenvalues of the pair $(M,\ L)$ lies on the unit circle, and if $\lambda \neq 0$ is a generalised eigenvalue of the pair, then $1/\lambda$ is also a generalised eigenvalue of the same multiplicity. In other words, the stable spectrum, consisting of n generalised eigenvalues lying in the open unit disc, is unique. Therefore, if there exists another triple $(V_1,\ V_2,\ T)$ satisfying (5.32), with V_1 being invertible and T stable, then there must exist an invertible R such that $T = R^{-1}SR$. Consequently,

$$\begin{bmatrix} U_1 \\ U_2 \end{bmatrix} = \begin{bmatrix} V_1 \\ V_2 \end{bmatrix} R^{-1} \qquad (5.33)$$

The solution, of course, remains the same, since $X = V_2V_1^{-1} = (U_2R)(U_1R)^{-1} = U_2U_1^{-1}$.

In the present case, we can accordingly define the three matrix pencils as

$$\begin{aligned} M_P &= \begin{bmatrix} \Phi & 0 \\ -C^TR_1^{-1}C & I \end{bmatrix} \\ L_P &= \begin{bmatrix} I & BR_2^{-1}B^T \\ 0 & \Phi^T \end{bmatrix} \end{aligned} \tag{5.34}$$

$$\begin{aligned} M_Q &= \begin{bmatrix} \Phi^T & 0 \\ -BR_2^{-1}B^T & I \end{bmatrix} \\ L_Q &= \begin{bmatrix} I & C^TR_1^{-1}C \\ 0 & \Phi \end{bmatrix} \end{aligned} \tag{5.35}$$

$$\begin{aligned} M_X &= \begin{bmatrix} A - \begin{bmatrix} -HZ_2^{-1} & BR_2^{-1/2} \end{bmatrix} R^{-1} \begin{bmatrix} Z_2^{-1}C \\ D^TR_1^{-1/2}C \end{bmatrix} & 0 \\ -C^TC + \begin{bmatrix} C^TZ_2^{-1} & C^TR_1^{-1/2}D \end{bmatrix} R^{-1} \begin{bmatrix} Z_2^{-1}C \\ D^TR_1^{-1/2}C \end{bmatrix} & I \end{bmatrix} \\ L_X &= \begin{bmatrix} I & \begin{bmatrix} -HZ_2^{-1} & BR_2^{-1/2} \end{bmatrix} R^{-1} \begin{bmatrix} -Z_2^{-1}H^T \\ R_2^{-1/2}B^T \end{bmatrix} \\ 0 & A^T - \begin{bmatrix} C^TZ_2^{-1} & C^TR_1^{-1/2}D \end{bmatrix} R^{-1} \begin{bmatrix} -Z_2^{-1}H^T \\ R_2^{-1/2}B^T \end{bmatrix} \end{bmatrix} \end{aligned} \tag{5.36}$$

With all the above properties of the DAREs and the notations, the following theorem may be proved ([58]).

Theorem 5.1. *Let P, Q and X_∞ be the stabilising solutions to the DAREs (5.17), (5.18) and (5.23), (or, (5.26), (5.27) and (5.28) when G is strictly proper), respectively, the following identity holds*

$$\begin{aligned} X_\infty &= P\left[\left(1-\gamma^{-2}\right)I_n - \gamma^{-2}QP\right]^{-1} \\ &= \gamma^2P\left[\gamma^2I_n - (I_n+QP)\right]^{-1} \end{aligned} \tag{5.37}$$

■

A similar result can be found in [155] for the relationship between three discrete-time algebraic Riccati equations arising in the general $\mathcal{H}_\infty$ suboptimal design. The results concerning the discrete-time loop-shaping design procedure are with three different DAREs and the conclusion is of a slightly different form.

5.4 A Mixed Optimisation Design Method with LSDP

It is well known that control-system design problems can be naturally formulated as constrained optimisation problems, the solutions of which will characterise acceptable designs. The numerical optimisation approach to controller design can directly tackle design specifications in both the frequency domain and the time domain. The optimisation problems derived, however, are usually very complicated with many unknowns, many nonlinearities, many constraints, and in most cases, they are multiobjective with several conflicting design aims that need to be simultaneously achieved. It is also known that a direct parameterisation of the controller will increase the complexity of the optimisation problem.

On the other hand, the $\mathcal{H}_\infty$ loop-shaping design procedure (LSDP) discussed in this chapter is an efficient, robust design method that just depends on appropriate choice of weighting functions to shape the open-loop system. In general, the weighting functions are of low orders. This indicates if we take the weighting functions as design parameters it may well combine the advantages of both the numerical optimisation approach and the $\mathcal{H}_\infty$ LSDP in terms of flexible formulation of design objectives, mature numerical algorithms, tractable optimisation problems, guaranteed stability and a certain degree of robustness. Such a design method is proposed in [161]. The numerical optimisation component of the method is based on the *method of inequalities(MOI)*.

Performance specifications for control-systems design are frequently, and more naturally, given in terms of algebraic or functional inequalities, rather than in the minimisation of some objective function. For example, the system may be required to have a rise time of less than one second, a settling time of less than five seconds and an overshoot of less than 10%. In such cases, it is obviously more logical and convenient if the design problem is expressed explicitly in terms of such inequalities. The method of inequalities [169] is a computer-aided multiobjective design approach, where the desired performance is represented by such a set of algebraic inequalities and where the aim of the design is to simultaneously satisfy these inequalities. The design problem is expressed as

$$\phi_i(p) \leq \epsilon_i, \quad \text{for} \quad i = 1, \cdots, n \tag{5.38}$$

where ϵ_i are real numbers, $p \in P$ is a real vector $(p_1, p_2, \cdots, p_q)$ chosen from a given set P, and ϕ_i are real functions of p. The functions ϕ_i are performance indices, the components of p represent the design parameters, and ϵ_i are chosen by the designer and represent the largest tolerable values of ϕ_i. The aim is the satisfaction of the set of inequalities so that an acceptable design p is reached.

For control-systems design, the functions $\phi_i(p)$ may be functionals of the system step response, for example, the rise-time, overshoot or the integral absolute tracking error, or functionals of the frequency responses, such as the bandwidth. They can also represent measures of the system stability, such as the maximum real part of the closed-loop poles. Additional inequalities that arise from the physical constraints of the system can also be included, to restrict, for example, the maximum control signal. In practice, the constraints on the design parameters p that define the set P are also included in the inequality set, *e.g.* to constrain the possible values of some of the design parameters or to limit the search to stable controllers only.

Each inequality $\phi_i(p) \leq \epsilon_i$ of the set of inequalities (5.38) defines a set $\mathcal{S}_i$ of points in the q-dimensional space $\mathcal{R}^q$ and the coordinates of this space are $p_1, p_2, \cdots, p_q$, so

$$\mathcal{S}_i = \{p : \ \phi_i(p) \leq \epsilon_i\} \tag{5.39}$$

The boundary of this set is defined by $\phi_i(p) = \epsilon_i$. A point $p \in \mathcal{R}^q$ is a solution to the set of inequalities (5.38) if and only if it lies inside every set $\mathcal{S}_i$, $i = 1, 2, \cdots, n$, and hence inside the set $\mathcal{S}$ that denotes the intersection of all the sets $\mathcal{S}_i$

$$\mathcal{S} = \bigcap_{i=1}^{n} \mathcal{S}_i \tag{5.40}$$

$\mathcal{S}$ is called the admissible set and any point p in $\mathcal{S}$ is called an admissible point denoted p_s.

The objective is thus to find a point p such that $p \in \mathcal{S}$. Such a point satisfies the set of inequalities (5.38) and is said to be a solution. In general, a point p_s is not unique unless the set $\mathcal{S}$ is a singleton in the space $\mathcal{R}^q$. In some cases, there is no solution to the problem, *i.e.* $\mathcal{S}$ is an empty set. It is then necessary to relax the boundaries of some of the inequalities, *i.e.* increase some of the numbers ϵ_i, until an admissible point p_s is found.

The actual solution to the set of inequalities (5.38) may be obtained by means of numerical search algorithms, such as the moving-boundaries process (MBP); details of the MBP may be found in [159, 169]. The procedure for obtaining a solution is interactive, in that it requires supervision and intervention from the designer. The designer needs to choose the configuration of the design, which determines the dimension of the design parameter vector p, and initial values for the design parameters. The progress of the search algorithm should be monitored and, if a solution is not found, the designer may either change the starting point, relax some of the desired bounds ϵ_i, or change the design configuration. Alternatively, if a solution is found easily, to improve the quality of the design, the bounds could be tightened or additional design objectives could be included in (5.38). The design process is thus a two-way process, with the MOI providing information to the designer about conflicting requirements, and the designer making decisions about the "trade-offs" between design requirements based on this information as well as on the designer's knowledge, experience, and intuition about the particular problem. The designer can be supported in this role by various graphical displays [117] or expert systems [56].

A difficult problem in control-systems design using the MOI method is how to define the design parameter vector p. A straightforward idea is to directly parametrise the controller, *i.e.* let p be the system matrices of the controller, or the coefficients of the numerator and denominator polynomials of the controller. In doing so, the designer has to choose the order of the controller first. In general, the lower the dimension of the design vector p, the easier it is for numerical search algorithms to find a solution, if one exists. Choosing a low-order controller, say a PI controller $p_1 + \frac{p_2}{s}$, may reduce the dimension of p. However, it may not yield a solution and, in that case, a solution may exist with higher order controllers. A further limitation of using the MOI alone in the design is that an initial point that gives the stability of the closed-loop system must be located as a prerequisite to searching the parameter space in order to improve the index set of (5.38).

Two aspects of the $\mathcal{H}_\infty$ LSDP make it amenable to combine this approach with the MOI. First, unlike the standard $\mathcal{H}_\infty$-optimisation approaches, the $\mathcal{H}_\infty$-optimal controller for the weighted plant can be synthesised from the solutions of two algebraic Riccati equations (5.5) and (5.7) and does not require time-consuming γ-iteration. Second, in the LSDP described earlier in this chapter, the weighting functions are chosen by considering the open-loop response of the weighted plant, so effectively the weights W_1 and W_2 are the design parameters. This means that the design problem can be formulated as in the method of inequalities, with the parameters of the weighting functions used as the design parameters p to satisfy the set of closed-loop performance inequalities. Such an approach to the MOI overcomes the limitations of the MOI. The designer does not have to choose the order of the controller, but instead chooses the order of the weighting functions. With low-order weighting functions, high-order controllers can be synthesised that often lead to significantly better performance or robustness than if simple low-order controllers were used. Additionally, the problem of finding an initial point for system stability does not exist, because the stability of a closed-loop system is guaranteed through the solution to the robust stabilisation problem, provided that the weighting functions do not cause undesirable pole/zero cancellations.

The design problem is now stated as follows.

Problem: For the system of Figure 5.2, find a $(W_1,\ W_2)$ such that

$$\gamma_o(W_1, W_2) \le \epsilon_\gamma \tag{5.41}$$

and

$$\phi_i(W_1, W_2) \le \epsilon_i \quad \text{for} \quad i = 1, \cdots, n \tag{5.42}$$

where

$$\gamma_o = [1 + \lambda_{max}(ZX)]^{1/2}$$

with Z and X the solutions of the two AREs, (5.5) and (5.7), of the weighted plant.

In the above formulation, ϕ_i are functionals representing design objectives, ϵ_γ and ϵ_i are real numbers representing desired bounds on γ_o and ϕ_i, respectively, and (W_1, W_2) a pair of fixed order weighting functions with real parameters $w = (w_1, w_2, \cdots, w_q)$.

Consequently, a design procedure can be stated as follows.

Design Procedure: A design procedure to solve the above problem is

1. Define the plant G, and define the functionals ϕ_i.
2. Define the values of ϵ_γ and ϵ_i.
3. Define the form and order of the weighting functions W_1 and W_2. Bounds should be placed on the values of w_j to ensure that W_1 and W_2 are stable and minimum phase to prevent undesirable pole/zero cancellations. The order of the weighting functions, and hence the value of q, should be small initially.
4. Define initial values of w_j based on the open-loop frequency response of the plant.
5. Implement the MBP, or other appropriate algorithms, in conjunction with (5.12) or (5.13) to find a (W_1, W_2) that satisfies inequalities (5.41) and (5.42). If a solution is found, the design is satisfactory; otherwise, either increase the order of the weighting functions, or relax one or more of the desired bounds ϵ, or try again with different initial values of w.
6. With satisfactory weighting functions W_1 and W_2, a satisfactory feedback controller is obtained from (5.12) or (5.13).

This mixed optimisation design approach has been applied in some design cases. Examples of application and some extensions can be found in [115, 147, 161, 160, 162, 158, 163].

6

μ-Analysis and Synthesis

As discussed earlier in the book, the $\mathcal{H}_\infty$ optimisation approach may achieve robust stabilisation against unstructured system perturbations and nominal performance requirements. It is though possible that by applying appropriate weighting functions some robust performance requirements can be obtained. Satisfactory designs have been reported, in particular when using the $\mathcal{H}_\infty$ loop-shaping design methods. In order to achieve robust stability **and** robust performance, design methods based on the structured singular value μ can be used. In this chapter, we first show how a robust performance design specification for a control system with unstructred/structured perturbations can be transformed into a robust stabilisation problem with regard to structured perturbation. We then discuss how these design specifications are related to some indications of this new setting, followed by introduction of two iterative μ-synthesis methods, the D-K iteration and μ-K iteration methods.

We recall, as defined in Chapter 3, for $M \in \mathcal{C}^{n\times n}$ and a known structure of Δ (usually representing uncertainties),

$$\mathbf{\Delta} = \{\mathrm{diag}[\delta_1 I_{r_1}, \cdots, \delta_s I_{r_s}, \Delta_1, \cdots, \Delta_f] : \delta_i \in \mathcal{C}, \Delta_j \in \mathcal{C}^{m_j\times m_j}\} \tag{6.1}$$

where $\sum_{i=1}^{s} r_i + \sum_{j=1}^{f} m_j = n$ with n is the dimension of the block Δ, the structure singular value μ is defined by

$$\mu_\Delta^{-1}(M) := \min_{\Delta\in\mathbf{\Delta}} \{\overline{\sigma}(\Delta) : \det(I - M\Delta) = 0\} \tag{6.2}$$

If there is no $\Delta \in \mathbf{\Delta}$ such that $\det(I - M\Delta) = 0$, then $\mu_\Delta(M) := 0$.

Some important properties of μ have been listed in Chapter 3.

6.1 Consideration of Robust Performance

When M is an interconnected transfer function matrix $M(s)$ formed with respect to the uncertainty set $\mathbf{\Delta}$, the structured singular value of $M(s)$

$$\mu_\Delta(M(s)) := \sup_{\omega \in \mathcal{R}} \mu_\Delta(M(j\omega)) \tag{6.3}$$

indicates the robust stability of the perturbed system. Without loss of generality, we assume that the uncertainties have been normalised in the rest of this chapter, *i.e.* $\Delta \in \mathbf{B\Delta}$. From Theorem 3.4, the standard configuration Figure 6.1 is robustly stable if $M(s)$ is stable and $\mu_\Delta(M(s)) < 1$ (or, $\|M\|_\mu < 1$).

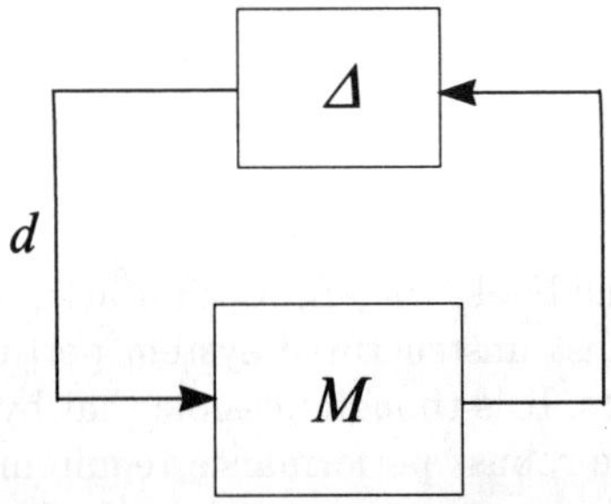

Fig. 6.1. Standard M-Δ configuration

As discussed in Chapter 3 in design of control systems, in addition to the stability the performance of the system must be taken into account. The designed system should perform well (for instance, good tracking) against exogenous disturbances, which requires a closed-loop structure. The feedback controller is usually designed based on a nominal plant model. In the case of the existence of plant perturbations, the closed-loop system may well possibly perform badly, even degrade to an unsatisfactory level. The *robust performance* requires that a designed control system maintains a satisfactory performance level even in the presence of plant dynamic uncertainties.

We now expand Figure 6.1 to include input and output signals, as depicted in Figure 6.2.

In Figure 6.2, w, z, v and d are usually vector signals. w denotes the exogenous input typically including command signals, disturbances, noises, *etc.*; z denotes the error output usually consisting of regulator output, tracking errors, filtered actuator signals, *etc.*; v and d are the input and output signals of the dynamic uncertainties. System-performance specifications can usually be interpreted as reduction of z with respect of w. With the assumption that w and z are both energy-bounded signals, the performance requirement is equivalent to the minimisation of the $\mathcal{H}_\infty$ norm of the transfer function from w to z. Let M be partitioned accordingly as

$$M = \begin{bmatrix} M_{11} & M_{12} \\ M_{21} & M_{22} \end{bmatrix}$$

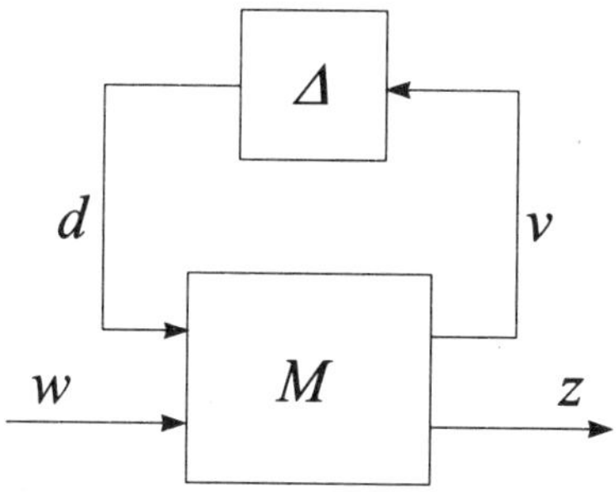

Fig. 6.2. Standard M-Δ configuration for RP analysis

it can be easily derived that

$$\begin{aligned} z &= \left[M_{22} + M_{21}\Delta(I - M_{11}\Delta)^{-1}M_{12}\right] w \\ &= F_u(M, \Delta)w \end{aligned} \tag{6.4}$$

Using normalisation, a satisfactory level of performance requirement can be set as

$$\|F_u(M, \Delta)\|_\infty < 1 \tag{6.5}$$

Equation (6.5) implies the stability of $F_u(M, \Delta)$ that denotes the robust stability with respect to the plant perturbations Δ.

Condition (6.5) is equivalent to the system loop in Figure 6.3 to be robustly stable with regard to a fictitious uncertainty block Δ_p. Δ_p is unstructured with appropriate dimensions, satisfies $\|\Delta_p\|_\infty \leq 1$ and is usually called the performance uncertainty block. The part enclosed in the dotted line is, of course, $F_u(M, \Delta)$.

For *robust performance*, (6.5) should hold for all $\Delta \in \mathbf{B\Delta}$. Based on Figure 6.3, the robust performance design as well as robust stabilisation against Δ, can be equivalently considered as a robust stabilisation problem in Figure 6.1 with the uncertainty block to be replaced by $\tilde{\Delta}$ where

$$\tilde{\Delta} \in \tilde{\mathbf{\Delta}} := \{\text{diag}\{\Delta,\ \Delta_p\} : \Delta \in \mathbf{B\Delta},\ \|\Delta_p\|_\infty \leq 1\}$$

This is thus a robust stabilisation problem with respect to a structured uncertainty $\tilde{\mathbf{\Delta}}$. And, we have the following indications.

- *Robust Performance (RP)* $\Longleftrightarrow \|M\|_\mu < 1$, for structured uncertainty $\tilde{\mathbf{\Delta}}$
- *Robust Stability (RS)* $\Longleftrightarrow \|M_{11}\|_\mu < 1$, for structured uncertainty $\mathbf{B\Delta}$
- *Nominal Performance (NP)* $\Longleftrightarrow \|M_{22}\|_\infty < 1$
- *Nominal Stability (NS)* $\Longleftrightarrow$ M is internally stable

Of course, if the uncertainty Δ is unstructured, then the robust stability requirement corresponds to $\|M_{11}\|_\infty < 1$.

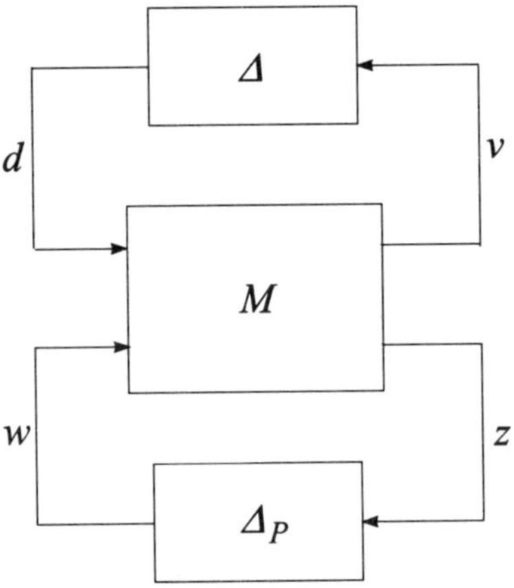

Fig. 6.3. Standard M-Δ configuration with Δ_p analysis

It is easy to see that, by setting $\Delta_p \equiv 0$, $\|M\|_\mu < 1$ implies $\|M_{11}\|_\mu < 1$. Hence, the former condition yields a ***robust stability and robust performance(RSRP)*** design.

6.2 μ-Synthesis: *D*-*K* Iteration Method

The transfer function matrix $M(s)$ in Figure 6.2 usually contains the controller K, hence a closed-loop interconnected matrix. For the purpose of controller design, we may rearrange the configuration and explicitly show the dependence of the closed-loop system on the controller K. That is, Figure 6.2 is reshown in Figure 6.4.

The nominal, open-loop interconnected transfer function matrix $P(s)$ in Figure 6.4 does not include the controller $K(s)$ nor any perturbations that are under consideration. $P(s)$ may be partitioned as

$$P(s) = \begin{bmatrix} P_{11} & P_{12} & P_{13} \\ P_{21} & P_{22} & P_{23} \\ P_{31} & P_{32} & P_{33} \end{bmatrix}$$

The signals y and u represent the feedback signals (measured output, tracking error, *etc.*, the input to the controller) and the control signal (the output from the controller), respectively.

The relationship between M in Figure 6.2 and P can be obtained by straightforward calculations as

$$\begin{aligned} M(P,K) &= F_l(P,K) \\ &= \begin{bmatrix} P_{11} & P_{12} \\ P_{21} & P_{22} \end{bmatrix} + \begin{bmatrix} P_{13} \\ P_{23} \end{bmatrix} K(I - P_{33}K)^{-1} \begin{bmatrix} P_{31} & P_{32} \end{bmatrix} \end{aligned} \quad (6.6)$$

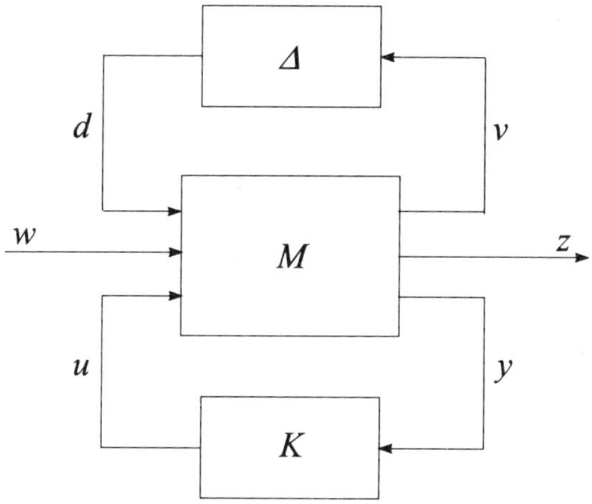

Fig. 6.4. Standard M-Δ configuration with K

where $M(s)$ is explicitly written as $M(P, K)$ to show that M is formed by P and K.

For the RSRP design, it is required to find a stabilising controller K such that

$$\sup_{\omega \in \mathcal{R}} \mu[M(P, K)(j\omega)] < 1 \tag{6.7}$$

where the subscript $\tilde{\Delta}$ has been suppressed for the sake of simplicity.

Or, for the "optimal" RSRP design, the objective is to solve for K

$$\inf_{K(s)} \sup_{\omega \in \mathcal{R}} \mu[M(P, K)(j\omega)] \tag{6.8}$$

An iterative method was proposed to solve (6.8) in [24]. The method is called the D-K iteration μ-synthesis method, and is based on solving the following optimisation problem, for a stabilising controller K and a diagonal constant scaling matrix D,

$$\inf_{K(s)} \sup_{\omega \in \mathcal{R}} \inf_{D \in \mathbf{D}} \overline{\sigma}[DM(P, K)D^{-1}(j\omega)] \tag{6.9}$$

where the scaling matrix set $\mathbf{D}$ is defined in Section 3.3. The justification for using (6.9) is obvious, from Theorem 3.7 and the discussion afterwards, with $\mu[M(P, K)]$ replaced by its upper bound $\inf_{D \in \mathbf{D}} \overline{\sigma}[DM(P, K)D^{-1}]$.

Corresponding to the case of (6.7), a stabilising controller is to be found such that

$$\sup_{\omega \in \mathcal{R}} \inf_{D \in \mathbf{D}} \overline{\sigma}[DM(P, K)D^{-1}(j\omega)] < 1 \tag{6.10}$$

The D-K iteration method is to alternately minimise (6.9), or to reduce the left-hand-side value of (6.10), for K and D in turn while keeping the other one fixed.

For a given matrix D, either constant or transfer function, (6.9) is a standard $\mathcal{H}_\infty$ optimisation problem

$$\inf_{K(s)} \|DM(P,K)D^{-1}\|_\infty \tag{6.11}$$

that can be further written as

$$\inf_K \|DF_l(P,K)D^{-1}\|_\infty = \inf_K \|F_l(\tilde{P},K)\|_\infty \tag{6.12}$$

with $\tilde{P} = \begin{bmatrix} D & 0 \\ 0 & I \end{bmatrix} P \begin{bmatrix} D^{-1} & 0 \\ 0 & I \end{bmatrix}$ compatible with the partition of P.

On the other hand, for a fixed $K(s)$, $\inf_{D\in\mathbf{D}} \overline{\sigma}[DM(P,K)D^{-1}(j\omega)]$ is a convex optimisation problem at each frequency ω. After the minimisation over a range of frequency of interest, the resultant diagonal (constant) matrices Ds can be approximated, via curve fitting, by a stable, minimum phase, rational transfer function matrix $D(s)$, which will be used in the next iteration for K.

The D-K iterative μ-synthesis algorithm is thus:

Step 1: Start with an initial guess for D, usually set $D = I$.

Step 2: Fix D and solve the $\mathcal{H}_\infty$-optimisation for K,

$$K = \arg\inf_K \|F_l(\tilde{P},K)\|_\infty$$

Step 3: Fix K and solve the following convex optimisation problem for D at each frequency over a selected frequency range,

$$D(j\omega) = \arg\inf_{D\in\mathbf{D}} \overline{\sigma}[DF_l(P,K)D^{-1}(j\omega)]$$

Step 4: Curve fit $D(j\omega)$ to get a stable, minimum-phase $D(s)$; goto Step 2 and repeat, until a prespecified convergence tolerance or (6.10) is achieved, or a prespecified maximum iteration number is reached.

Successful applications of the D-K iteration method have been reported. This method has also been applied in the case studies in Part II of this book. It should be noticed, however, that the algorithm may not converge in some cases. The resultant controller may be very conservative, particularly in the case of real and/or mixed perturbations, due to the possible gap between $\mu(M)$ and $\inf_{D\in\mathbf{D}} \overline{\sigma}[DMD^{-1}]$ in general cases, and due to the lack of powerful algorithms to compute the real and mixed μ values and to solve the optimisation problem for D. Another adverse effect in practical designs is that the order of resultant μ-controllers is usually very high, and hence controller-order reduction must be applied.

6.3 μ-Synthesis: μ-K Iteration Method

As discussed above, the structured singular value μ plays an important role in the robust stability and robust performance designs. The μ-design is to seek a stabilising controller that either minimises or reduces the μ value over a frequency range. There is another design method proposed in the literature that can be applied to find a μ-controller. That is the so-called μ-K iteration method ([89, 90, 127]).

It is shown in [64] that in many optimisation-based control-system designs, the resultant, optimal controller will make a "well-posed" cost function (*i.e.* satisfying certain assumptions ([64], Theorem 4.1)) *constant* in the frequency ω almost everywhere. This feature is obvious in the $\mathcal{H}_\infty$ optimisations and can also be observed in the μ-designs using the D-K iterations where the controller $K(s)$ obtained in the iteration gradually flattens the curve of $\mu(M)$. The μ-K iteration method is motivated by the above. In the μ-K iterations, the obtained μ curves are used as weighting functions on the $\mathcal{H}_\infty$ cost function $\|F_l(\tilde{P}, K)\|_\infty$ with the aim to suppress the peak values of the μ curves in the successive iterations. An algorithm to implement the μ-K iteration method is described below.

Step 1: Solve the $\mathcal{H}_\infty$-optimisation problem for K_0,

$$K_0 := \arg \inf_K \|F_l(\tilde{P}, K)\|_\infty$$

Step 2: Compute the μ curve corresponding to K_0 over a chosen frequency range,

$$\mu_0(j\omega) := \mu[F_l(P, K_0)(j\omega)]$$

Step 3: Normalise μ_0 by its maximum value, *i.e.*

$$\tilde{\mu}_0 := \frac{\mu_0}{\max_\omega \mu_0}$$

Step 4: Curve fit $\tilde{\mu}_0(j\omega)$ to get a stable, minimum-phase, rational $\tilde{\mu}_0(s)$.

Step 5: Solve for the $\mathcal{H}_\infty$ optimal controller $K_1(s)$,

$$K_1 := \arg \inf_K \|\tilde{\mu}_0 F_l(\tilde{P}, K)\|_\infty \tag{6.13}$$

goto Step 2, multiply the newly obtained μ curve function onto the previous cost function in (6.13) (*e.g.* $\|\tilde{\mu}_1 \tilde{\mu}_0 F_l(\tilde{P}, K)\|_\infty$); repeat until the μ curve is sufficiently flat or until a desired level of performance (measured by the peak value of μ) has been achieved.

7

Lower-order Controllers

There is a dilemma concerning design of control systems. Due to increasing demands on quality and productivity of industrial systems and with deeper understanding of these systems, mathematical models derived to represent the system dynamics are more complete, usually of multi-input-multi-output form, and are of high orders. Consequently, the controllers designed are complex. The order of such controllers designed using, for instance, the $\mathcal{H}_\infty$ optimisation approach or the μ-method, is higher than, or at least similar to, that of the plant. On the other hand, in the implementation of controllers, high-order controllers will lead to high cost, difficult commissioning, poor reliability and potential problems in maintenance. Lower-order controllers are always welcomed by practising control engineers. Hence, how to obtain a low-order controller for a high-order plant is an important and interesting task, and is the subject of the present chapter.

In general, there are three directions to obtain a lower-order controller for a relatively high-order plant, as depicted in Figure 7.1,

(1) plant model reduction followed by controller design;
(2) controller design followed by controller-order reduction; and,
(3) direct design of low-order controllers.

Approaches (1) and (2) are widely used and can be used together. When a controller is designed using a robust design method, Approach (1) would usually produce a stable closed loop, though the reduction of the plant order is likely to be limited. In Approach (2), there is freedom in choosing the final order of the controller, but the stability of the closed-loop system should always be verified.

The third approach usually would heavily depend on some properties of the plant, and require numerous computations. This approach will not be introduced in this book. Interested readers are referred to [10, 17, 55, 68, 71].

From the viewpoint of being a control system, model reductions (usually referred to the reduction of orders of the original plant models) and controller

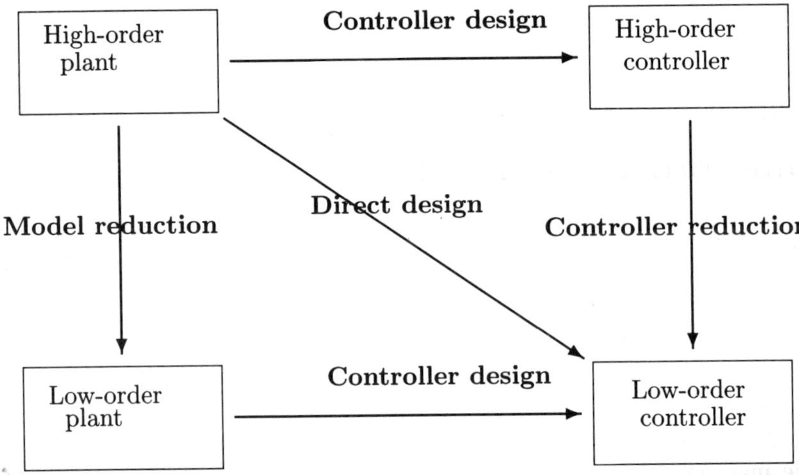

Fig. 7.1. Diagram for design of low-order controllers

reductions are similar. Indeed, most methods introduced in this chapter are applicable in both cases, though some methods may be more suitable for one of them. This will be discussed along with the introduction of methods.

In this book, only details of continuous-time case reduction will be discussed. Where the method is applicable in the discrete-time case as well references will be given. When we say order reductions of plants/systems, we actually refer to the order reduction of a model of the plant/system.

In the rest of the chapter, let the original system have a state-space representation

$$G(s) := C(sI - A)^{-1}B + D \tag{7.1}$$

$$=: \left[\begin{array}{c|c} A & B \\ \hline C & D \end{array}\right] \tag{7.2}$$

where $A : n \times n$, $B : n \times m$, $C : p \times n$, $D : p \times m$, and $[A, B, C]$ is assumed to be a minimal realisation. A reduced-order system $G_r(s)$ is represented by

$$G_r(s) := C_r(sI - A_r)^{-1}B_r + D_r \tag{7.3}$$

$$=: \left[\begin{array}{c|c} A_r & B_r \\ \hline C_r & D_r \end{array}\right] \tag{7.4}$$

with $A_r : r \times r$, $B_r : r \times m$, $C_r : p \times r$, $D_r : p \times m$, and $r < n$.

7.1 Absolute-error Approximation Methods

The aim of the following methods is to find a reduced-order system $G_r(s)$ such that some norm of the error system is small, *i.e.* to minimise

$$\|G(s) - G_r(s)\| \tag{7.5}$$

Three methods are to be introduced: Balanced Truncation, Singular Perturbation Approximation (Balanced Residualisation), and Hankel-Norm Approximation. These methods are for *stable* systems only. The case of unstable systems will be discussed at the end of this section.

7.1.1 Balanced Truncation Method

The general idea of truncation methods is to neglect those parts of the original system that are less observable or/and less controllable. Hopefully, this would lead to a system that is of lower order and retains the important dynamic behaviour of the original system. However, in some systems, a mode would be weakly observable but highly controllable, or *vice versa*. To delete such a mode may be inappropriate with regard to the whole characteristics of the system. Hence, in the balanced truncation method ([30, 29, 108, 126]), a state similarity transformation is applied first to "balance" the controllability and observability features of the system.

A stable system $G(s)$ is called *balanced* if the solutions P and Q to the following Lyapunov equations

$$AP + PA^T + BB^T = 0 \tag{7.6}$$
$$A^TQ + QA + C^TC = 0 \tag{7.7}$$

are such that $P = Q = \mathrm{diag}(\sigma_1, \sigma_2, \cdots, \sigma_n) := \Sigma$, with $\sigma_1 \geq \sigma_2 \geq \cdots \geq \sigma_n > 0$. P and Q are called the controllability gramian and observability gramian, respectively. When the system is balanced, both gramians are diagonal and equal. σ_i, $i = 1, \cdots, n$, is the ith Hankel singular value of the system.

For a general system not in the balanced realisation form, the following algorithm ([81]) can be applied.

Balanced Realisation Algorithm:

Step 1 : Calculate the gramians P and Q from (7.6) and (7.7), respectively
Step 2 : Calculate a Cholesky factor R of Q, *i.e.* $Q = R^TR$
Step 3 : Form a positive-definite matrix RPR^T and diagonalise it,

$$RPR^T = U\Sigma^2U^T$$

where U is an orthonormal matrix, $U^TU = I$, and $\Sigma = \mathrm{diag}(\sigma_1, \sigma_2, \cdots, \sigma_n)$, $\sigma_1 \geq \sigma_2 \geq \cdots \geq \sigma_n > 0$

Step 4 : Let $T = \Sigma^{-\frac{1}{2}}U^TR$. $[T, T^{-1}]$ is a required state similarity transformation (balancing transformation). That is, $[TAT^{-1}, TB, CT^{-1}]$ is a balanced realisation. ■

We now assume that the state-space model of the original system $G(s)$, $[A, B, C, D]$, is already in the balanced realisation form. Assume $\Sigma = \text{diag}(\Sigma_1, \Sigma_2)$, with $\Sigma_1 = \text{diag}(\sigma_1, \cdots, \sigma_r)$, $\Sigma_2 = \text{diag}(\sigma_{r+1}, \cdots, \sigma_n)$, where $\sigma_r > \sigma_{r+1}$. The matrices A, B and C can be partitioned compatibly as $A = \begin{bmatrix} A_{11} & A_{12} \\ A_{21} & A_{22} \end{bmatrix}$, $B = \begin{bmatrix} B_1 \\ B_2 \end{bmatrix}$, and $C = \begin{bmatrix} C_1 & C_2 \end{bmatrix}$. Then, a reduced-order system $G_r(s)$ can be defined by

$$G_r(s) = C_1(sI - A_{11})^{-1}B_1 + D = \left[\begin{array}{c|c} A_{11} & B_1 \\ \hline C_1 & D \end{array}\right] \tag{7.8}$$

Such a $G_r(s)$ is of rth order and is called a balanced truncation of the full order (nth) system $G(s)$. It can be shown that $G_r(s)$ is stable, in the balanced realisation form, and

$$\|G(s) - G_r(s)\|_\infty \leq 2tr(\Sigma_2) \tag{7.9}$$

where $tr(\Sigma_2)$ denotes the trace of the matrix Σ_2, *i.e.* $tr(\Sigma_2) = \sigma_{r+1} + \cdots + \sigma_n$, the sum of the last $(n - r)$ Hankel singular values ([30, 46]).

In most applications, to reduce the original system into an rth-order system there should be a large gap between σ_r and σ_{r+1}, *i.e.* $\sigma_r \gg \sigma_{r+1}$.

7.1.2 Singular Perturbation Approximation

In many engineering systems, the steady-state gain of a system, usually called *dc-gain* (the system gain at infinitive time, equivalent to $G(0)$), plays an important role in assessing system performances. It is thus better to maintain the dc gain in a reduced-order model, *i.e.* $G_r(0) = G(0)$. The Balanced Truncation method introduced in the last subsection does not keep the dc gain unchanged in the reduced-order system. The singular perturbation approximation method [93] (or, balanced residualisation method [136]) presented below does retain the dc gain.

Assume that $[A, B, C, D]$ is a minimal and balanced realisation of a stable system $G(s)$, and partitioned compatibly as in the previous subsection. It can be shown that A_{22} is stable (*e.g.* see Theorem 4.2 of [46]), and thus invertible. Define

$$A_r = A_{11} - A_{12}A_{22}^{-1}A_{21} \tag{7.10}$$
$$B_r = B_1 - A_{12}A_{22}^{-1}B_2 \tag{7.11}$$
$$C_r = C_1 - C_2A_{22}^{-1}A_{21} \tag{7.12}$$
$$D_r = D - C_2A_{22}^{-1}B_2 \tag{7.13}$$

A reduced-order system $G_r(s)$ defined by

$$G_r(s) = C_r(sI - A_r)^{-1}B_r + D_r \tag{7.14}$$

is called a singular perturbation approximation (or, balanced residualisation) of $G(s)$. It is a straightforward computation to show that the dc-gain remains unchanged, *i.e.*

$$-CA^{-1}B + D = -C_r A_r^{-1} B_r + D_r \tag{7.15}$$

by noting that

$$\begin{bmatrix} I & 0 \\ -A_{22}^{-1}A_{21} & I \end{bmatrix} \begin{bmatrix} I & 0 \\ A_{22}^{-1}A_{21} & I \end{bmatrix} = I \tag{7.16}$$

$$\begin{bmatrix} I & A_{12}A_{22}^{-1} \\ 0 & I \end{bmatrix} \begin{bmatrix} I & -A_{12}A_{22}^{-1} \\ 0 & I \end{bmatrix} = I \tag{7.17}$$

and

$$\left(\begin{bmatrix} I & -A_{12}A_{22}^{-1} \\ 0 & I \end{bmatrix} \begin{bmatrix} A_{11} & A_{12} \\ A_{21} & A_{22} \end{bmatrix} \begin{bmatrix} I & 0 \\ -A_{22}^{-1}A_{21} & I \end{bmatrix} \right)^{-1} = \begin{bmatrix} A_r^{-1} & 0 \\ 0 & A_{22}^{-1} \end{bmatrix} \tag{7.18}$$

It can also be shown that such a reduction $G_r(s)$ is a stable and balanced realisation [136] and enjoys the same error bound as the balanced truncation method, *i.e.*

$$\|G(s) - G_r(s)\|_\infty \leq 2(\sigma_{r+1} + \cdots + \sigma_n)$$

It can be seen that instead of discarding the "less important" part totally as in the balanced truncation method, the derivative of x_2 in the following equation is set to zero, in the singular perturbation approximation (balanced residualisation) method,

$$\dot{x_2} = A_{21}X_1 + A_{22}x_2 + B_2 u \tag{7.19}$$

x_2 is then solved in (7.19) in terms of x_1 and u, and is substituted as residual into the state equation of x_1 and output equation to obtain the reduced-order system $G_r(s)$ as given above.

This idea resembles what happens in analysis of singular perturbation systems with

$$\epsilon\dot{x_2} = A_{21}X_1 + A_{22}x_2 + B_2 u$$

where $0 < \epsilon \ll 1$, and hence the term of singular perturbation approximation.

7.1.3 Hankel-norm Approximation

For a stable system $G(s)$ with Hankel singular values defined in Section 7.1.1, the largest Hankel singular value σ_1 is defined as the Hankel-norm of $G(s)$ ([46]). The Hankel-norm denotes the largest possible $\mathcal{L}_2$-gain from past inputs to future outputs. In some order-reduction cases, minimisation of the Hankel-norm of the error system is more appropriate and thus required.

Approximation 1: Let $G(s)$ represent a stable and square system with a state-space model $[A, B, C, D]$ of minimal and balanced realisation. Let the gramians be $P = Q = \mathrm{diag}(\Sigma_1, \sigma I_l)$, where σ is the smallest Hankel singular value with multiplicity l and every diagonal element of Σ_1 is larger than σ. Let $[A, B, C]$ be partitioned compatibly. An $(n-l)$th-order system $G_h(s)$ can be constructed as follows. Define

$$\hat{A} = \Gamma^{-1}(\sigma^2 {A_{11}}^T + \Sigma_1 A_{11} \Sigma_1 - \sigma {C_1}^T U {B_1}^T) \quad (7.20)$$
$$\hat{B} = \Gamma^{-1}(\Sigma_1 B_1 + \sigma {C_1}^T U) \quad (7.21)$$
$$\hat{C} = C_1 \Sigma_1 + \sigma U {B_1}^T \quad (7.22)$$
$$\hat{D} = D - \sigma U \quad (7.23)$$

where U is an orthonormal matrix satisfying

$$B_2 = -{C_2}^T U \quad (7.24)$$

and

$$\Gamma = {\Sigma_1}^2 - \sigma^2 I \quad (7.25)$$

The reduced-order system $G_h(s)$ is defined as

$$G_h(s) = \hat{C}(sI - \hat{A})^{-1}\hat{B} + \hat{D} \quad (7.26)$$

It is shown in [46] that the $(n-l)$th-order $G_h(s)$ is stable and is an optimal approximation of $G(s)$ satisfying

$$\|G(s) - G_h(s)\|_H = \sigma \quad (7.27)$$

And, it is also true that $G(s) - G_h(s)$ is *all-pass* with the inf-norm

$$\|G(s) - G_h(s)\|_\infty = \sigma \quad (7.28)$$

Approximation 2: It can be shown that the Hankel singular values of $G_h(s)$ defined in (7.26) are correspondingly equal to those first $(n-l)$ Hankel singular values of $G(s)$. Hence, the above reduction formula can be repeatedly applied to get further reduced-order systems with known error bounds.

Let the Hankel singular values of $G(s)$ be $\sigma_1 > \sigma_2 > \cdots > \sigma_r$ with multiplicities m_i, $i = 1, \cdots, r$, *i.e.* $m_1 + m_2 + \cdots + m_r = n$. By repeatedly applying the formulae (7.20)–(7.26), we may have

$$G(s) = D_0 + \sigma_1 E_1(s) + \sigma_2 E_2(s) + \cdots + \sigma_r E_r(s) \quad (7.29)$$

where D_0 is a constant matrix and $E_i(s)$, $i = 1, \cdots, r$, are stable, norm-1, all-pass transfer function matrices. $E_i(s)s$ are the differences at each approximation. Consequently, we may define reduced-order models, for $k = 1, \cdots, r-1$,

$$\hat{G}_k(s) = D_0 + \Sigma_{i=1}^{k} \sigma_i E_i(s) \tag{7.30}$$

Such a $\hat{G}_k(s)$ is stable, with the order $m_1 + \cdots + m_k$, and satisfies

$$\|G(s) - \hat{G}_k(s)\|_\infty \leq (\sigma_{k+1} + \cdots + \sigma_r) \tag{7.31}$$

However, $\hat{G}_k(s)$ is not an optimal Hankel approximation, for $k < r-1$. The method to obtain an optimal Hankel approximation with "general" order is given below.

Approximation 3: Let the Hankel singular values of $G(s)$ be $\sigma_1 \geq \sigma_2 \geq \cdots \geq \sigma_k > \sigma_{k+1} = \cdots = \sigma_{k+l} > \sigma_{k+l+1} \geq \cdots \sigma_n$. Apply appropriate state similarity transformations to make the gramians of $G(s)$ be arranged as $\Sigma = \text{diag}(\sigma_1, \sigma_2, \cdots, \sigma_k, \sigma_{k+l+1}, \cdots, \sigma_n, \sigma_{k+1}, \cdots, \sigma_{k+l})$. Define the last l Hankel singular values to be σ. Following the formulae (7.20)–(7.26), define an $(n-l)$th-order $\hat{G}(s)$. This $\hat{G}(s)$ is not stable but has exactly k stable poles. The kth-order stable part $G_{h,k}(s)$ of $\hat{G}(s)$, obtained by using modal decompositions say, is an kth-order Hankel optimal approximation of $G(s)$ and satisfies

$$\|G(s) - G_{h,k}(s)\|_H = \sigma \tag{7.32}$$

Nonsquare plants can be augmented with zero columns/rows and then be applied by the above procedures.

7.1.4 *Remarks*

1. The three methods introduced in the last 3 subsections can be applied to original systems (plants) as well as to controllers. However, most reported cases are on plant reductions. This may be due to robust controller design methods used subsequently that leads to better closed-loop performance even with a reduced-order plant. Examples of application on controller-size reduction can be found in [136]. In [136] it is also observed that the Balanced Truncation method and the Hankel-norm Approximation usually perform better at high frequency, while the singular perturbation approximation (balanced residualisation) method performs better in the low- and medium-frequency ranges.
2. Glover shows in [46] that any stable, rth-order approximation G_r of $G(s)$ can never achieve $\|G(s) - G_r(s)\|_\infty \leq \sigma_{r+1}$. This lower error bound may serve as a yardstick to compare with the actual error obtained in practice.
3. All the three methods are applicable for *stable* systems only. If a system is unstable, modal decomposition can be applied first. That is, find a stable $G_s(s)$ and an antistable $G_{us}(s)$ (with all the poles in the closed right-half complex plane) such that

$$G(s) = G_s(s) + G_{us}(s) \tag{7.33}$$

Then, $G_s(s)$ can be reduced to $G_{sr}(s)$, by using any of the three methods, and a reduced-order system of the original $G(s)$ can be formed as

$$G_r(s) = G_{sr}(s) + G_{us}(s) \tag{7.34}$$

The routines to calculate a modal decomposition can be found in software packages such as MATLAB® or Slicot.

4. The formulae introduced here are for continuous-time systems. In the case of discrete-time systems, the gramians are calculated from the discrete Lyapunov equations instead,

$$APA^T - P + BB^T = 0 \tag{7.35}$$
$$A^TQA - Q + C^TC = 0 \tag{7.36}$$

The balanced truncation method can then be applied similar to the case of continuous time. However, it should be noted that the reduced-order system is no longer in a balanced realisation form ([126, 34]), though the same error bound still holds ([8]).
For using the singular perturbation approximation (balanced residualisation) on a system with zero D-matrix, the reduced-order system $G_r(s) = [A_r, B_r, C_r, D_r]$ can be instead defined by

$$\begin{aligned} A_r &= A_{11} + A_{12}(I - A_{22})^{-1}A_{21} \\ B_r &= B_1 + A_{12}(I - A_{22})^{-1}B_2 \\ C_r &= C_1 + C_2(I - A_{22})^{-1}A_{21} \\ D_r &= C_2(I - A_{22})^{-1}B_2 \end{aligned} \tag{7.37}$$

Such a reduced-order system is still in a balanced realisation and enjoys the same error bound ([8, 35, 93, 111]).
The discrete-time case of Hankel-norm approximation has been studied for a long time and is also called the Hankel-matrix approximation. Details can be found in [6, 78, 79, 116, 173].

5. Research has been conducted on numerical implementations of the above reduction methods. For instance, in balanced transformation in order to avoid numerical instability of forming products BB^T and C^TC, algorithms for direction calculation of the Choleski factors and improved balanced truncation scheme have been proposed ([62, 82, 148]). Also, to avoid ill-conditioned computational problems, balancing-free approaches can be considered ([133, 134, 151]). It is recommended that the model-reduction subroutines developed in the software package Slicot be used because of their superior numerical properties.

7.2 Reduction via Fractional Factors

The modal decomposition can be used to reduce the order of a general, unstable system as discussed in the last section. However, the order of the antistable part will not be reduced at all. Another reduction method applicable in the case of unstable systems is via reduction of normalised coprime factors of the system introduced in Chapter 5.

For a minimal realisation model $G(s) = \left[\begin{array}{c|c} A & B \\ \hline C & 0 \end{array}\right]$, recall that a normalised left coprime factorisation is defined by $G(s) = \tilde{M}(s)^{-1}\tilde{N}(s)$, where $\tilde{M}(s)$ and $\tilde{N}(s)$ are stable and satisfy (5.2) and (5.3). Note that we assume $G(s)$ has a zero D-matrix and model reduction is conducted with regard to such a strictly proper system. In the case of a nonzero feedthrough term in $G(s)$, the nonzero D-matrix should be added to the reduced-order model. Such a treatment greatly simplifies formulae. It will keep the high-frequency gain of the original system in the *fractional balanced truncation(FBT)* method, or maintain the dc-gain in the *fractional singular perturbation approximation(FSPA)* method, both introduced below.

The factors $\tilde{M}(s)$ and $\tilde{N}(s)$ have the following state-space models

$$\left[\tilde{N}\ \tilde{M}\right] = \left[\begin{array}{c|c} A + HC & B\ H \\ \hline C & 0\ I \end{array}\right] \tag{7.38}$$

where $H = -ZC^T$ with $Z > 0$ solves the following algebraic Riccati equation

$$AZ + ZA^T - ZC^TCZ + BB^T = 0 \tag{7.39}$$

$\left[\tilde{N}\ \tilde{M}\right]$ in (7.38) is stable and a balanced realisation transformation can be found. Following the balanced realisation algorithm in Section 7.1.1, a state similarity transformation $[T, T^{-1}]$ can be obtained such that $[T(A + HC)T^{-1}, T\left[B\ H\right], CT^{-1}]$ is a balanced realisation with the gramian $\Sigma = \mathrm{diag}(\sigma_1, \sigma_2, \cdots, \sigma_n)$. To reduce the system to rth order, where $\sigma_r > \sigma_{r+1}$, let

$$T(A + HC)T^{-1} =: \begin{bmatrix} \tilde{A}_{11} & \tilde{A}_{12} \\ \tilde{A}_{21} & \tilde{A}_{22} \end{bmatrix} = \begin{bmatrix} A_{11} + H_1C_1 & A_{12} + H_1C_2 \\ A_{21} + H_2C_1 & A_{22} + H_2C_2 \end{bmatrix} \tag{7.40}$$

$$T\left[B\ H\right] =: \begin{bmatrix} \tilde{B}_1 \\ \tilde{B}_2 \end{bmatrix} = \begin{bmatrix} B_1 & H_1 \\ B_2 & H_2 \end{bmatrix} \tag{7.41}$$

$$CT^{-1} =: \left[\tilde{C}_1\ \tilde{C}_2\right] = \left[C_1\ C_2\right] \tag{7.42}$$

$$\tilde{D} := \left[0\ I\right] \tag{7.43}$$

where

$$TAT^{-1} =: \begin{bmatrix} A_{11} & A_{12} \\ A_{21} & A_{22} \end{bmatrix} \tag{7.44}$$

$$TB =: \begin{bmatrix} B_1 \\ B_2 \end{bmatrix} \tag{7.45}$$

$$CT^{-1} =: \left[C_1\ C_2\right] \tag{7.46}$$

$$TH =: \begin{bmatrix} H_1 \\ H_2 \end{bmatrix} \tag{7.47}$$

Accordingly, the gramian is divided as

$$\Sigma = \begin{bmatrix} \Sigma_1 & 0 \\ 0 & \Sigma_2 \end{bmatrix} \tag{7.48}$$

with $\Sigma_1 = \mathrm{diag}(\sigma_1, \cdots, \sigma_r)$ and $\Sigma_2 = \mathrm{diag}(\sigma_{r+1}, \cdots, \sigma_n)$.

Now, the reduction methods introduced in Section 7.1 can be applied to obtain a rth-order $\begin{bmatrix} \tilde{N}_r & \tilde{M}_r \end{bmatrix}$ which leads to a reduced order (rth-order) model $G_r(s) = \tilde{M}_r(s)^{-1}\tilde{N}_r(s)$.

Fractional Balanced Truncation(FBT) Method

In this direct truncation method, we define

$$\begin{bmatrix} \tilde{N}_r & \tilde{M}_r \end{bmatrix} := \left[\begin{array}{c|c} \tilde{A}_{11} & \tilde{B}_1 \\ \hline \tilde{C}_1 & [0\ I] \end{array}\right] = \left[\begin{array}{c|cc} A_{11} + H_1C_1 & B_1 & H_1 \\ \hline C_1 & 0 & I \end{array}\right] \tag{7.49}$$

It is easy to check that the above realisation of $\begin{bmatrix} \tilde{N}_r & \tilde{M}_r \end{bmatrix}$ is still a balanced realisation with the gramian Σ_1. Define

$$G_r(s) := \tilde{M}_r(s)^{-1}\tilde{N}_r(s) \tag{7.50}$$

By direct manipulations, we have

$$G_r(s) = \left[\begin{array}{c|c} A_{11} & B_1 \\ \hline C_1 & 0 \end{array}\right] \tag{7.51}$$

An error bound directly follows the result in Section 7.1.1 and is shown in [107], that is

$$\| \begin{bmatrix} \tilde{N} - \tilde{N}_r & \tilde{M} - \tilde{M}_r \end{bmatrix} \|_\infty \leq 2 \sum_{i=r+1}^{n} \sigma_i \tag{7.52}$$

Fractional Singular Perturbation Approximation(FSPA) Method

Naturally, the singular perturbation approximation method (or, the balanced residualisation method) in Section 7.1.2 can be used to reduce the order of $\begin{bmatrix} \tilde{N} & \tilde{M} \end{bmatrix}$.

Define

$$A_r = \tilde{A}_{11} - \tilde{A}_{12}\tilde{A}_{22}^{-1}\tilde{A}_{21} \tag{7.53}$$

$$B_r = \tilde{B}_1 - \tilde{A}_{12}\tilde{A}_{22}^{-1}\tilde{B}_2 \tag{7.54}$$

$$C_r = \tilde{C}_1 - \tilde{C}_2\tilde{A}_{22}^{-1}\tilde{A}_{21} \tag{7.55}$$

$$D_r = \tilde{D} - \tilde{C}_2\tilde{A}_{22}^{-1}\tilde{B}_2 \tag{7.56}$$

Furthermore, B_r and D_r can be compatibly partitioned as

$$B_r = \begin{bmatrix} B_{r,1} & B_{r,2} \end{bmatrix} := \begin{bmatrix} B_1 - \tilde{A}_{12}\tilde{A}_{22}^{-1}B_2 & H_1 - \tilde{A}_{12}\tilde{A}_{22}^{-1}H_2 \end{bmatrix} \tag{7.57}$$

$$D_r = \begin{bmatrix} D_{r,1} & D_{r,2} \end{bmatrix} := \begin{bmatrix} -C_2\tilde{A}_{22}^{-1}B_2 & I - C_2\tilde{A}_{22}^{-1}H_2 \end{bmatrix} \tag{7.58}$$

Hence, let

$$\begin{bmatrix} \tilde{N}_r & \tilde{M}_r \end{bmatrix} := \left[\begin{array}{c|cc} A_r & B_{r,1} & B_{r,2} \\ \hline C_r & D_{r,1} & D_{r,2} \end{array}\right] \tag{7.59}$$

which is of balanced realisation form with the gramian Σ_1. An rth-order model $G_r(s)$ is then obtained by

$$G_r(s) = \tilde{M}_r(s)^{-1}\tilde{N}_r(s) = \left[\begin{array}{c|c} A_r - B_{r,2}D_{r,2}^{-1}C_r & B_{r,1} - B_{r,2}D_{r,2}^{-1}D_{r,1} \\ \hline D_{r,2}^{-1}C_r & D_{r,2}^{-1}D_{r,1} \end{array}\right] \tag{7.60}$$

In the case that the original model $G(s)$ is not strictly proper, the nonzero feedthrough term should be added in the D-matrix in (7.60).

The error bound (7.52) obviously holds here as well, from the result in Section 7.1.2.

Remarks:

1. Meyer [107] shows that $\| \begin{bmatrix} \tilde{N} & \tilde{M} \end{bmatrix} \|_H < 1$, *i.e.* $\Sigma < I$. Also, the infinitive norm $\| \begin{bmatrix} \tilde{N} & \tilde{M} \end{bmatrix} \|_\infty = 1$, because of the normalised coprime factorisation.
2. It can be obtained that, in either the FBT method or FSPA method, we have

$$\|G(s) - G_r(s)\|_\infty \leq \|\tilde{M}_r^{-1}\|_\infty \| \begin{bmatrix} \tilde{N} - \tilde{N}_r & \tilde{M} - \tilde{M}_r \end{bmatrix} \|_\infty \|\tilde{M}^{-1}\|_\infty \tag{7.61}$$

by writing

$$\begin{aligned} G - G_r &= \tilde{M}^{-1}\tilde{N} - \tilde{M}_r^{-1}\tilde{N}_r \\ &= \tilde{M}_r^{-1}\left(\begin{bmatrix} \tilde{N} - \tilde{N}_r & \tilde{M} - \tilde{M}_r \end{bmatrix} \begin{bmatrix} I \\ -\tilde{M}^{-1}\tilde{N} \end{bmatrix}\right) \\ &= \tilde{M}_r^{-1}\left(\begin{bmatrix} \tilde{N} - \tilde{N}_r & \tilde{M} - \tilde{M}_r \end{bmatrix} \tilde{M}^{-1} \begin{bmatrix} \tilde{M} \\ -\tilde{N} \end{bmatrix}\right) \end{aligned} \tag{7.62}$$

and by the fact that $\| \begin{bmatrix} \tilde{M} \\ -\tilde{N} \end{bmatrix} \|_\infty = 1$.

3. Note that the methods introduced above are based on the left coprime factorisation. Similarly, model reductions can be done with regard to the normalised *right* coprime factorisation.
4. The model reduction in the discrete-time case using the fractional balanced truncation method is straightforward, and using the fractional singular perturbation approximation method can be found in [112, 111].

7.3 Relative-error Approximation Methods

As discussed in Section 7.1.4, the balanced truncation method gives a good approximation over high-frequency ranges, while the singular perturbation approximation performs better over low- and medium-frequency ranges. If a reduced-order system is required in some practical problems to approximate equally well over the whole frequency range, then the method called the *balanced stochastic truncation(BST)* method may be considered [19, 52, 51, 92, 152]). The effect of this method may be viewed as, for a stable, square and invertible $G(s)$, a minimisation of

$$\|G^{-1}(s)(G(s) - G_r(s))\|_\infty \tag{7.63}$$

Hence, the reduced-order system $G_r(s)$ approximates the original system in the sense of making $G^{-1}G_r$ nearer to identity. Problem (7.63) represents a minimisation of a relative error and is one of several relative-error approximation methods (*e.g.*, see [47, 49]).

The idea of the BST method is the following. First, a spectral factor $W(s)$ of $G(s)G^{-}(s)$ is to be found. That is,

$$W^{-}(s)W(s) = G(s)G^{-}(s)$$

where $W^{-}(s) := W^T(-s)$, similarly for $G^{-}(s)$; and $W(s)$ is stable, square and of minimum phase (*i.e.* $(W(s))^{-1} \in \mathcal{H}_\infty$). $W(s)$ contains the "magnitude" part of $G(s)$. Correspondingly, a "phase" matrix of $G(s)$ can be defined as $F(s) = (W^{-}(s))^{-1}G(s)$. $F(s)$ is an all-pass transfer function matrix and contains both stable and unstable modes. The BST method is then to apply a balanced truncation on the stable part of $F(s)$ (which is of the same order as $G(s)$), with the same state similarity transformation and partition on the state space model of $G(s)$ to obtain a reduced-order $G_r(s)$.

For a given nth-order, stable and square $G(s) = \left[\begin{array}{c|c} A & B \\ \hline C & D \end{array}\right]$ we assume this is a minimal realisation and the invertibility of $G(s)$ implies the existence of D^{-1}.

The computational steps of the BST method can be described as:

Step 1: Solve the Lyapunov equation

$$AP + PA^T + BB^T = 0 \tag{7.64}$$

where the solution $P > 0$.

Step 2: Let

$$B_W = PC^T + BD^T \tag{7.65}$$
$$C_W = D^{-1}(C - B_W^T Q_W) \tag{7.66}$$

where Q_W is the stabilising solution to the following Riccati equation

$$A^T Q_W + Q_W A + C_W^T C_W = 0 \tag{7.67}$$

Remark: (A, B_W, C_W) forms the stable part of $F(s)$.

Step 3: Decide a balanced realisation transformation with regard to (P, Q_W) and apply the transformation onto $G(s)$. Let the balanced gramian matrix be $\Sigma = \mathrm{diag}(\sigma_1, \cdots, \sigma_n)$ in descending order.

Remarks: (1) After applying the above transformation, $G(s)$ is, in general, not in the balanced realisation form, but with its controllability gramian being Σ; (2) $\Sigma \leq I$, due to $F(s)$ being an all-pass matrix.

Step 4: Partition Σ as

$$\Sigma = \begin{bmatrix} \Sigma_1 & 0 \\ 0 & \Sigma_2 \end{bmatrix} \tag{7.68}$$

where $\Sigma_1 = \mathrm{diag}(\sigma_1, \cdots, \sigma_r)$, $\Sigma_2 = \mathrm{diag}(\sigma_{r+1}, \cdots, \sigma_n)$, with $\sigma_r > \sigma_{r+1}$. Partition compatibly the matrices A, B and C (of the transformed state-space model of $G(s)$) $A = \begin{bmatrix} A_{11} & A_{12} \\ A_{21} & A_{22} \end{bmatrix}$, $B = \begin{bmatrix} B_1 \\ B_2 \end{bmatrix}$, and $C = \begin{bmatrix} C_1 & C_2 \end{bmatrix}$. Then, a reduced-order system $G_r(s)$ can be defined by

$$G_r(s) = C_1(sI - A_{11})^{-1}B_1 + D \tag{7.69}$$

■

For this reduced-order system, a relative error bound can be derived ([52]) as

$$\|G^{-1}(G - G_r)\|_\infty \leq \prod_{i=r+1}^{n} \frac{1 + \sigma_i}{1 - \sigma_i} - 1 \tag{7.70}$$

The errors between the phase matrices, with the same antistable and constant parts, are bounded by

$$\|F(s) - F_r(s)\|_\infty \leq 4(\sigma_{r+1} + \cdots + \sigma_n) \tag{7.71}$$
$$\|F(s) - F_r(s)\|_H \leq 2(\sigma_{r+1} + \cdots + \sigma_n) \tag{7.72}$$

Remarks:

1. The BST method can be applied to nonsquare $G(s)$ as well, with slight modifications. The invertibility of $G(s)$ is changed to the assumption that

D is of full row rank. The constant matrix of the square spectral factor $W(s)$ would be D_W, with $D_W^T D_W = DD^T$, and the output matrix $C_W = D_W(DD^T)^{-1}(C - B_W^T Q_W)$. However, in the nonsquare case, there would be no explicit explanation of the relative error format (7.63). The reduction just shows an approximation with respect to phase.

2. In the above method, instead of balanced truncation, the Hankel-norm approximation can be used ([47]). That is, the balanced realisation of the stable part, $F_s(s)$, of the phase matrix $F(s)$ is to be approximated by a Hankel-norm approximant, $F_{s,r}(s)$, calculated using the formula in Section 7.1.3. The reduced model, G_r, can then be defined as

$$G_r = G - W^-(F_s - F_{s,r}) \tag{7.73}$$

It can be shown ([47, 52]) that G_r is stable and satisfies the following error bound

$$\|G^{-1}(G - G_r)\|_\infty \leq \prod_{i=r+1}^{n} (1 + \sigma_i) - 1 \tag{7.74}$$

7.4 Frequency-weighted Approximation Methods

The model-order reduction approaches introduced above can be in theory applied to plants (the original system models) as well as to controllers. However, to reduce the order of a designed controller, it is necessary to take into account the plant that is being compensated and other design specifications of the closed-loop system. With such considerations, the controller-order reduction problem would be better formulated as a frequency-weighted model reduction, and suitable approaches have been suggested.

Assume that a controller $K(s)$ has been designed for a plant with a model $G(s)$, and denote a reduced-order controller by $K_r(s)$. The configuration with $K(s)$ replaced by $K_r(s)$ in the closed-loop system can be depicted by Figure 7.2.

From the Small-Gain Theorem (Theorem 3.1), it is easy to obtain that the closed-loop system with the reduced-order controller $K_r(s)$ remains stable if $K(s)$ and $K_r(s)$ have the same number of unstable poles and if

$$\|[K(s) - K_r(s)]G(s)[I + K(s)G(s)]^{-1}\|_\infty < 1 \tag{7.75}$$

or

$$\|[I + G(s)K(s)]^{-1}G(s)[K(s) - K_r(s)]\|_\infty < 1 \tag{7.76}$$

Let $W_i(s) := G(s)[I+K(s)G(s)]^{-1}$ and $W_o(s) := [I+G(s)K(s)]^{-1}G(s)$. Then in order to reserve the stability of the closed-loop system, a reduced-order controller $K_r(s)$ is sought to minimise the frequency-weighted cost functions

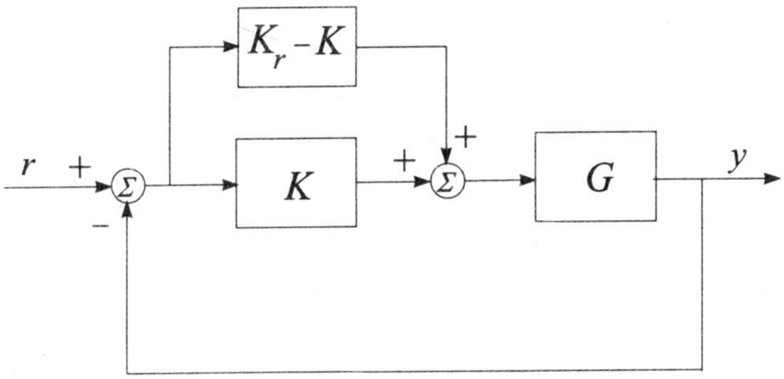

Fig. 7.2. Closed-loop system with reduced-order controllers

$$\|[K(s) - K_r(s)]W_i(s)\|_\infty \tag{7.77}$$

or

$$\|W_o(s)[K(s) - K_r(s)]\|_\infty \tag{7.78}$$

Note in this case the input frequency weight function $W_i(s)$ equals the output frequency weight function $W_o(s)$.

Another consideration is about the performance of the closed-loop system. The performance is closely related to the transfer function of the closed-loop system. Naturally, to maintain the performance of the designed, closed-loop system, it requires the transfer function of the closed-loop system with reduced-order controller to be as near as possible to that with the original controller. The two transfer functions are, respectively,

$$G(s)K(s)[I + G(s)K(s)]^{-1}$$
$$G(s)K_r(s)[I + G(s)K_r(s)]^{-1}$$

The difference between these two transfer functions, by neglecting terms of second and higher orders in $K - K_r$, is

$$G(s)[I + K(s)G(s)]^{-1}[K(s) - K_r(s)][I + G(s)K(s)]^{-1}$$

Hence, a reduced-order controller $K_r(s)$ should try to minimise

$$\|W_o(s)[K(s) - K_r(s)]W_i(s)\|_\infty \tag{7.79}$$

where $W_i(s) := [I + G(s)K(s)]^{-1}$ and $W_o(s) := G(s)[I + K(s)G(s)]^{-1}$.

Let us now concentrate on the general form of frequency-weighted model reduction of (7.79), but replacing $K(s)$ and $K_r(s)$ by $G(s)$ and $G_r(s)$, respectively. Assume that $G(s)$is stable and has the minimal realisation as defined in (7.2). The input weight function $W_i(s)$ and the output weight $W_o(s)$ are also stable with minimal realisations: $W_i(s) = \left[\begin{array}{c|c} A_i & B_i \\ \hline C_i & D_i \end{array}\right]$ and $W_o(s) = \left[\begin{array}{c|c} A_o & B_o \\ \hline C_o & D_o \end{array}\right]$, respectively.

Remark:

Obviously the stability assumption of $G(s)$ would be a restriction in the case of controller-order reduction. In the case of unstable $G(s)$, the modal decomposition discussed in Section 7.1.4 can be considered. ■

The augmented systems $G(s)W_i(s)$ and $W_o(s)G(s)$ have the state-space models

$$G(s)W_i(s) = \left[\begin{array}{c|c} \hat{A}_i & \hat{B}_i \\ \hline \hat{C}_i & \hat{D}_i \end{array}\right] = \left[\begin{array}{cc|c} A & BC_i & BD_i \\ 0 & A_i & B_i \\ \hline C & DC_i & DD_i \end{array}\right] \tag{7.80}$$

$$W_oG(s) = \left[\begin{array}{c|c} \hat{A}_o & \hat{B}_o \\ \hline \hat{C}_o & \hat{D}_o \end{array}\right] = \left[\begin{array}{cc|c} A & 0 & B \\ B_oC & A_o & B_oD \\ \hline D_oC & C_o & D_oD \end{array}\right] \tag{7.81}$$

Let $\hat{P}$ and $\hat{Q}$ be two non-negative matrices satisfying the following two Lyapunov equations, respectively,

$$\hat{A}\hat{P} + \hat{P}\hat{A}^T + \hat{B}\hat{B}^T = 0 \tag{7.82}$$

$$\hat{A}^T\hat{Q} + \hat{Q}\hat{A} + \hat{C}^T\hat{C} = 0 \tag{7.83}$$

Furthermore, partition $\hat{P}$ and $\hat{Q}$ as

$$\hat{P} = \begin{bmatrix} P & P_{12} \\ P_{12}^T & P_{22} \end{bmatrix} \tag{7.84}$$

$$\hat{Q} = \begin{bmatrix} Q & Q_{12} \\ Q_{12}^T & Q_{22} \end{bmatrix} \tag{7.85}$$

where, P and Q are of n-dimension, and are called the input weighted gramian and output weighted gramian, respectively.

Several frequency-weighted model-reduction algorithms use balanced realisation transformations on P and Q or are related with truncations. Three such methods are introduced below.

Frequency-weighted Balanced Truncation(FWBT)

Enns [30, 29] proposes to find a balanced realisation on P and Q, *i.e.* to find an invertible $n \times n$ matrix T (see Section 7.1.1) such that

$$TPT^T = (T^{-1})^TQT^{-1} = \text{diag}(\sigma_1, \cdots, \sigma_r, \sigma_{r+1}, \cdots, \sigma_n)$$

with $\sigma_1 \geq \sigma_2 \geq \cdots \geq \sigma_r > \sigma_{r+1} \geq \cdots \geq \sigma_n \geq 0$. Apply such a state similarity transformation (T, T^{-1}) on $G(s)$ and partition it accordingly,

$$\left[\begin{array}{c|c} TAT^{-1} & TB \\ \hline CT^{-1} & D \end{array}\right] = \left[\begin{array}{cc|c} A_{11} & A_{12} & B_1 \; B_2 \\ A_{21} & A_{22} & \\ \hline C_1 & C_2 & D \end{array}\right] \tag{7.86}$$

where A_{11} is of $r \times r$ dimension. A reduced-order $G_r(s)$ can then be defined by

$$G_r(s) = \left[\begin{array}{c|c} A_{11} & B_1 \\ \hline C_1 & D \end{array}\right] \tag{7.87}$$

$G_r(s)$ obtained in (7.87) is not necessarily stable, except in the cases where either $W_i = I$ or $W_o = I$ (one-side weight only). There is an error bound derived ([75]) for

$$\|W_o(s)[G(s) - G_r(s)]W_i(s)\|_\infty$$

However, this bound has to be computed iteratively, depending on reduced-order models of $n-1, \cdots, r+1$, and is not practically useful.

Frequency-weighted Singular Perturbation Approximation(FWSPA)

Lin and Chiu [88] introduce another truncation method to obtain a frequency-weighted reduced model. Assume that P_{22} and Q_{22} in (7.84) and (7.85), respectively, are nonsingular. This condition is guaranteed, for example, in the case that the realisations (7.80) and (7.81) are minimal, *i.e.* if there are no pole/zero cancellations between $G(s)$ and $W_i(s)$, nor between $W_o(s)$ and $G(s)$. Instead of applying a balanced realisation transformation on P and Q as in the Enns method, a balanced realisation transformation is to be found with regard to $P - P_{12}P_{22}^{-1}P_{12}^T$ and $Q - Q_{12}^TQ_{22}^{-1}Q_12$. This balanced realisation is then applied onto the original model $G(s)$ and truncation taken in the same way as in (7.86) and (7.87).

Apparently, this method is so named because the matrices $P - P_{12}P_{22}^{-1}P_{12}^T$ and $Q - Q_{12}^TQ_{22}^{-1}Q_12$ are in the form of the reduced-order state matrix used in the Singular Perturbation Approximation method (Section 7.1.2). It is observed that, by pre/postmultiplying $\begin{bmatrix} I & -P_{12}P_{22}^{-1} \\ 0 & I \end{bmatrix}$ and $\begin{bmatrix} I & 0 \\ -P_{22}^{-1}P_{12}^T & I \end{bmatrix}$ on (7.82), and similar multiplications with regard to Q on (7.83), matrices $P - P_{12}P_{22}^{-1}P_{12}^T$ and $Q - Q_{12}^TQ_{22}^{-1}Q_12$ satisfy two Lyapunov equations, respectively. Hence, the diagonalised matrices of these two after the balanced realisation transformations satisfy the Lyapunov equations, too. This indicates the reduced-order system is guaranteed to be stable.

There is an error bound available for this method ([144]). However, it suffers the same weakness as the error bound for the Enns method. The error bound cannot be simply calculated from the original model data.

Frequency-weighted Modulii Truncation Method(FWMT)

The error bounds for the above two methods are not practically useful. In [157], Wang *et al.* propose another truncation method with *a priori* computable error bound.

Quoting the upper-left blocks of (7.82) and (7.83) gives the following two matrix equations:

$$AP + PA^T + BC_iP_{12} + P_{12}^TC_i^TB^T + BD_iD_i^TB^T = 0 \tag{7.88}$$
$$A^TQ + QA + Q_{12}B_oC + C^TB_o^TQ_{12}^T + C^TD_o^TD_oC = 0 \tag{7.89}$$

Let

$$X = BC_iP_{12} + P_{12}^TC_i^TB^T + BD_iD_i^TB^T \tag{7.90}$$
$$Y = Q_{12}B_oC + C^TB_o^TQ_{12}^T + C^TD_o^TD_oC \tag{7.91}$$

Note that X and Y defined above are symmetric but not sign-definite in general. Apply congruent transformations on X and Y to obtain orthogonal matrices U and V such that

$$X = U\Theta U^T \tag{7.92}$$
$$Y = V\Gamma V^T \tag{7.93}$$

where $\Theta = \text{diag}(\theta_1, \cdots, \theta_i, 0, \cdots, 0)$, $\Gamma = \text{diag}(\gamma_1, \cdots, \gamma_o, 0, \cdots, 0)$, with $|\theta_1| \geq \cdots \geq |\theta_i| > 0$ and $|\gamma_1| \geq \cdots \geq |\gamma_o| > 0$. Now, define

$$\tilde{B} = U\text{diag}(|\theta_1|^{\frac{1}{2}}, \cdots, |\theta_i|^{\frac{1}{2}}, 0, \cdots, 0) \tag{7.94}$$
$$\tilde{C} = \text{diag}(|\gamma_1|^{\frac{1}{2}}, \cdots, |\gamma_o|^{\frac{1}{2}}, 0, \cdots, 0)V^T \tag{7.95}$$

Solve the following two Lyapunov equations:

$$A\tilde{P} + \tilde{P}A^T + \tilde{B}\tilde{B}^T = 0 \tag{7.96}$$
$$\tilde{Q}A + A^T\tilde{Q} + \tilde{C}^T\tilde{C} = 0 \tag{7.97}$$

It can be shown ([157]) that $(A, \tilde{B}, \tilde{C})$ is a minimal realisation and hence the solutions $\tilde{P}$ and $\tilde{Q}$ to (7.96) and (7.97), respectively, are positive definite. Similar to FWBT and FWSPA methods, a balanced realisation is found with regard to $\tilde{P}$ and $\tilde{Q}$ and the transformation is applied to the original model (A, B, C) to yield a reduced-order model $G_r(s)$. Such a $G_r(s)$ is stable, following the same reasoning as in the FWSPA method. Furthermore, the following error bound can be derived ([157]).

Define

$$K = \text{diag}(|\theta_1|^{\frac{1}{2}}, \cdots, |\theta_i|^{\frac{1}{2}}, 0, \cdots, 0)U^TB \tag{7.98}$$
$$L = CV\text{diag}(|\gamma_1|^{\frac{1}{2}}, \cdots, |\gamma_o|^{\frac{1}{2}}, 0, \cdots, 0) \tag{7.99}$$

Then, it is shown in ([157]) that

$$\|W_o(s)(G(s) - G_r(s))W_i(s)\|_\infty \leq k \sum_{j=r+1}^{n} \sigma_j \tag{7.100}$$

where

$$k = 2\|W_o(s)L\|_\infty \|KW_i(s)\|_\infty$$

and $(\sigma_1, \cdot, \sigma_r, \sigma_{r+1}, \cdots, \sigma_n)$ are the diagonal elements of the balanced form of $\tilde{P}$ $(\tilde{Q})$.

Part II

Design Examples

8

Robust Control of a Mass-Damper-Spring System

In this chapter we consider the design of a robust control system for a simple, second-order, mechanical system, namely a *mass-damper-spring system.* The mass-damper-spring system is a common control experimental device frequently seen in an undergraduate teaching laboratory. As the first case study considered in this book, we show in detail the design of three different controllers for this system and present robust stability and robust performance analysis of the corresponding closed-loop systems, respectively. In order to keep the designs simple we take into account only the structured (parametric) perturbations in the plant coefficients. In this design example we give some information for several basic commands from the Robust Control Toolbox that are used in the analysis and design in this and subsequent case studies. To illuminate in a clear way the implementation of the most important Robust Control Toolbox commands, we include in the chapter all files used in computations related to the analysis and design of the mass-damper-spring control system. It is hoped that this chapter may serve as a tutorial introduction not only to robust control systems analysis and design but also to the use of the Robust Control Toolbox.

8.1 System Model

The *one-degree-of-freedom(1DOF)* mass-damper-spring system is depicted in Figure 8.1.

The dynamics of such a system can be described by the following 2nd-order differential equation, by Newton's Second Law,

$$m\ddot{x} + c\dot{x} + kx = u$$

where x is the displacement of the mass block from the equilibrium position and $u = F$ is the force acting on the mass, with m the mass, c the damper constant and k the spring constant. A block diagram of such a system is shown in Figure 8.2.

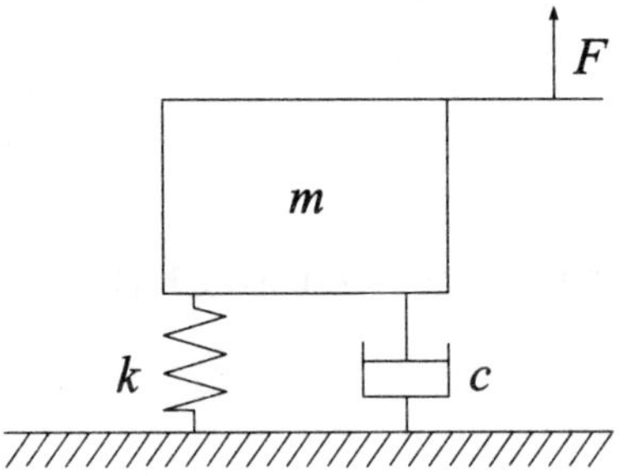

Fig. 8.1. Mass-damper-spring system

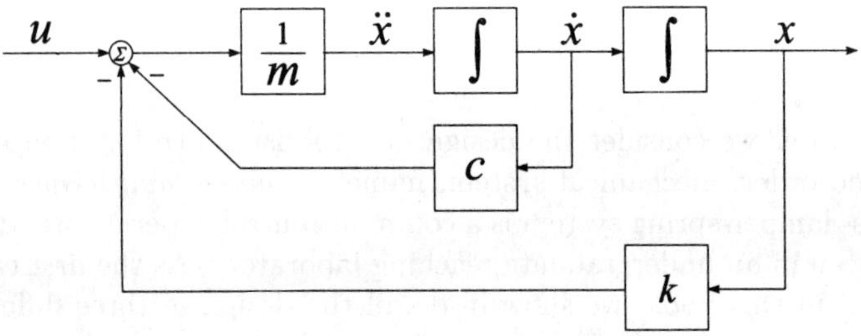

Fig. 8.2. Block diagram of the mass-damper-spring system

In a realistic system, the three physical parameters m, c and k are not known exactly. However, it can be assumed that their values are within certain, known intervals. That is,

$$m = \overline{m}(1 + p_m\delta_m), \;\; c = \overline{c}(1 + p_c\delta_c), \;\; k = \overline{k}(1 + p_k\delta_k)$$

where $\overline{m} = 3$, $\overline{c} = 1$, $\overline{k} = 2$ are the so-called nominal values of m, c and k. p_m, p_c and p_k and δ_m, δ_c and δ_k represent the possible (relative) perturbations on these three parameters. In the present study, we let $p_m = 0.4$, $p_c = 0.2$, $p_k = 0.3$ and $-1 \leq \delta_m, \; \delta_c, \; \delta_k \leq 1$. Note that this represents up to 40% uncertainty in the mass, 20% uncertainty in the damping coefficient and 30% uncertainty in the spring stiffness.

The three constant blocks in Figure 8.2 can be replaced by block diagrams in terms of $\overline{m}$, p_m, δ_m, *etc.*, in a unified approach. We note that the quantity $\frac{1}{m}$ may be represented as a linear fractional transformation (LFT) in δ_m

$$\begin{aligned} \frac{1}{m} &= \frac{1}{\overline{m}(1 + p_m\delta_m)} = \frac{1}{\overline{m}} - \frac{p_m}{\overline{m}}\delta_m(1 + p_m\delta_m)^{-1} \\ &= F_U(M_{mi}, \delta_m) \end{aligned}$$

with

$$M_{mi} = \begin{bmatrix} -p_m & \frac{1}{\overline{m}} \\ -p_m & \frac{1}{\overline{m}} \end{bmatrix}$$

Similarly, the parameter $c = \bar{c}(1 + p_c\delta_c)$ may be represented as an upper LFT in δ_c

$$c = F_U(M_c, \delta_c)$$

with

$$M_c = \begin{bmatrix} 0 & \bar{c} \\ p_c & \bar{c} \end{bmatrix}$$

and the parameter $k = \bar{k}(1 + p_k\delta_k)$ may be represented as an upper LFT in δ_k,

$$k = F_U(M_k, \delta_k)$$

with

$$M_k = \begin{bmatrix} 0 & \bar{k} \\ p_k & \bar{k} \end{bmatrix}$$

All these LFTs are depicted by block diagrams in Figure 8.3.

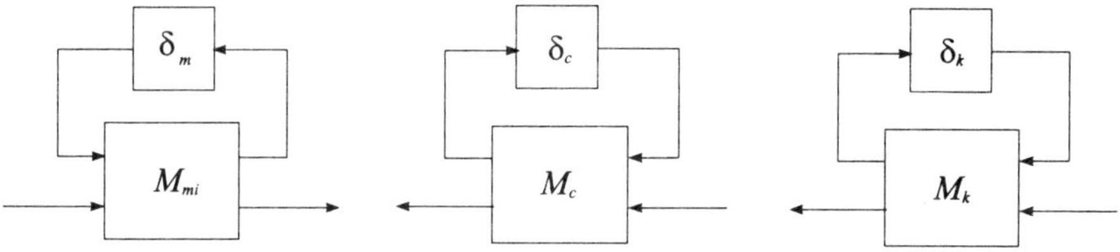

Fig. 8.3. Representation of uncertain parameters as LFTs

To further represent the system model as an LFT of the unknown, real perturbations δ_m, δ_c and δ_k, we use the block diagrams in Figure 8.3 and denote the inputs and outputs of δ_m, δ_c and δ_k as y_m, y_c, y_k and u_m, u_c, u_k, respectively, as shown in Figure 8.4.

With the above substitutions, the equations relating all "inputs" to corresponding "outputs" around these perturbed parameters can now be obtained as

$$\begin{bmatrix} y_m \\ \ddot{x} \end{bmatrix} = \begin{bmatrix} -p_m & \frac{1}{\overline{m}} \\ -p_m & \frac{1}{\overline{m}} \end{bmatrix} \begin{bmatrix} u_m \\ u - v_c - v_k \end{bmatrix}$$

$$\begin{bmatrix} y_c \\ v_c \end{bmatrix} = \begin{bmatrix} 0 & \bar{c} \\ p_c & \bar{c} \end{bmatrix} \begin{bmatrix} u_c \\ \dot{x} \end{bmatrix}$$

$$\begin{bmatrix} y_k \\ v_k \end{bmatrix} = \begin{bmatrix} 0 & \bar{k} \\ p_k & \bar{k} \end{bmatrix} \begin{bmatrix} u_k \\ x \end{bmatrix}$$

$$u_m = \delta_m y_m$$
$$u_c = \delta_c y_c$$
$$u_k = \delta_k y_k$$

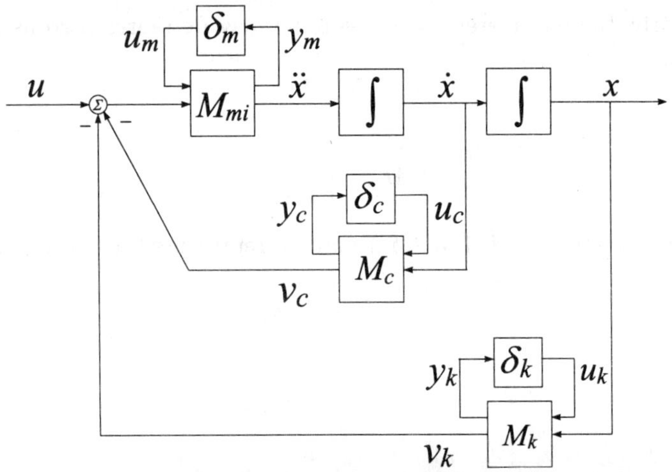

Fig. 8.4. Block diagram of the mass-damper-spring system with uncertain parameters

Let us set

$$x_1 = x, \; x_2 = \dot{x} = \dot{x}_1, \; y = x_1$$

such that

$$\dot{x}_2 = \ddot{x} = \ddot{x}_1$$

As a result, we obtain the following equations

$$\begin{aligned}
\dot{x}_1 &= x_2 \\
\dot{x}_2 &= -p_m u_m + \tfrac{1}{\overline{m}}(u - v_c - v_k) \\
y_m &= -p_m u_m + \tfrac{1}{\overline{m}}(u - v_c - v_k) \\
y_c &= \overline{c} x_2 \\
y_k &= \overline{k} x_1 \\
v_c &= p_c u_c + \overline{c} x_2 \\
v_k &= p_k u_k + \overline{k} x_1 \\
y &= x_1 \\
u_m &= \delta_m y_m \\
u_c &= \delta_c y_c \\
u_k &= \delta_k y_k
\end{aligned}$$

By eliminating the variables v_c and v_k, the equations governing the system dynamic behaviour are given by

$$
\begin{bmatrix} \dot{x}_1 \\ \dot{x}_2 \\ -- \\ y_m \\ y_c \\ y_k \\ -- \\ y \end{bmatrix} = \left[\begin{array}{cc|ccc|c} 0 & 1 & 0 & 0 & 0 & 0 \\ -\frac{\overline{k}}{\overline{m}} & -\frac{\overline{c}}{\overline{m}} & -p_m & -\frac{p_c}{\overline{m}} & -\frac{p_k}{\overline{m}} & \frac{1}{\overline{m}} \\ \hline -\frac{\overline{k}}{\overline{m}} & -\frac{\overline{c}}{\overline{m}} & -p_m & -\frac{p_c}{\overline{m}} & -\frac{p_k}{\overline{m}} & \frac{1}{\overline{m}} \\ 0 & \overline{c} & 0 & 0 & 0 & 0 \\ \overline{k} & 0 & 0 & 0 & 0 & 0 \\ \hline 1 & 0 & 0 & 0 & 0 & 0 \end{array}\right] \begin{bmatrix} x_1 \\ x_2 \\ -- \\ u_m \\ u_c \\ u_k \\ -- \\ u \end{bmatrix}
$$

$$
\begin{bmatrix} u_m \\ u_c \\ u_k \end{bmatrix} = \begin{bmatrix} \delta_m & 0 & 0 \\ 0 & \delta_c & 0 \\ 0 & 0 & \delta_k \end{bmatrix} \begin{bmatrix} y_m \\ y_c \\ y_k \end{bmatrix}
$$

Let G_{mds} denote the input/output dynamics of the mass-damper-spring system, which takes into account the uncertainty of parameters as shown in Figure 8.5. G_{mds} has four inputs $(u_m,\ u_c,\ u_k,\ u)$, four outputs $(y_m,\ y_c,\ y_k,\ y)$ and two states $(x_1,\ x_2)$.

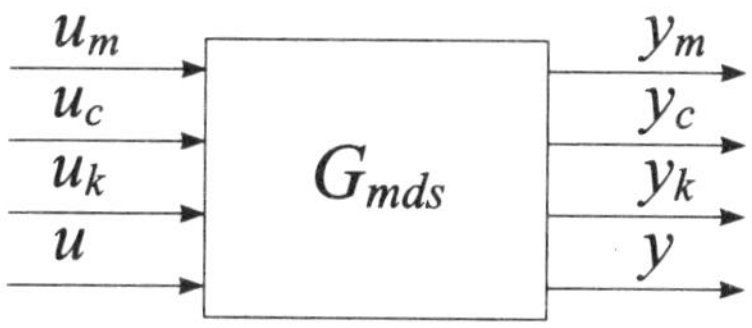

Fig. 8.5. Input/output block diagram of the mass-damper-spring system

The state space representation of G_{mds} is

$$
G_{\text{mds}} = \left[\begin{array}{c|cc} A & B_1 & B_2 \\ \hline C_1 & D_{11} & D_{12} \\ C_2 & D_{21} & D_{22} \end{array}\right]
$$

where

$$
A = \begin{bmatrix} 0 & 1 \\ -\frac{\overline{k}}{\overline{m}} & -\frac{\overline{c}}{\overline{m}} \end{bmatrix}, \quad B_1 = \begin{bmatrix} 0 & 0 & 0 \\ -p_m & -\frac{p_c}{\overline{m}} & -\frac{p_k}{\overline{m}} \end{bmatrix}, \quad B_2 = \begin{bmatrix} 0 \\ \frac{1}{\overline{m}} \end{bmatrix}
$$

$$
C_1 = \begin{bmatrix} -\frac{\overline{k}}{\overline{m}} & -\frac{\overline{c}}{\overline{m}} \\ 0 & \overline{c} \\ \overline{k} & 0 \end{bmatrix}, \quad D_{11} = \begin{bmatrix} -p_m & -\frac{p_c}{\overline{m}} & -\frac{p_k}{\overline{m}} \\ 0 & 0 & 0 \\ 0 & 0 & 0 \end{bmatrix}, \quad D_{12} = \begin{bmatrix} \frac{1}{\overline{m}} \\ 0 \\ 0 \end{bmatrix}
$$

$$
C_2 = \begin{bmatrix} 1 & 0 \end{bmatrix}, \quad D_{21} = \begin{bmatrix} 0 & 0 & 0 \end{bmatrix}, \quad D_{22} = 0
$$

Note that G_{mds} depends only on $\overline{m}$, $\overline{c}$, $\overline{k}$, p_m, p_c, p_k and on the original differential equation connecting y with u. Hence, G_{mds} is known and contains no uncertain parameters.

Below, we give the M-file mod_mds.m, which can be used to compute the system matrix G_{mds} and to save it in the MATLAB® variable G.

File mod_mds.m

```
m = 3;
c = 1;
k = 2;
pm = 0.4;
pc = 0.2;
pk = 0.3;
%
A = [ 0      1
     -k/m  -c/m];
B1 = [ 0    0       0
      -pm  -pc/m   -pk/m];
B2 = [ 0
      1/m];
C1 = [-k/m  -c/m
       0      c
       k      0 ];
C2 = [ 1 0 ];
D11 = [-pm  -pc/m  -pk/m
        0     0      0
        0     0      0 ];
D12  = [1/m
         0
         0 ];
D21 = [0 0 0];
D22 = 0;
G  = pck(A,[B1,B2],[C1;C2],[D11 D12;D21 D22]);
```

The uncertain behaviour of the original system can be described by an upper LFT representation

$$y = F_U(G_{\text{mds}}, \Delta)u$$

with diagonal uncertainty matrix $\Delta = \text{diag}(\delta_m, \delta_c, \delta_k)$, as shown in Figure 8.6. It should be noted that the unknown matrix Δ, which will be called the *uncertainty matrix*, has a fixed *structure*. It is a diagonal matrix. It could, in general, be block diagonal. Such uncertainty is thus called *structured uncertainty*.

Apart from the method presented above in which system equations are derived first followed by explicitly defining all those coefficient matrices, the system matrix G_{mds} may also be obtained by using the MATLAB® command sysic. The following file sys_mds.m shows how this can be done.

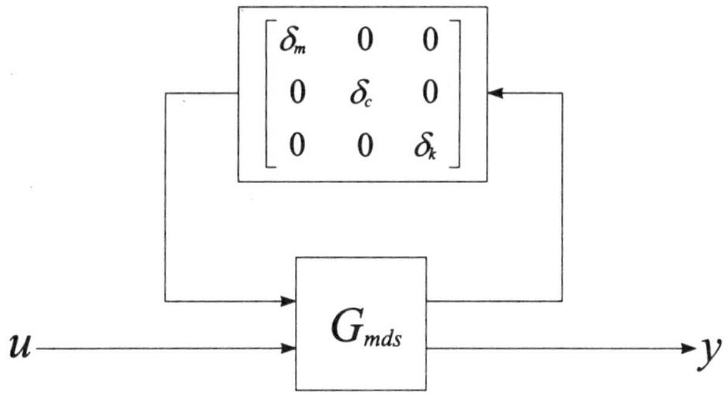

Fig. 8.6. LFT representation of the mass-damper-spring system with uncertainties

File sys_mds.m

```
m_nom = 3; c_nom = 1; k_nom = 2;
p_m = 0.4; p_c = 0.2; p_k = 0.3;
mat_mi = [-p_m 1/m_nom; -p_m 1/m_nom];
mat_c = [0 c_nom; p_c c_nom];
mat_k = [0 k_nom; p_k k_nom];
int1 = nd2sys([1],[1 0]);
int2 = nd2sys([1],[1 0]);
systemnames = 'mat_mi mat_c mat_k int1 int2';
sysoutname = 'G';
inputvar = '[um;uc;uk;u]';
input_to_mat_mi = '[um;u-mat_c(2)-mat_k(2)]';
input_to_mat_c = '[uc;int1]';
input_to_mat_k = '[uk;int2]';
input_to_int1 = '[mat_mi(2)]';
input_to_int2 = '[int1]';
outputvar = '[mat_mi(1);mat_c(1);mat_k(1);int2]';
sysic;
```

8.2 Frequency Analysis of Uncertain System

The frequency responses of the perturbed open-loop system may be computed by using the command **starp** at a few different values of the perturbation parameters δ_m, δ_c, δ_k. In the M-file **pfr_mds** below, three values of each perturbation are chosen, the corresponding open-loop transfer function matrices generated and frequency responses calculated and plotted.

File pfr_mds.m

```
%
% Frequency responses of the perturbed plants
%
mod_mds
omega = logspace(-1,1,100);
[delta1,delta2,delta3] = ndgrid([-1 0 1],[-1 0 1], ...
                                 [-1 0 1]);
for j = 1:27
    delta = diag([delta1(j),delta2(j),delta3(j)]);
    olp = starp(delta,G);
    olp_ic = sel(olp,1,1);
    olp_g = frsp(olp_ic,omega);
    figure(1)
    vplot('bode',olp_g,'c-')
    subplot(2,1,1)
    hold on
    subplot(2,1,2)
    hold on
end
subplot(2,1,1)
olp_ic = sel(G,4,4);
olp_g = frsp(olp_ic,omega);
vplot('bode',olp_g,'r--')
subplot(2,1,1)
title('BODE PLOTS OF PERTURBED PLANTS')
hold off
subplot(2,1,2)
hold off
```

The Bode plots of the family of perturbed systems for $-1 \leq \delta_m,\ \delta_c,\ \delta_k \leq 1$ are shown in Figure 8.7.

8.3 Design Requirements of Closed-loop System

The design objective for the mass-damper-spring system in this study is to find a linear, output feedback control $u(s) = K(s)y(s)$, which ensures the following properties of the closed-loop system.

Nominal stability and performance:

The controller designed should make the closed-loop system internally stable. Further, the required closed-loop system performance should be achieved for

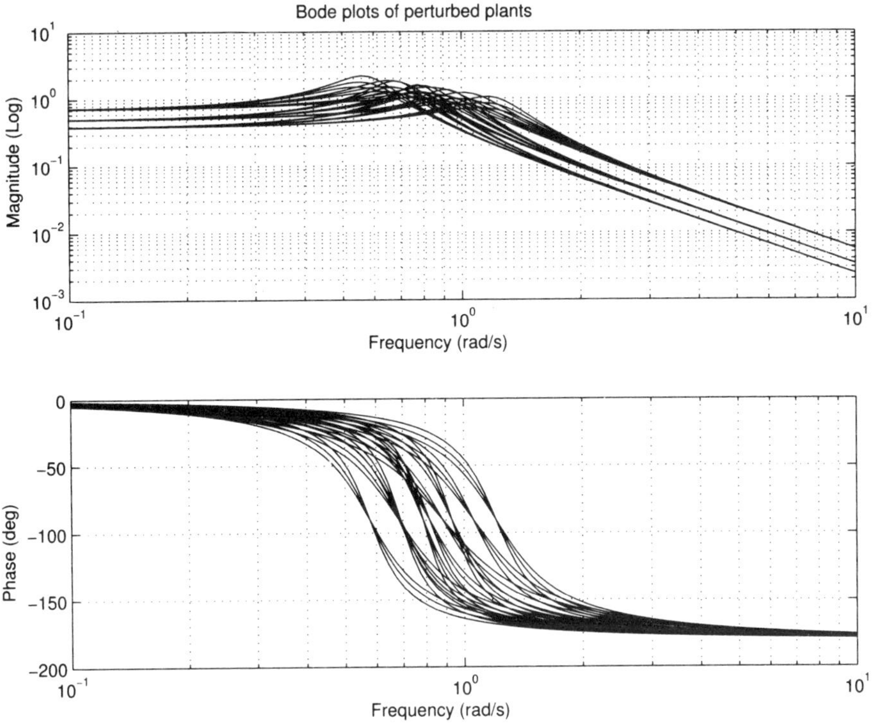

Fig. 8.7. Bode plots of perturbed open-loop systems

the nominal plant model G_{mds}. In this case study, the performance criterion for the closed-loop system is the so-called S *over* KS design and is described by

$$\left\| \begin{bmatrix} W_p S(G_{\mathrm{mds}}) \\ W_u K S(G_{\mathrm{mds}}) \end{bmatrix} \right\|_\infty < 1 \tag{8.1}$$

where $S(G_{\mathrm{mds}}) = (I + G_{\mathrm{mds}} K)^{-1}$ is the output sensitivity function of the nominal system, and W_p, W_u are weighting functions chosen to represent the frequency characteristics of some external (output) disturbance d and performance requirement (including consideration of control-effort constraint) level. Satisfaction of the above norm inequality indicates that the closed-loop system successfully reduces the effect of the disturbance to an acceptable level, and achieves the required performance. It should also be noted that the sensitivity function S denotes the transfer function of the reference tracking error.

Robust stability:

The closed-loop system achieves robust stability if the closed-loop system is internally stable for all possible plant models $G = F_U(G_{\text{mds}}, \Delta)$. In the present case this means that the system must remain stable for any $1.8 \leq m \leq 4.2,\ 0.8 \leq c \leq 1.2,\ 1.4 \leq k \leq 2.6$.

Robust performance:

In addition to the robust stability, the closed-loop system, for all $G = F_U(G_{\text{mds}}, \Delta)$, must satisfy the performance criterion

$$\left\| \begin{bmatrix} W_p(I+GK)^{-1} \\ W_u K(I+GK)^{-1} \end{bmatrix} \right\|_\infty < 1$$

Also, it is desirable that the complexity of the controller is acceptable, *i.e.*, it is of sufficiently low order.

The block diagram of the closed-loop system showing the feedback structure and including the elements reflecting model uncertainty and performance requirements, is given in Figure 8.8.

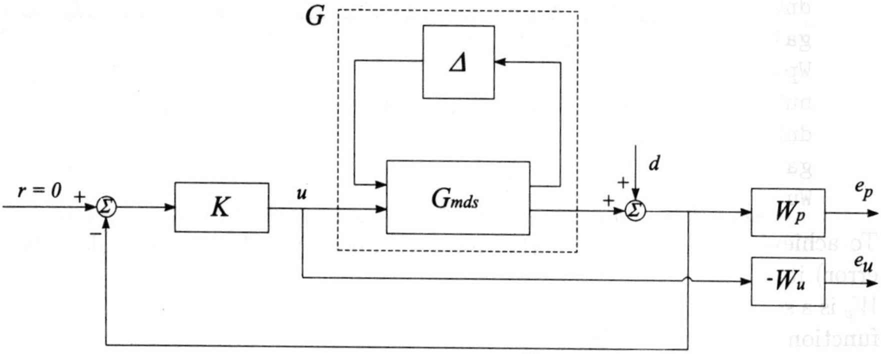

Fig. 8.8. Closed-loop system structure

The rectangle with dashed lines in Figure 8.8 represents the transfer function matrix G. Inside the rectangle is the nominal model G_{mds} of the mass-damper-spring system and the uncertainty matrix Δ that includes the model uncertainties. In general, the matrix Δ could be a transfer function matrix and is assumed to be stable. Δ is unknown but satisfies the norm condition $\|\Delta\|_\infty < 1$. The variable d is the disturbance on the system, at the system output. It is easy to work out that

$$\begin{bmatrix} e_p \\ e_u \end{bmatrix} = \begin{bmatrix} W_p(I+GK)^{-1} \\ W_u K(I+GK)^{-1} \end{bmatrix} d$$

Hence, the performance criterion is that the transfer functions from d to e_p and e_u should be small in the sense of $\|.\|_\infty$, for all possible uncertain transfer matrices Δ. The weighting functions W_p and W_u are used to reflect the relative significance of the performance requirement over different frequency ranges.

In the given case, the performance weighting function is a scalar function $W_p(s) = w_p(s)$ and is chosen as

$$w_p(s) = 0.95\frac{s^2 + 1.8s + 10}{s^2 + 8.0s + 0.01}$$

which ensures, apart from good disturbance attenuation, good transient response (settling time less than 10 s and overshoot less than 20% for the nominal system). The control weighting function W_u is chosen simply as the scalar $w_u = 10^{-2}$. Note that finding appropriate weighting functions is a crucial step in robust controller design and usually needs a few trials. For complex systems, significant efforts may be required.

To define the chosen weighting functions in MATLAB®, the following file `wts_mds.m` is used.

File `wts_mds.m`

```
nuWp = [1  1.8  10];
dnWp = [1  8    0.01];
gainWp = 0.95;
Wp = nd2sys(nuWp,dnWp,gainWp);
nuWu = 1;
dnWu = 1;
gainWu = 10^(-2);
Wu = nd2sys(nuWu,dnWu,gainWu);
```

To achieve the desired performance of disturbance rejection (or, of tracking error) it is necessary to satisfy the inequality $\|W_p(I+GK)^{-1}\|_\infty < 1$. Since W_p is a scalar function in the present case, the singular values of the sensitivity function $(I+GK)^{-1}$ over the frequency range must lie below that of $\frac{1}{w_p}$. This indicates $\|W_p(I+GK)^{-1}\|_\infty < 1$ if and only if for all frequencies $\sigma[(I+GK)^{-1}(j\omega)] < |1/w_p(j\omega)|$. The inverse weighting function $\frac{1}{w_p}$ is calculated by the commands

```
omega = logspace(-4,4,100);
Wp_g = frsp(Wp,omega);
Wpi_g = minv(Wp_g);
vplot('liv,lm',Wpi_g)
title('Inverse of Performance Weighting Function')
xlabel('Frequency (rad/sec)')
ylabel('Magnitude')
```

The singular values of $\frac{1}{w_p}$ over the frequency range $[10^{-4},\ 10^4]$ are shown in Figure 8.9.

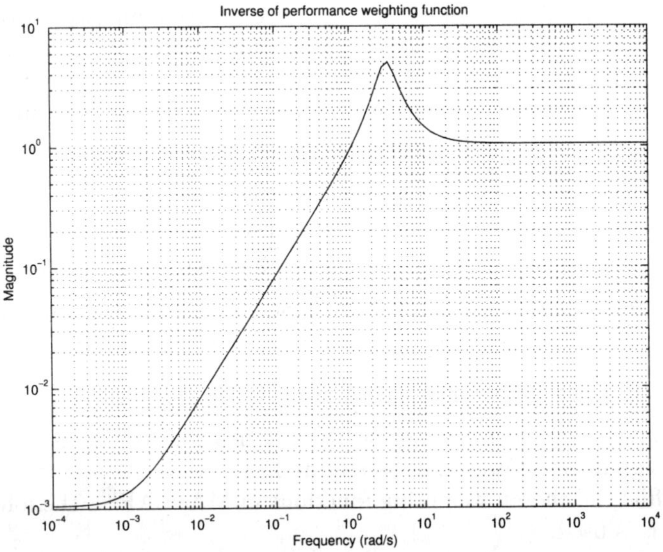

Fig. 8.9. Singular values of $\frac{1}{w_p}$

This weighting function shows that for low frequencies the closed-loop system (the nominal as well as perturbed) must attenuate the output disturbance in the ratio of 10 to 0.01. In other words, The effect of a unit disturbance on the steady-state output should be of the order 10^{-3} or less. (The same is also valid for the reference tracking error since in the given case the corresponding transfer function coincides with the sensitivity transfer function.) This performance requirement becomes less stringent with increasing frequency. It is seen in Figure 8.9 that from the frequency 1 rad/s the disturbance is no longer to be "attenuated". This shows in the time response the effect of disturbance will not be alleviated until some time later and will then be reduced to a scale of 10^{-3} or even less.

8.4 System Interconnections

The structure of the open-loop system is represented in Figure 8.10. The variable `y_c` (the controller input) is taken with negative sign since the commands intended for design in the Robust Control Toolbox produce controllers with positive feedback.

The variables `pertin` and `pertout` have three elements, and the variables `control`, `dist`, `e_p`, `e_u` and `y_c` have one element.

The command `sysic` can be used to create the structure of open-loop systems in MATLAB®. In the present case study, the open-loop system is

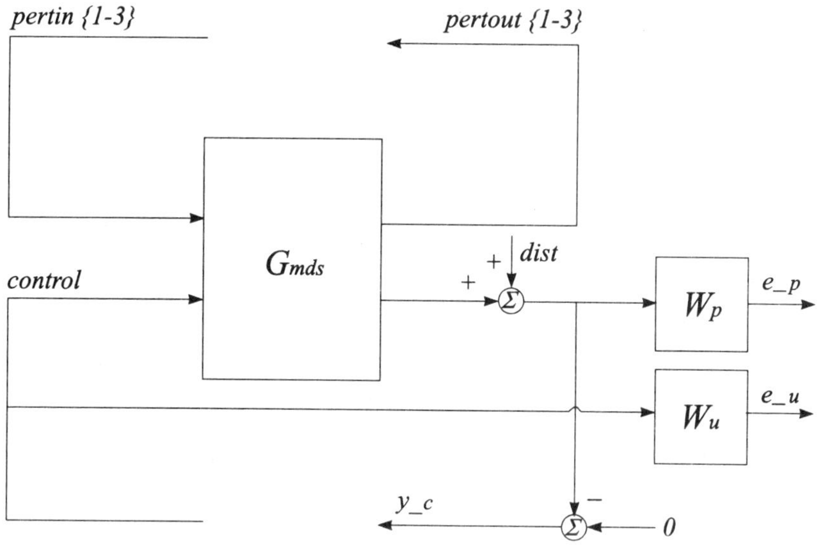

Fig. 8.10. Structure of open-loop system

saved as the variable `sys_ic`. `sys_ic` has five inputs and six outputs, as shown in Figure 8.11, and is a variable of the type SYSTEM (denoted also as P).

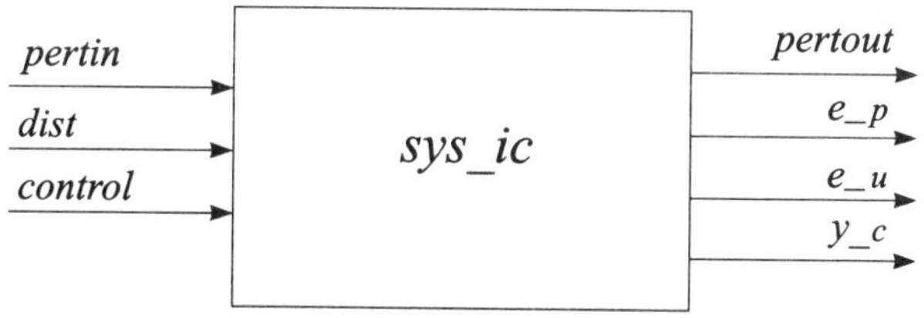

Fig. 8.11. Generalised block diagram of open-loop system

The following M-file `olp_mds.m` is used to create the variable `sys_ic`.

File `olp_mds.m`

```
systemnames = ' G Wp Wu ';
inputvar = '[ pert{3}; dist; control ]';
outputvar = '[ G(1:3); Wp; -Wu; -G(4)-dist ]';
input_to_G = '[ pert; control ]';
input_to_Wp = '[ G(4)+dist ]';
input_to_Wu = '[ control ]';
sysoutname = 'sys_ic';
```

```
cleanupsysic = 'yes';
sysic
```

To analyse the open-loop system, the following commands can be used.

```
minfo(sys_ic)
spoles(sys_ic)
spoles(W_p)
```

To assess the performance of designed systems, a unified simulation file of codes, `clp_mds.m`, is used and will be listed in Section 8.6. This simulation of closed-loop systems with designed controllers is based on the structure shown in Figure 8.12. Note that the weighting functions W_p and W_u are not included in the block-diagram for obvious reasons.

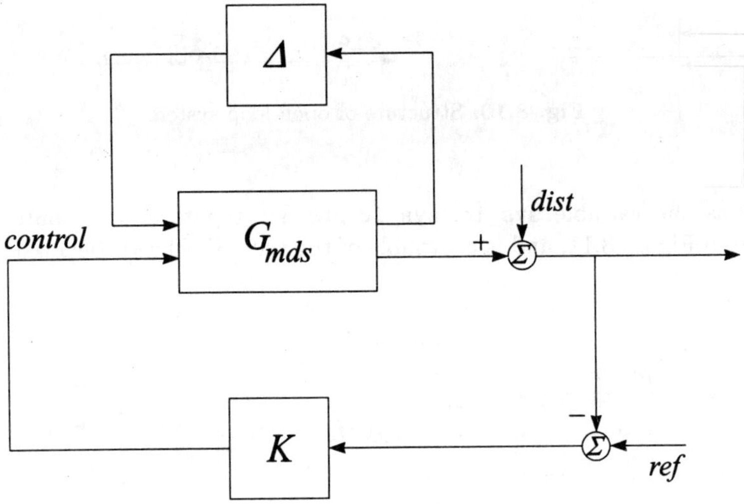

Fig. 8.12. Structure of the closed-loop system

The model of the open-loop system with uncertainties is set by the M-file `sim_mds.m` below.

File `sim_mds.m`

```
systemnames = ' G ';
inputvar = '[ pert{3}; ref; dist; control ]';
outputvar = '[ G(1:3); G(4)+dist; ref - G(4) - dist ]';
input_to_G = '[ pert; control ]';
sysoutname = 'sim_ic';
cleanupsysic = 'yes';
sysic
```

8.5 Suboptimal $\mathcal{H}_\infty$ Controller Design

The first controller to be designed for the connection of type SYSTEM, shown in Figure 8.11, is an $\mathcal{H}_\infty$ (sub)optimal controller. This controller minimizes the infinite-norm of $F_L(P, K)$ over all stabilising controllers K. Remember that $F_L(P, K)$ is the transfer function matrix of the nominal closed-loop system from the disturbance (the variable `dist`) to the errors (`e`), as shown in Figure 8.13, where $e = \begin{bmatrix} e_p \\ e_u \end{bmatrix}$. For this purpose, we first extract from `sys_ic` the corresponding transfer function matrix P using the following command and save it in the variable `hin_ic`.

```
hin_ic = sel(sys_ic,[4:6],[4:5])
```

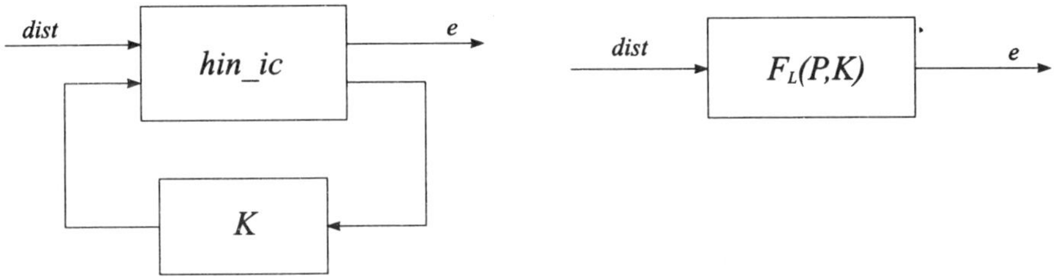

Fig. 8.13. Closed-loop LFTs in $\mathcal{H}_\infty$ design

The design uses the command `hinfsyn` that computes a suboptimal $\mathcal{H}_\infty$ controller, based on the given open-loop structure. The syntax, input and output arguments of `hinfsyn` are

```
[k,clp] = hinfsyn(p,nmeas,ncon,glow,ghigh,tol)
```

The arguments have the following meanings.

Input arguments

open-loop interconnection (matrix of type SYSTEM)	`p`
number of measurements	`nmeas`
number of controls	`ncons`
lower bound of bisection	`glow`
upper bound of bisection	`ghigh`
absolute tolerance for the bisection method	`tol`

Output arguments

controller (matrix of type SYSTEM)	`k`
closed-loop system (matrix of type SYSTEM)	`clp`

In the present exercise, the open-loop interconnection is saved in the variable `hin_ic`. It consists of one measurement (obtained by a sensor), two error signals, one control input, one disturbance and four states (two states of the plant plus two of the weighting function W_p). *Note that for the given structure, the open-loop system whose norm is to be minimised has 1 input and 2 outputs.* The interval for γ iteration is chosen between 0.1 and 10 with tolerance $tol = 0.001$. At each iteration the program displays the current value of γ and the results of five tests for existence of a suboptimal controller. At the end of each iteration the symbol `p` or `f` is displayed, which indicates the value of the current γ is either accepted or rejected. The symbol # is used to denote which of the five conditions for the existence of $\mathcal{H}_\infty$ (sub)optimal controllers is violated for the γ used. When the iteration procedure succeeds, the achievable minimum value of γ is given. The transfer function matrix of the closed-loop system from `dist` to `e` is saved in the variable `clp`. Below, we give the file `hin_mds.m` used to design an $\mathcal{H}_\infty$ (sub)optimal controller K_{hin}, followed by the display of results obtained in the exercise.

File `hin_mds.m`

```
nmeas = 1;
ncon = 1;
gmin = 1;
gmax = 10;
tol = 0.001;
hin_ic = sel(sys_ic,4:6,4:5);
[K_hin,clp] = hinfsyn(hin_ic,nmeas,ncon,gmin,gmax,tol);
```

```
Resetting value of Gamma min based on D_11, D_12, D_21 terms

Test bounds:        0.9500 <  gamma  <=     10.0000

  gamma  hamx_eig  xinf_eig  hamy_eig  yinf_eig   nrho_xy  p/f
 10.000  8.9e-001  6.2e-003  1.3e-003  0.0e+000   0.0000   p
  5.475  8.9e-001  6.2e-003  1.3e-003  0.0e+000   0.0000   p
  3.212  8.9e-001  6.3e-003  1.3e-003  -1.1e-020  0.0000   p
  2.081  8.8e-001  6.3e-003  1.3e-003  0.0e+000   0.0000   p
  1.516  8.8e-001  6.4e-003  1.3e-003  -2.2e-014  0.0000   p
  1.233  8.8e-001  6.5e-003  1.3e-003  -7.9e-021  0.0000   p
  1.091  8.8e-001  6.7e-003  1.3e-003  0.0e+000   0.0000   p
  1.021  8.8e-001  6.8e-003  1.3e-003  0.0e+000   0.0000   p
  0.985  8.8e-001  6.9e-003  1.3e-003  -7.9e-021  0.0000   p
```

```
0.968  8.8e-001  7.0e-003  1.3e-003  0.0e+000   0.0000   p
0.959  8.8e-001  7.1e-003  1.3e-003  -7.9e-021  0.0000   p
0.954  8.8e-001  7.2e-003  1.3e-003  0.0e+000   0.0000   p
0.952  8.8e-001  7.2e-003  1.3e-003  0.0e+000   0.0000   p
0.951  8.8e-001  7.2e-003  1.3e-003  -7.9e-021  0.0000   p
0.951  8.8e-001  7.2e-003  1.3e-003  0.0e+000   0.0000   p

Gamma value achieved:     0.9506
```

The controller obtained is of 4th order. To check the achieved $\mathcal{H}_\infty$-norm of the closed-loop system that is found to be 0.95 in the iteration, the following command line can be used.

```
hinfnorm(clp)
```

8.6 Analysis of Closed-loop System with K_{hin}

From the last section, it is obvious that the required condition

$$\left\| \begin{bmatrix} W_p(I + G_{\text{mds}}K)^{-1} \\ W_u K(I + G_{\text{mds}}K)^{-1} \end{bmatrix} \right\|_\infty < 1$$

is satisfied and thus the closed-loop system achieves nominal performance requirements. Further, we use the following commands to form the transfer function of Figure 8.13, with the designed (sub)optimal $\mathcal{H}_\infty$ controller K_{hin}, to analyse the behaviour of the closed-loop system.

```
minfo(K_hin)
spoles(K_hin)
omega = logspace(-2,6,100);
clp_g = frsp(clp,omega);
vplot('liv,lm',vsvd(clp_g))
title('Singular Value Plot of clp')
xlabel('Frequency (rad/sec)')
ylabel('Magnitude')
```

Figure 8.14 shows the singular values of the closed-loop system `clp`. Since `clp` is of dimensions 2×1, there is just one (nonzero) singular value at each frequency.

Since the $\mathcal{H}_\infty$-norm of the closed-loop system is less than one, the condition $\|W_p(I + G_{\text{mds}}K)^{-1}\|_\infty < 1$ is satisfied in the given case. This may be checked by computing the sensitivity function of the closed-loop system and comparing it with the inverse of the performance weighting function. The following file `sen_mds.m` can be used for this purpose.

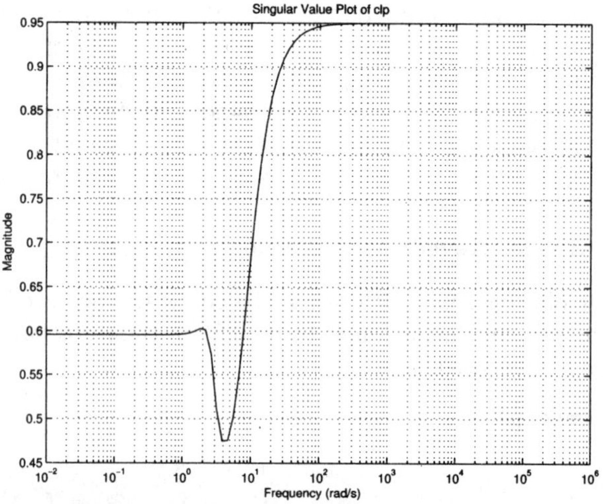

Fig. 8.14. Singular values of the closed-loop system with K_{hin}

File `sen_mds.m`

```
sim_mds
K = K_hin;
clp = starp(sim_ic,K);
%
% inverse performance weighting function
wts_mds
omega = logspace(-4,2,100);
Wp_g = frsp(Wp,omega);
Wpi_g = minv(Wp_g);
%
% sensitivity function
sen_loop = sel(clp,4,5);
sen_g = frsp(sen_loop,omega);
vplot('liv,lm',Wpi_g,'m--',vnorm(sen_g),'y-')
title('CLOSED-LOOP SENSITIVITY FUNCTION')
xlabel('Frequency (rad/sec)')
ylabel('Magnitude')
```

The result of the comparison is shown in Figure 8.15. It is seen that in the low-frequency range the sensitivity function lies below $\frac{1}{w_p}$.

The test for robust stability is conducted on the leading 3×3 diagonal block of the transfer function matrix `clp` and the test for nominal performance is tested on its bottom-right (the $(4-4)$ element), 2×1 transfer function. These transfer functions may be obtained by using the command `sel` from `clp`, and their frequency responses can be calculated afterwards. Or, it is

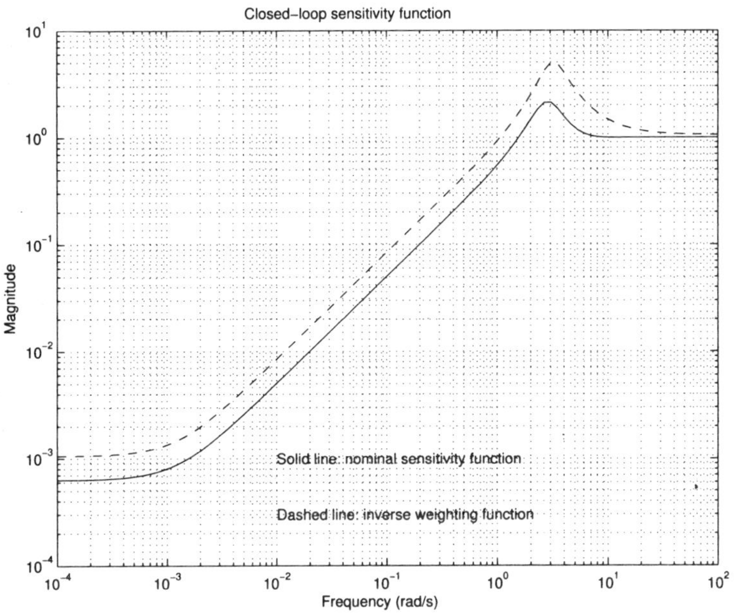

Fig. 8.15. Sensitivity function with K_{hin}

possible to calculate the frequency response `clp_g` of the whole system `clp` first, and then use `sel` to obtain corresponding frequency responses. The closed-loop transfer function matrix `clp_ic` is obtainable from the open-loop interconnection `sys_ic` together with the controller `K` by implementing the function `starp`.

Since the uncertainty considered is *structured*, verification of the robust stability and robust performance needs the frequency response in terms of μ values. The syntax of the command `mu` is as the following.

```
[bnds,dvec,sens,pvec] = mu(matin,deltaset)
```

The function `mu` for μ-analysis computes upper and lower bounds for the structured singular value of the matrix `matin` with respect to the block structure `deltaset`. The matrix `matin` may be a matrix of the type CONSTANT or a matrix of the type VARYING, such as the frequency response of the closed-loop system. The command `mu` finds the upper and lower bounds for μ values in the 1×2 matrix `bnds` of the type VARYING, the frequency-dependent D-scaling matrices in `dvec`, the frequency-dependent perturbation, related to the lower bound, in `pvec` and the sensitivity of the upper bound to the D-scalings in `sens`. To achieve robust stability it is necessary that the upper bound of μ values is less than 1 over the frequency range.

In the present example, the frequency response in terms of μ values is denoted by `rob_stab` and the block structure is set by

```
blksR = [-1 1;-1 1;-1 1]
```

This means that in the robust stability analysis the parametric perturbations are assumed to be real. However, for better convergence of the algorithm that computes the lower bound of μ, we include 1% complex perturbations. This may, of course, produce some conservativeness in the design.

The file `rob_mds.m` below analyses the robust stability of the designed system, in which

```
K = K_hin;
```

has been defined in advance. The same file may be used for robust stability analysis of other controllers K.

File `rob_mds.m`

```
clp_ic = starp(sys_ic,K);
omega = logspace(-1,2,100);
clp_g = frsp(clp_ic,omega);
blkrsR = [-1 1;-1 1;-1 1];
rob_stab = sel(clp_g,[1:3],[1:3]);
pdim = ynum(rob_stab);
fixl = [eye(pdim); 0.1*eye(pdim)]; % 1% Complex
fixr = fixl';
blkrs = [blkrsR; abs(blkrsR)];
clp_mix = mmult(fixl,rob_stab,fixr);
[rbnds,rowd,sens,rowp,rowg] = mu(clp_mix,blkrs);
disp(' ')
disp(['mu-robust stability: ' ...
      num2str(pkvnorm(sel(rbnds,1,1)))])
disp(' ')
vplot('liv,lm',sel(rbnds,1,1),'y--',sel(rbnds,1,2),'m-', ...
      vnorm(rob_stab),'c-.')
title('ROBUST STABILITY')
xlabel('Frequency (rad/s)')
ylabel('mu')
```

The frequency responses of the upper and lower bounds of μ are shown in Figure 8.16. It is clear from the figure that the closed-loop system with K_{hin} achieves robust stability. The maximum value of μ is 0.764 that shows that structured perturbations with norm less than $\frac{1}{0.764}$ are allowable, *i.e.*, the stability maintains for $\|\Delta\|_\infty < \frac{1}{0.764}$. In the same figure we also plot the frequency response of the maximum singular value of the leading 3×3 transfer function matrix, which characterises the robust stability with respect to unstructured perturbations. It is seen that the latter is greater than 1 over the frequency interval [0.5, 1] rad/s roughly. Thus, the robust stability

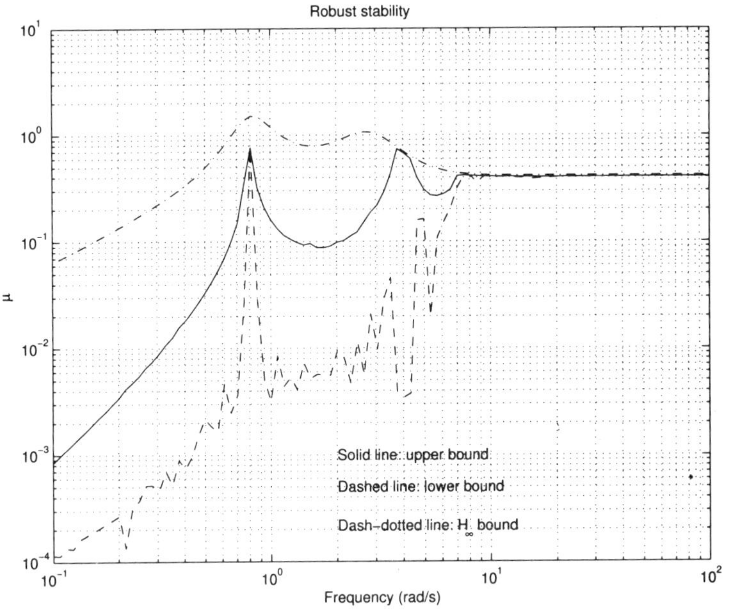

Fig. 8.16. Robust stability analysis of $K_{\rm hin}$

is not preserved if the uncertainty is unstructured, which shows that the μ values give less conservative results if further information is known about the uncertainty.

The nominal performance of the closed-loop system (with respect to the weighting performance function) is analysed by means of the frequency response of the lower-right 2×1 transfer function block of `clp`. The nominal performance is achieved, if and only if for the frequency range considered the response magnitude is less than 1.

The robust performance of the closed-loop system with $K_{\rm hin}$ may be tested by means of the μ-analysis. The closed-loop transfer function `clp` have four inputs and five outputs. The first three inputs/outputs correspond to the three channels of perturbations Δ, while the 4th input/4th and 5th outputs pair corresponds to the weighted mixed sensitivity function. Hence, for μ-analysis of the robust performance the block structure must consist of a 3×3 uncertainty block and an 1×2 performance block as

$$\Delta_P := \left\{ \begin{bmatrix} \Delta & 0 \\ 0 & \Delta_F \end{bmatrix} : \Delta \in \mathcal{R}^{3\times 3},\ \Delta_F \in \mathcal{C}^{1\times 2} \right\}$$

The robust performance of the designed system is achieved, if and only if $\mu_{\Delta_P}(\cdot)$ is less than 1 for each frequency.

The nominal and robust performance of the closed-loop system with the controller $K_{\rm hin}$ is analysed using the file `nrp_mds.m`, which may be used for other controllers as well by defining, accordingly, the controller K.

File `nrp_mds.m`

```
clp_ic = starp(sys_ic,K);;
omega = logspace(-1,2,100);
clp_g = frsp(clp_ic,omega);
blkrsR = [-1 1;-1 1;-1 1];
%
% nominal performance
nom_perf = sel(clp_g,4,4);
%
% robust performance
rob_perf = clp_g;
blkrp = [blkrsR;[1 2]];
bndsrp = mu(rob_perf,blkrp);
vplot('liv,lm',vnorm(nom_perf),'y-',sel(bndsrp,1,1),'m--',...
      sel(bndsrp,1,2),'c--')
tmp1 = 'NOMINAL PERFORMANCE (solid) and';
tmp2 = ' ROBUST PERFORMANCE (dashed)';
title([tmp1 tmp2])
xlabel('Frequency (rad/s)')
disp(' ')
disp(['mu-robust performance: ' ...
     num2str(pkvnorm(sel(bndsrp,1,1)))])
disp(' ')
```

The frequency responses showing the nominal and robust performance are given in Figure 8.17. It is seen from this figure that the system with $K_{\rm hin}$ achieves nominal performance but fails to satisfy the robust performance criterion. This conclusion follows from the fact that the frequency response of the nominal performance has a maximum of 0.95, while the μ curve for the robust performance has a maximum of 1.67.

With respect to the robust performance, this means in the present case that the size of the perturbation matrix Δ must be limited to $\|\Delta\|_\infty \leq \frac{1}{1.67}$, to ensure the (perturbed) performance function satisfying

$$\left\| \begin{bmatrix} W_p(I + F_U(G_{\rm mds}, \Delta_G)K)^{-1} \\ W_uK(I + F_U(G_{\rm mds}, \Delta_G)K)^{-1} \end{bmatrix} \right\|_\infty \leq 1$$

The closed-loop simulation with the file `clp_mds.m` includes determination of the transient responses in respect to the reference and disturbance, in which the command `trsp` is used.

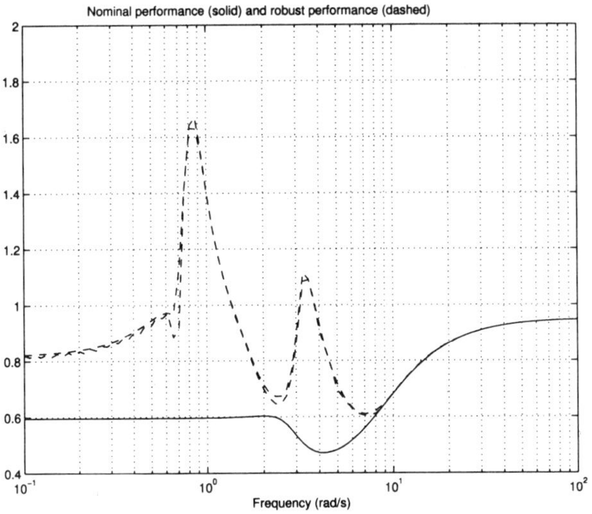

Fig. 8.17. Nominal and robust performance of K_{hin}

File clp_mds.m

```
% response to the reference
sim_mds
clp = starp(sim_ic,K);
timedata = [0 20 40];
stepdata = [1 0 1];
dist = 0;
ref = step_tr(timedata,stepdata,0.1,60);
u = abv(0,0,0,ref,dist);
y = trsp(clp,u,60,0.1);
figure(1)
vplot(sel(y,4,1),'y-',ref,'r--')
title('CLOSED-LOOP TRANSIENT RESPONSE')
xlabel('Time (secs)')
ylabel('y (m)')
%
% response to the disturbance
timedata = [0 20 40];
stepdata = [1 0 1];
dist = step_tr(timedata,stepdata,0.1,60);
ref = 0;
u = abv(0,0,0,ref,dist);
y = trsp(clp,u,60,0.1);
figure(2)
```

```
vplot(sel(y,4,1),'y-',dist,'r--')
title('TRANSIENT RESPONSE TO THE DISTURBANCE')
xlabel('Time (secs)')
ylabel('y (m)')
```

The transient responses to the reference input and to the disturbance input are shown in Figures 8.18 and 8.19, respectively. The transient responses are relatively slow and have slight overshoots.

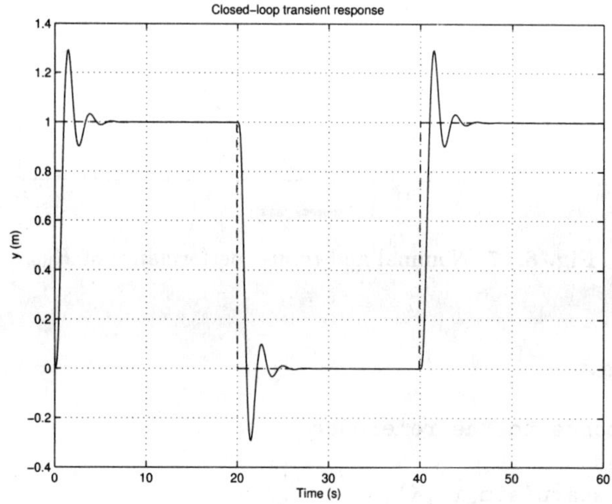

Fig. 8.18. Transient response to reference input (K_{hin})

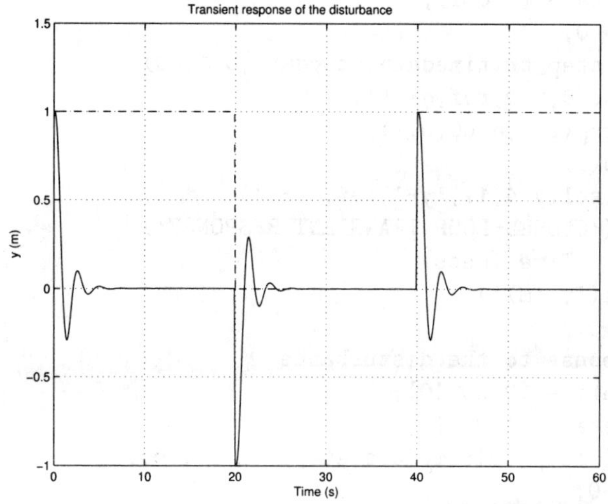

Fig. 8.19. Transient response to disturbance input (K_{hin})

8.7 $\mathcal{H}_\infty$ Loop-shaping Design

Let us consider now using the $\mathcal{H}_\infty$ loop-shaping design procedure (LSDP) for the mass-damper-spring system. For this aim the pre- and postcompensators (see Chapter 5) are taken as

$$W_1(s) = 2\frac{8s+1}{0.9}, \quad W_2(s) = 1$$

The precompensator W_1 is chosen so as to introduce an integrating effect in the low-frequency range that leads to good attenuation of disturbances.

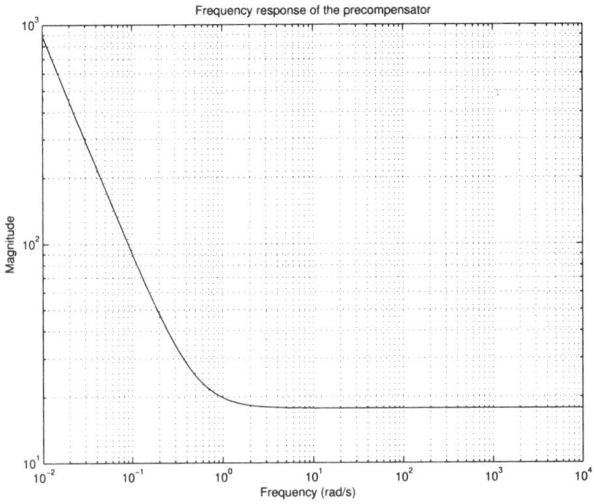

Fig. 8.20. Frequency response of the precompensator

The frequency response of the precompensator W_1 is shown in Figure 8.20, and the frequency responses of the plant and of the shaped plant are given in Figure 8.21.

The $\mathcal{H}_\infty$ LSDP controller is computed by calling the function `ncfsyn` from the Robust Control Toolbox. The syntax, input and output arguments of `ncfsyn` are

```
[sysk,emax,sysobs] = ncfsyn(sysgw,factor,opt)
```

The arguments have the following meanings:

Input arguments

`sysgw` the shaped plant
(matrix of type SYSTEM);

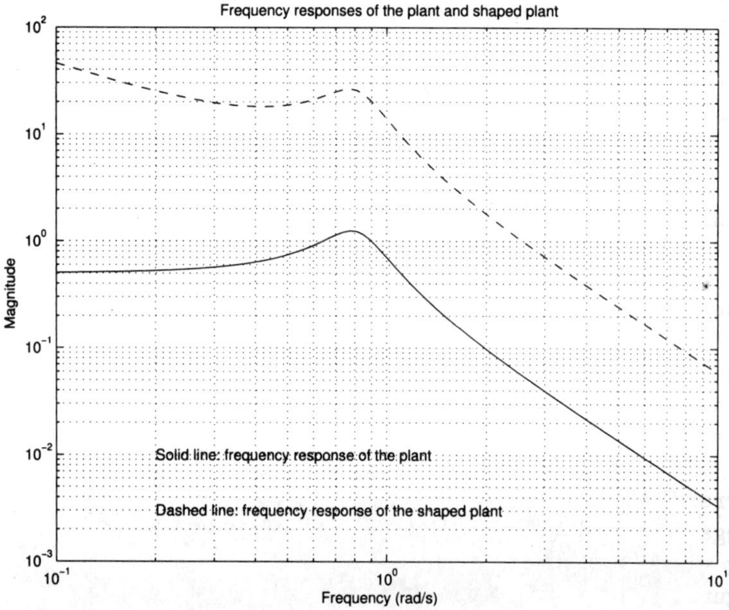

Fig. 8.21. Frequency responses of the plant and shaped plant

`factor` = 1 implies that optimal controller is required.
> 1 implies that suboptimal controller is required achieving a performance that is `FACTOR` times less than the optimal one;

`opt` `'ref'` the controller includes an additional set of reference signals and should be implemented as a two-degree-of-freedom controller (optional).

Output arguments

`sysk` $\mathcal{H}_\infty$ loop-shaping controller;

`emax` Stability margin that shows the robustness to unstructured perturbations. `emax` is always less than 1 and values of `emax` larger than 0.3 generally indicate good robustness margins;

`sysobs` $\mathcal{H}_\infty$ controller with state observer. This variable is created only if `factor>1` and `opt = 'ref'`.

The loop-shaping design of the mass-damper-spring system is executed by the file `lsh_mds.m` below. The parameter `factor` is chosen as 1.1, which means that the resulted suboptimal controller will be close to the optimal one. The designed controller is named $K_{\rm lsh}$ in the program.

File `lsh_mds.m`

```
%
% set the precompensator
nuw1 = [2 1];
dnw1 = [0.9 0];
gainw1 = 8;
w1 = nd2sys(nuw1,dnw1,gainw1);
%
% frequency response of w1
omega = logspace(-2,4,100);
w1g = frsp(w1,omega);
figure(1)
vplot('liv,lm',w1g,'r-')
title('Frequency response of the precompensator')
xlabel('Frequency (rad/sec)')
ylabel('Magnitude')
%
% form the shaped plant
G_l = sel(G,4,4);
sysGW = mmult(G_l,w1);
%
% frequency responses of the plant and shaped plant
omega = logspace(-1,1,100);
G_lg = frsp(G_l,omega);
sysGWg = frsp(sysGW,omega);
figure(2)
vplot('liv,lm',G_lg,'c-',sysGWg,'m--')
title('Frequency responses of the plant and shaped plant')
xlabel('Frequency (rad/sec)')
ylabel('Magnitude')
%
% compute the suboptimal positive feedback controller
[sysK,emax] = ncfsyn(sysGW,1.1);
disp(['Stability margin emax = ' num2str(emax)]);
K = mmult(w1,sysK);
%
% construct the negative feedback controller
[ak,bk,ck,dk] = unpck(K);
```

```
ck = -ck;
dk = -dk;
K_lsh = pck(ak,bk,ck,dk);
```

8.8 Assessment of $\mathcal{H}_\infty$ Loop-shaping Design

The $\mathcal{H}_\infty$ LSDP controller K_{lsh} ensures a stability margin `emax = 0.395`, which is a good indication with respect to the robust stability. As in the case of the previous $\mathcal{H}_\infty$ design, the order of the resulting controller is equal to 4.

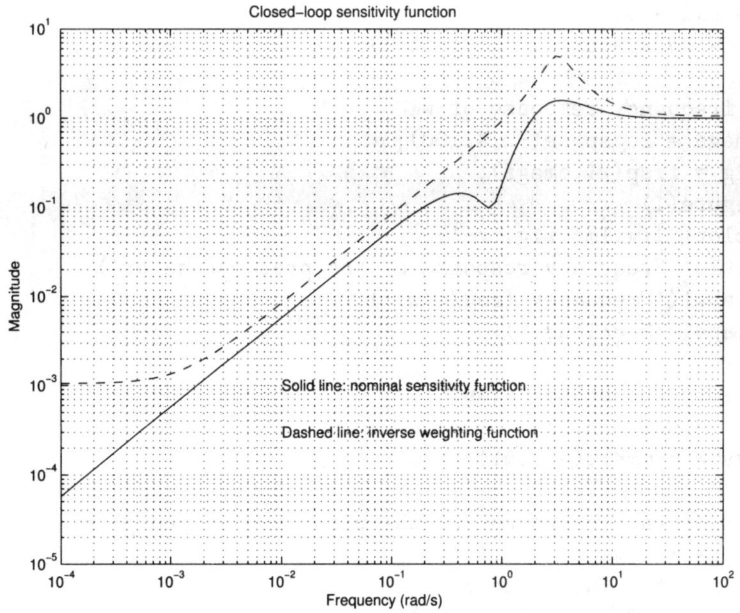

Fig. 8.22. Sensitivity function of K_{lsh}

The sensitivity function of the closed-loop system of K_{lsh} is shown in Figure 8.22. It is clear that the requirement for disturbance attenuation is satisfied.

To check the robust stability of the designed closed-loop system with K_{lsh}, the upper- and low-bounds of μ values of the corresponding transfer function matrix (the leading 3×3 diagonal block of `clp` with K_{lsh}) are shown in Figure 8.23. It is clear from that figure that the closed-loop system achieves robust stability for the parametric perturbations under consideration.

The nominal and robust performance of the closed-loop system are assessed in Figure 8.24. It is seen in the figure that with the designed $\mathcal{H}_\infty$

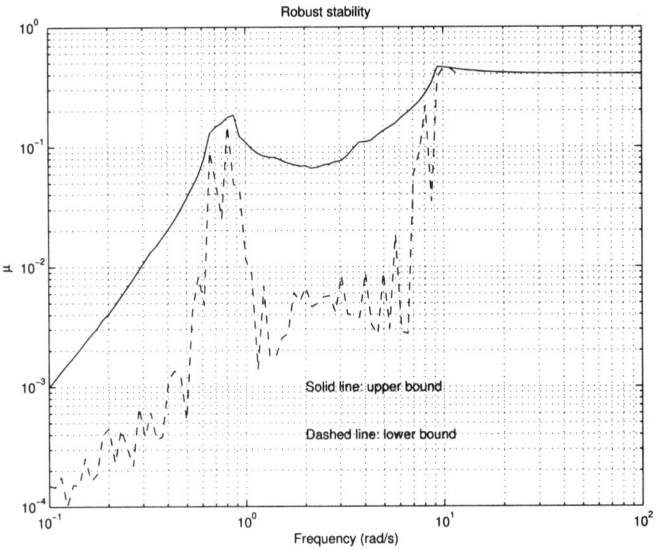

Fig. 8.23. Robust stability of closed-loop system with $K_{\rm lsh}$

LSDP controller the closed-loop system achieves both nominal and robust performance.

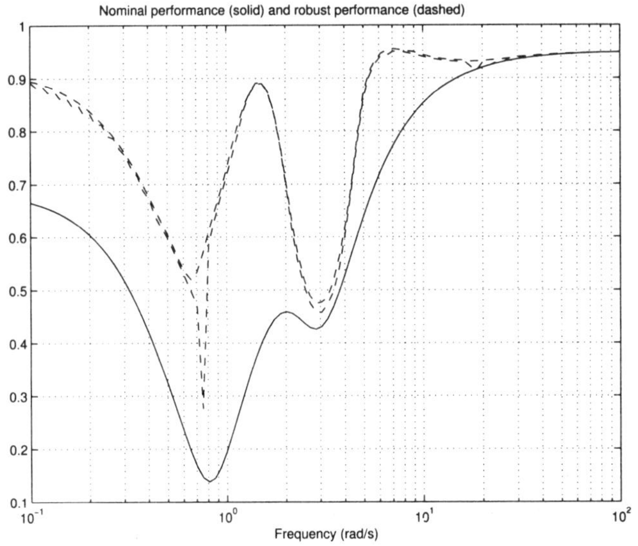

Fig. 8.24. Nominal and robust performance of $K_{\rm lsh}$

Further, the transient responses of the closed-loop system are obtained, once more using the file `clp_mds.m`, and are shown in Figures 8.25 and 8.26. The time responses are slower than in the case of the $\mathcal{H}_\infty$ controller but with smaller overshoots (less than 25%).

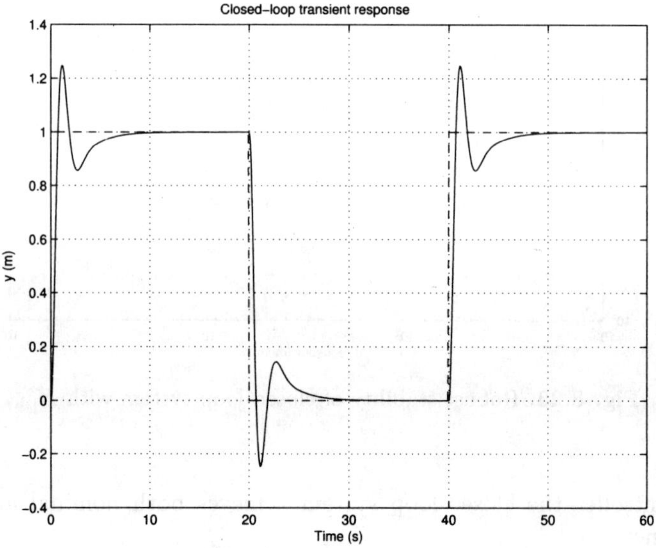

Fig. 8.25. Transient response to reference input (K_{lsh})

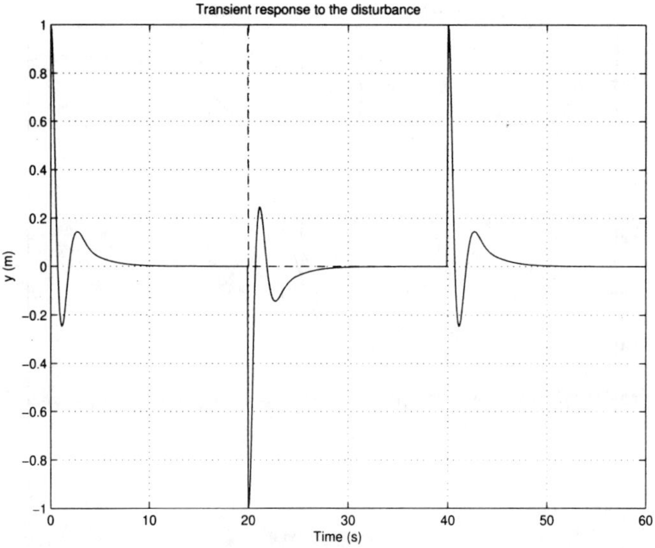

Fig. 8.26. Transient response to disturbance input (K_{lsh})

8.9 μ-Synthesis and D-K Iterations

The block diagram of the closed-loop system used in the μ-synthesis is given in Figure 8.8 and is reproduced for convenience in Figure 8.27.

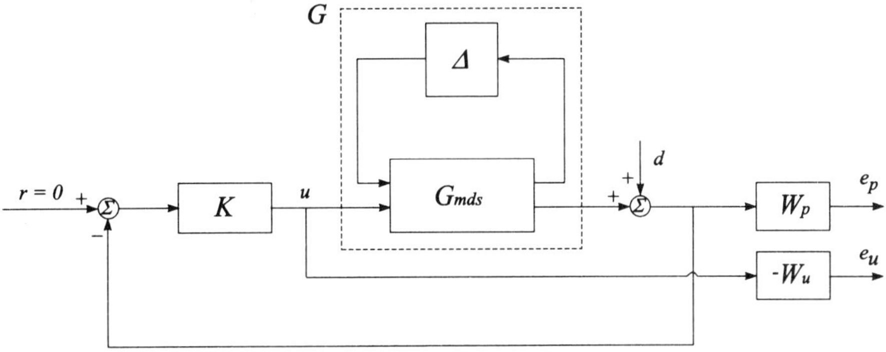

Fig. 8.27. Structure of closed-loop system

Let $P(s)$ denote the transfer function matrix of the five inputs, six outputs open-loop system sys_ic in Figure 8.11 and let the block structure Δ_P be defined as in the case of the robust performance analysis in previous sections as

$$\Delta_P := \left\{ \begin{bmatrix} \Delta & 0 \\ 0 & \Delta_F \end{bmatrix} : \Delta \in \mathcal{R}^{3\times 3},\ \Delta_F \in \mathcal{C}^{1\times 2} \right\}$$

The first uncertainty block Δ of this structured matrix is diagonal and corresponds to the uncertainties used in the modelling of the mass-damper-spring system. The second block Δ_F is a fictitious uncertainty block that is introduced to represent the performance requirements in the framework of the μ-approach.

The following optimisation problem is formed to minimize the upper bound of μ values which in turn reduces the maximum value of μ.

$$\min_{\substack{K \\ \text{stabilizing}}} \quad \min_{\substack{D_\ell(s),\, D_r(s) \\ \text{stable,} \\ \text{min. phaze}}} \left\| D_\ell(s) F_L(P,K) D_r^{-1}(s) \right\|_\infty$$

$$D_\ell(s) = \begin{bmatrix} d_1(s) & 0 & 0 & 0 \\ 0 & d_2(s) & 0 & 0 \\ 0 & 0 & d_3(s) & 0 \\ 0 & 0 & 0 & d_4(s)I_2 \end{bmatrix}$$

$$D_r(s) = \begin{bmatrix} d_1(s) & 0 & 0 & 0 \\ 0 & d_2(s) & 0 & 0 \\ 0 & 0 & d_3(s) & 0 \\ 0 & 0 & 0 & d_4(s) \end{bmatrix}$$

where $d_1(s)$, $d_2(s)$, $d_3(s)$, $d_4(s)$ are scaling transfer functions. The finding of a minimum value of the cost function and construction of a controller K, which would achieve performance arbitrary close to the optimal level, is called the μ-synthesis.

In other words, the aim of μ-synthesis is to find a stabilising controller K, such that for each frequency $\omega \in [0, \infty]$ the structured singular value is to satisfy the condition

$$\mu_{\Delta_P}[F_L(P, K)(j\omega)] < 1$$

Fulfilment of the above condition guarantees robust performance of the closed-loop system, *i.e.*,

$$\left\| \begin{bmatrix} W_p(I + F_U(G_{\text{mds}}, \Delta)K)^{-1} \\ W_u K(I + F_U(G_{\text{mds}}, \Delta)K)^{-1} \end{bmatrix} \right\|_\infty < 1$$

The μ-synthesis is executed by the M-file **dkit** from the Robust Control Toolbox, which automates the procedure by using D-K iterations. To implement the function **dkit** it is necessary to prepare a file such as the following **dk_mds.m**, in which the necessary variables are assigned. The file below can be easily modified for the μ-synthesis of other systems.

File dk_mds.m

```
% dk_mds
%
%  This script file contains the USER DEFINED VARIABLES
%  for the mutools DKIT script file. The user MUST define
%  the 5 variables below.
%----------------------------------------------------%
%          REQUIRED USER DEFINED VARIABLES           %
%----------------------------------------------------%
% Nominal plant interconnection structure
NOMINAL_DK = sys_ic;

% Number of measurements
NMEAS_DK = 1;

% Number of control inputs
NCONT_DK = 1;

% Block structure for mu calculation
```

```
BLK_DK = [-1 1;-1 1;-1 1;1 2];

% Frequency response range
OMEGA_DK = logspace(-2,4,100);

AUTOINFO_DK = [1 4 1 4*ones(1,size(BLK_DK,1))];

NAME_DK = 'mds';
%------------------ end of dk_mds -------------------- %
```

The variables, set in the above file, have the following meanings.

NOMINAL_DK	A variable containing the nominal open-loop system
NMEAS_DK	Number of measurements (number of inputs of K)
NCONT_DK	Number of control inputs (number of outputs of K)
BLK_DK	Block structure for computation of μ (involves the uncertainty blocks as well as the performance block)
OMEGA_DK	Frequency response range
AUTINFO_DK	Variable, which is used for full automation of the D-K-iteration. It has the following components:
AUTOINFO_DK(1)	Initial iteration
AUTOINFO_DK(2)	Final iteration
AUTOINFO_DK(3)	Visualisation flag (1 - the results appear on the display, 2 - the results are not displayed) The rest elements in **AUTOINFO_DK** (their number is equal to the number of blocks in BLK_DK) set the maximum dynamic order of the transfer functions during the D-scaling
NAME_DK	Suffix to the names of the saved variables

After preparing the file **dk_mds**, it is necessary to assign the name of the string variable DK_DEF_NAME, in the MATLAB® workspace, to **DK_MDS**. Now the command **dkit** can be called, with which begins the D-K-iteration procedure. The whole design is completed by the file **ms_mds.m** below.

File ms_mds.m

```
DK_DEF_NAME = 'dk_mds';
dkit
```

```
K_mu = k_dk4mds;
```

Note that the μ-controller named as K_μ is obtained after four iterations in this exercise.

The results from the first iteration are listed below, in which the numbers have been shortened for display.

```
Iteration Number:  1
--------------------

Resetting value of Gamma min based on D_11, D_12, D_21 terms

Test bounds:       0.9500 <  gamma  <=     100.0000

gamma    hamx_eig  xinf_eig  hamy_eig  yinf_eig   nrho_xy p/f
100.000  6.4e-001  2.2e-001  1.3e-003  -1.6e-013  0.0002  p
 50.475  6.3e-001  2.2e-001  1.3e-003  0.0e+000   0.0010  p
 25.713  6.3e-001  2.2e-001  1.3e-003  -2.1e-014  0.0038  p
 13.331  6.3e-001  2.2e-001  1.3e-003  -1.9e-013  0.0143  p
  7.141  6.1e-001  2.2e-001  1.3e-003  0.0e+000   0.0522  p
  4.045  5.5e-001  2.3e-001  1.3e-003  -7.3e-018  0.1924  p
  2.498  3.8e-001  2.4e-001  1.3e-003  2.9e-019   1.6946# f
  3.097  4.8e-001  2.3e-001  1.3e-003  0.0e+000   0.4322  p

 Gamma value achieved:      3.0966
Calculating MU of closed-loop system:
points completed....
1.2.3.4.5.6.7.8.9.10.11.12.13.14.15.16.17.
18.19.20.21.22.23.24.25.26.27.28.29.30.31.32.33.34.35.
36.37.38.39.40.41.42.43.44.45.46.47.48.49.50.51.52.53.
54.55.56.57.58.59.60.61.62.63.64.65.66.67.68.69.70.71.
72.73.74.75.76.77.78.79.80.81.82.83.84.85.86.87.88.89.
90.91.92.93.94.95.96.97.98.99.100.
points completed....
1.2.3.4.5.6.7.8.9.10.11.12.13.14.15.16.17.
18.19.20.21.22.23.24.25.26.27.28.29.30.31.32.33.34.35.
36.37.38.39.40.41.42.43.44.45.46.47.48.49.50.51.52.53.
54.55.56.57.58.59.60.61.62.63.64.65.66.67.68.69.70.71.
72.73.74.75.76.77.78.79.80.81.82.83.84.85.86.87.88.89.
90.91.92.93.94.95.96.97.98.99.100.
Iteration Summary
-----------------------------
Iteration #                 1
Controller Order            4
Total D-Scale Order         0
Gamma Acheived          3.097
```

```
Peak mu-Value            2.340

Auto Fit in Progress
     Block 1, MaxOrder=4, Order = 0 1 2 3 4
     Block 2, MaxOrder=4, Order = 0 1 2 3
     Block 3, MaxOrder=4, Order = 0 1 2
     Block 4, MaxOrder=4, Order = 0
```

The results obtained may be interpreted in the following way.

First, the design of a fourth-order $\mathcal{H}_\infty$ controller is finished (initially the open-loop system is of fourth order), with the scaling matrix being set equal to the unit matrix. With this controller one achieves a value of γ equal to 3.097 and maximum value of μ equal to 2.34.

Then, an approximation (curve fitting) of the diagonal elements of the scaling matrix, obtained in the computation of μ, takes place. These elements are functions of the frequency and are approximated by stable minimum phase transfer functions whose orders do not exceed 4. It is seen from the results that the first diagonal element is approximated by a fourth-order transfer function, the second one by a third-order transfer function, the third one by a second-order transfer function, and the fourth one by a scalar. These transfer functions are "absorbed" into the transfer function P, which is to be used in the $\mathcal{H}_\infty$ design at the second iteration. Altogether, the order of the "absorbed" transfer functions is $9 + 9 = 18$. This increase of the open-loop system order leads to the increase of the controller order at the second iteration to 22.

The structured singular value μ of the closed-loop system with the $\mathcal{H}_\infty$ controller from the first step and the $\mathcal{H}_\infty$-norm of the scaled system are shown in Figure 8.28. The latter is an upper bound of the former.

The second iteration gives:

```
Iteration Number:  2
--------------------


Resetting value of Gamma min based on D_11, D_12, D_21 terms

Test bounds:        0.9500 <  gamma   <=       2.3873

 gamma  hamx_eig  xinf_eig  hamy_eig   yinf_eig   nrho_xy  p/f
 2.387  1.1e-002  2.6e-008  1.3e-003   -1.2e-017  0.0671    p
 1.669  1.1e-002  2.6e-008  1.3e-003   -1.1e-014  0.1720    p
 1.309  1.1e-002  2.6e-008  1.3e-003   -1.7e-017  1.5886#   f
 1.433  1.1e-002  2.6e-008  1.3e-003   1.4e-017   0.3344    p
 1.358  1.1e-002  2.6e-008  1.3e-003   -2.9e-020  0.5738    p
 1.333  1.1e-002  2.6e-008  1.3e-003   -1.9e-020  0.8201    p

 Gamma value achieved:     1.3334
Calculating MU of closed-loop system:
```

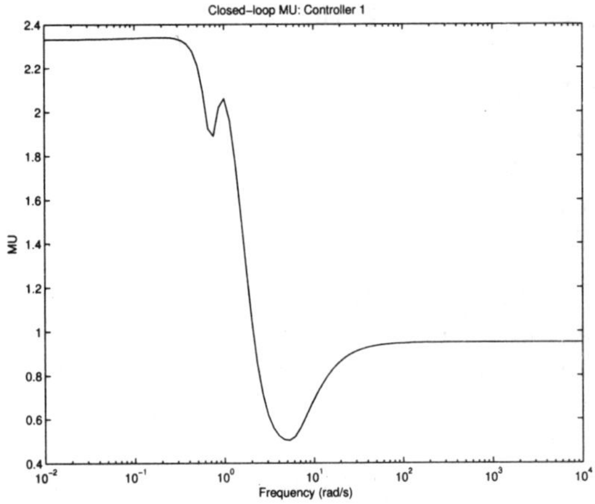

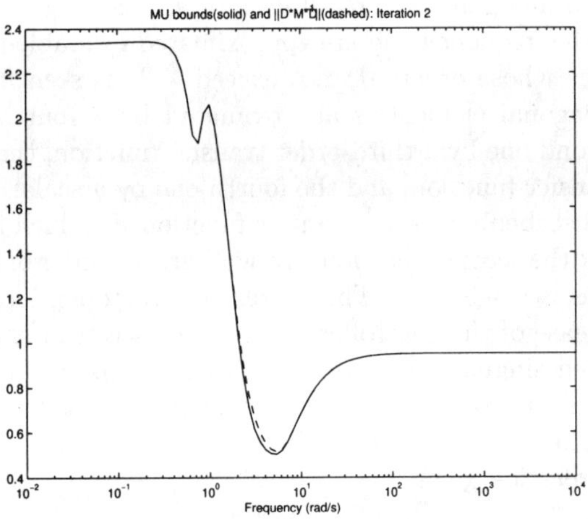

Fig. 8.28. μ values and D–scaling at first iteration

```
points completed....
1.2.3.4.5.6.7.8.9.10.11.12.13.14.15.16.17.
18.19.20.21.22.23.24.25.26.27.28.29.30.31.32.33.34.35.
36.37.38.39.40.41.42.43.44.45.46.47.48.49.50.51.52.53.
54.55.56.57.58.59.60.61.62.63.64.65.66.67.68.69.70.71.
72.73.74.75.76.77.78.79.80.81.82.83.84.85.86.87.88.89.
90.91.92.93.94.95.96.97.98.99.100.
points completed....
1.2.3.4.5.6.7.8.9.10.11.12.13.14.15.16.17.
18.19.20.21.22.23.24.25.26.27.28.29.30.31.32.33.34.35.
36.37.38.39.40.41.42.43.44.45.46.47.48.49.50.51.52.53.
54.55.56.57.58.59.60.61.62.63.64.65.66.67.68.69.70.71.
72.73.74.75.76.77.78.79.80.81.82.83.84.85.86.87.88.89.
```

```
90.91.92.93.94.95.96.97.98.99.100.
Iteration Summary
-----------------------------------------
Iteration #                  1        2
Controller Order             4       22
Total D-Scale Order          0       18
Gamma Acheived           3.097    1.333
Peak mu-Value            2.340    1.274

Auto Fit in Progress
     Block 1, MaxOrder=4, Order = 0 1 2 3 4
     Block 2, MaxOrder=4, Order = 0 1 2 3
     Block 3, MaxOrder=4, Order = 0 1 2 3
     Block 4, MaxOrder=4, Order = 0
```

After this iteration, the value of γ decreases to 1.333, and the value of μ reduces to 1.274. This is achieved, however, by a 22nd-order controller. The corresponding closed-loop system obtained is of 26th order. The diagonal elements of the scaling matrix are approximated by transfer functions of orders 4, 3, 3 and 0, respectively, which increases the order of the open-loop system by 20. Thus, the open-loop system is now of 24th order. Note that the increase of the open-loop system order is significant for the controller order only at the next step and there is no "accumulation" of order in the absorbed transfer functions in the sense that at each iteration new scaling functions are generated and old ones are discarded.

The μ values and the $\mathcal{H}_\infty$-norm of DMD^{-1} at Iteration 2 are shown in Figure 8.29. Note the approximation error in the range from 1 rad/s to 1000 rad/s.

The results after the third iteration are:

```
Iteration Number:  3
--------------------


Resetting value of Gamma min based on D_11, D_12, D_21 terms

Test bounds:      0.9500 <  gamma  <=      1.4889

gamma  hamx_eig  xinf_eig  hamy_eig   yinf_eig   nrho_xy  p/f
1.489  1.1e-002  4.4e-011  1.3e-003  -8.9e-019  0.2997    p
1.219  1.1e-002  4.4e-011  1.3e-003   1.8e-019  0.5799    p
1.085  1.1e-002  4.4e-011  1.3e-003  -4.0e-018  0.9993    p
1.017  1.1e-002  4.4e-011  1.3e-003  -1.9e-018  1.5626#   f
1.071  1.1e-002  4.4e-011  1.3e-003  -2.5e-019  1.0760#   f
1.082  1.1e-002  4.4e-011  1.3e-003  -3.4e-020  1.0138#   f

 Gamma value achieved:      1.0847
```

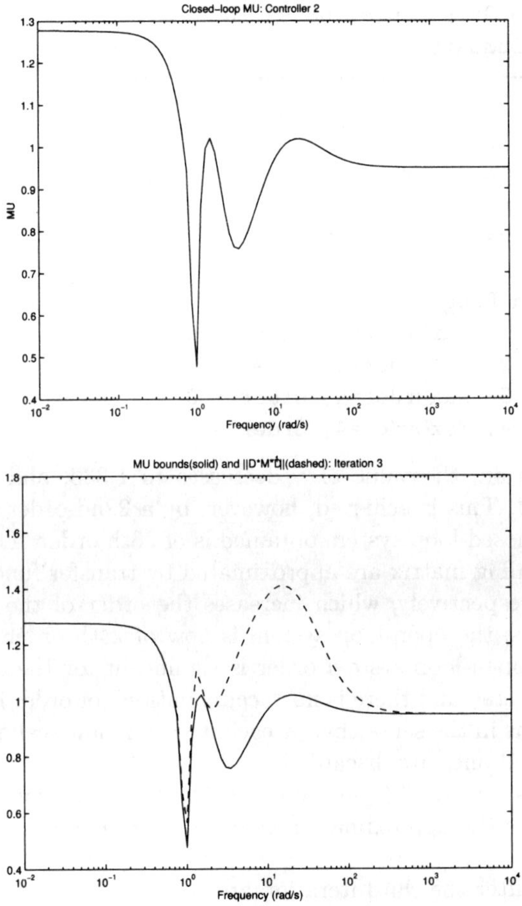

Fig. 8.29. μ and D–scaling at second iteration

```
Calculating MU of closed-loop system:
points completed....
1.2.3.4.5.6.7.8.9.10.11.12.13.14.15.16.17.
18.19.20.21.22.23.24.25.26.27.28.29.30.31.32.33.34.35.
36.37.38.39.40.41.42.43.44.45.46.47.48.49.50.51.52.53.
54.55.56.57.58.59.60.61.62.63.64.65.66.67.68.69.70.71.
72.73.74.75.76.77.78.79.80.81.82.83.84.85.86.87.88.89.
90.91.92.93.94.95.96.97.98.99.100.
points completed....
1.2.3.4.5.6.7.8.9.10.11.12.13.14.15.16.17.
18.19.20.21.22.23.24.25.26.27.28.29.30.31.32.33.34.35.
36.37.38.39.40.41.42.43.44.45.46.47.48.49.50.51.52.53.
```

```
54.55.56.57.58.59.60.61.62.63.64.65.66.67.68.69.70.71.
72.73.74.75.76.77.78.79.80.81.82.83.84.85.86.87.88.89.
90.91.92.93.94.95.96.97.98.99.100.
Iteration Summary
-------------------------------------------------
Iteration #                  1        2        3
Controller Order             4       22       24
Total D-Scale Order          0       18       20
Gamma Acheived           3.097    1.333    1.085
Peak mu-Value            2.340    1.274    1.079

Auto Fit in Progress
     Block 1, MaxOrder=4, Order = 0 1 2 3
     Block 2, MaxOrder=4, Order = 0 1 2 3
     Block 3, MaxOrder=4, Order = 0 1 2
     Block 4, MaxOrder=4, Order = 0
```

The value of γ after this iteration is 1.085, and the maximum value of μ is 1.079. The elements of the scaling matrix are approximated by transfer functions of 3rd, 3rd, 2nd and 0th orders, respectively, so that the order of the open-loop system for computation of a controller at the next iteration becomes 20.

The μ values of M and the $\mathcal{H}_\infty$-norm of DMD^{-1} at the third iteration are shown in Figure 8.30.

After the fourth iteration we obtain:

```
Iteration Number:  4
--------------------


Resetting value of Gamma min based on D_11, D_12, D_21 terms

Test bounds:       0.9500 <  gamma   <=      1.1153

gamma   hamx_eig  xinf_eig  hamy_eig   yinf_eig   nrho_xy  p/f
1.115  7.1e-003  1.3e-009  1.3e-003  -5.3e-021  0.3322    p
1.033  7.1e-003  1.4e-009  1.3e-003  -9.0e-019  0.4737    p
0.991  7.1e-003  1.4e-009  1.3e-003  -2.3e-019  0.6042    p
0.971  7.1e-003  1.4e-009  1.3e-003  7.7e-022   0.7069    p

 Gamma value achieved:      0.9707
Calculating MU of closed-loop system:
points completed....
1.2.3.4.5.6.7.8.9.10.11.12.13.14.15.16.17.
18.19.20.21.22.23.24.25.26.27.28.29.30.31.32.33.34.35.
36.37.38.39.40.41.42.43.44.45.46.47.48.49.50.51.52.53.
54.55.56.57.58.59.60.61.62.63.64.65.66.67.68.69.70.71.
```

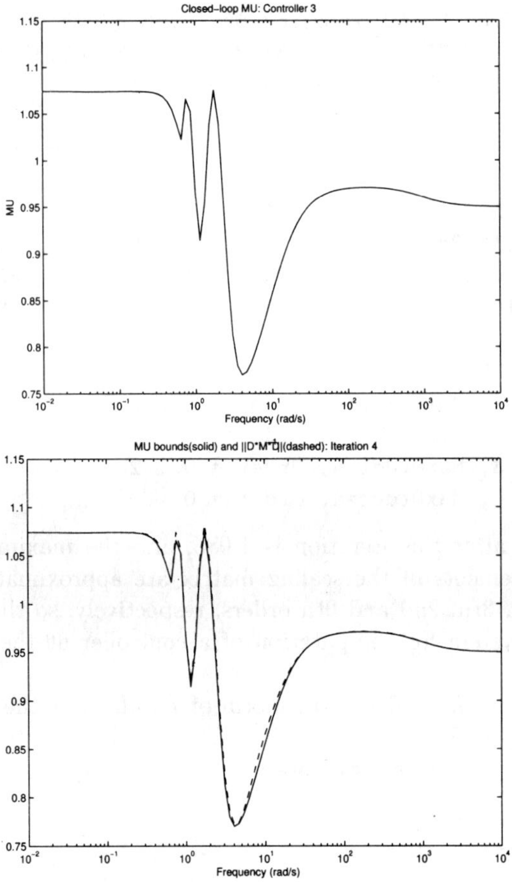

Fig. 8.30. μ and D–scaling at third iteration

```
72.73.74.75.76.77.78.79.80.81.82.83.84.85.86.87.88.89.
90.91.92.93.94.95.96.97.98.99.100.
points completed....
1.2.3.4.5.6.7.8.9.10.11.12.13.14.15.16.17.
18.19.20.21.22.23.24.25.26.27.28.29.30.31.32.33.34.35.
36.37.38.39.40.41.42.43.44.45.46.47.48.49.50.51.52.53.
54.55.56.57.58.59.60.61.62.63.64.65.66.67.68.69.70.71.
72.73.74.75.76.77.78.79.80.81.82.83.84.85.86.87.88.89.
90.91.92.93.94.95.96.97.98.99.100.
Iteration Summary
-----------------------------------------------------------
Iteration #                  1          2          3          4
Controller Order             4         22         24         20
```

```
Total D-Scale Order          0        18        20        16
Gamma Acheived           3.097     1.333     1.085     0.971
Peak mu-Value            2.340     1.274     1.079     0.965

  Next MU iteration number:  5
```

It is seen that at this iteration the value of γ decreases to 0.971 and the value of μ becomes equal to 0.965, which means that robust performance has been achieved.

The μ plot of the closed-loop system with the newly obtained controller K_{mu} at Iteration 4 is shown in Figure 8.31.

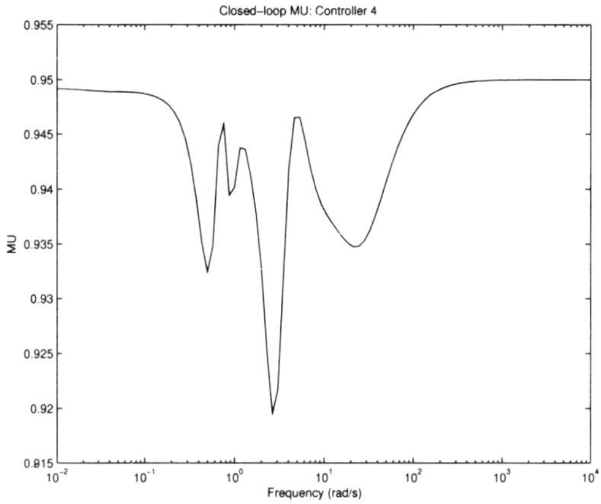

Fig. 8.31. μ plot after fourth iteration

8.10 Robust Stability and Performance of K_{mu}

In Figure 8.32 we show the sensitivity function of the closed-loop system with the 20th-order controller K_{mu}. It is clear that the sensitivity function is below the inverse of the performance weighting function, which shows that the nominal performance is achieved.

The robust stability of the closed-loop system with μ-controller is analysed by calling the file `rob_mds.m`, with defining in advance

```
K = K_mu;
```

The upper and lower bounds of μ are shown in Figure 8.33. It is seen that in this case the robust stability of the closed-loop system is achieved since the maximum value of μ is equal to 0.468, *i.e.*, the system stability is preserved for $\|\Delta\|_\infty < \frac{1}{0.468}$.

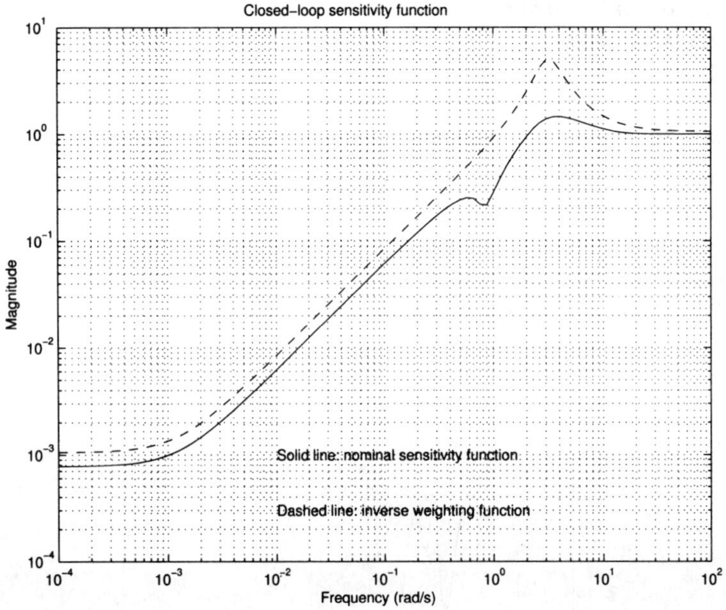

Fig. 8.32. Sensitivity function of K_{mu}

The frequency responses of the nominal and robust performance criteria are obtained by the commands from the file `nrp_mds.m`, also used in the analysis of the $\mathcal{H}_\infty$ controller, and are shown in Figure 8.34. The maximum value of μ in the robust performance analysis is 0.96. This means that the closed-loop system with μ-controller achieves robust performance since

$$\left\| \begin{bmatrix} W_p(I + F_U(G_{\text{mds}}, \Delta)K)^{-1} \\ W_u K(I + F_U(G_{\text{mds}}, \Delta)K)^{-1} \end{bmatrix} \right\|_\infty < 0.96$$

for every diagonal Δ with $\|\Delta\|_\infty < 1$.

To illustrate the robust properties of the system with the μ-controller, we show in Figure 8.35 the frequency response of the sensitivity functions of the perturbed closed-loop systems, obtained by using the file `psf_mds.m` below.

File `psf_mds.m`

```
%
% Sensitivity functions of perturbed systems
%
sim_mds
clp = starp(sim_ic,K);
```

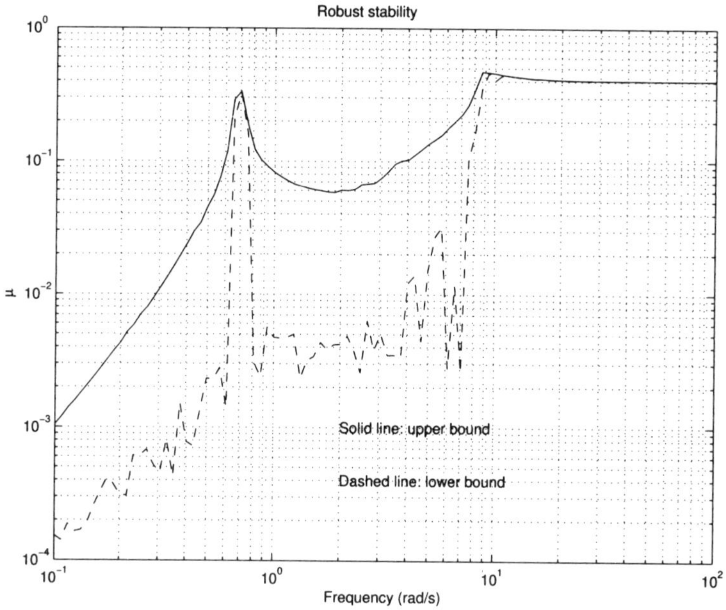

Fig. 8.33. Robust stability of K_{mu}

```
%
% inverse performance weighting function
wts_mds
omega = logspace(-4,2,100);
Wp_g = frsp(Wp,omega);
Wpi_g = minv(Wp_g);
figure(1)
vplot('liv,lm',Wpi_g,'m--')
hold on
[delta1,delta2,delta3] = ndgrid([-1 0 1],[-1 0 1], ...
                                [-1 0 1]);
for j = 1:27
    delta = diag([delta1(j),delta2(j),delta3(j)]);
    clp = starp(delta,starp(sim_ic,K));
    sen_loop = sel(clp,1,2);
    clp_g = frsp(sen_loop,omega);
    figure(1)
    vplot('liv,lm',clp_g,'c-')
    hold on
end
title('PERTURBED SENSITIVITY FUNCTIONS')
xlabel('Frequency (rad/s)')
```

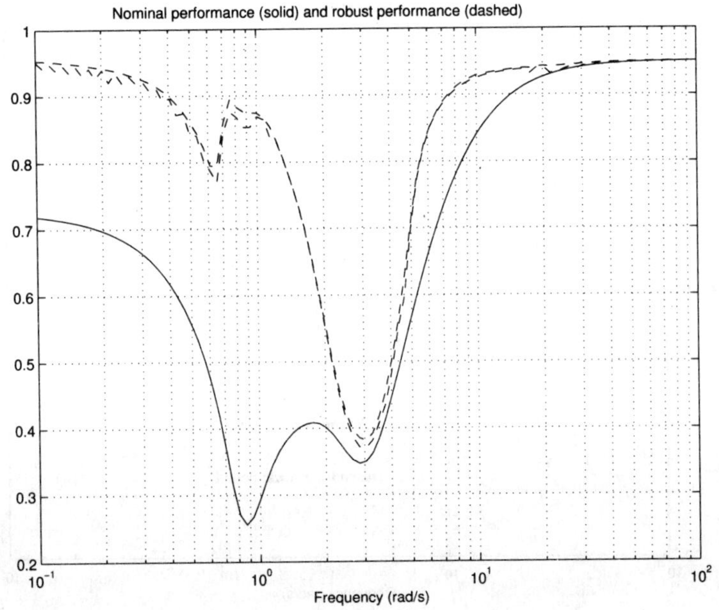

Fig. 8.34. Nominal and robust performance of K_{mu}

```
ylabel('Magnitude')
hold off
```

It is seen that the frequency responses of the perturbed sensitivity functions remain below the frequency response of the inverse of the performance weighting function.

In Figure 8.36 we show the magnitude responses of

$$\begin{bmatrix} W_p(I + F_U(G_{\mathrm{mds}}, \Delta)K)^{-1} \\ W_uK(I + F_U(G_{\mathrm{mds}}, \Delta)K)^{-1} \end{bmatrix}$$

(the weighted, mixed sensitivity function) of the perturbed system, using the following file `ppf_mds.m`. It is clear from Figure 8.36 that for all perturbed systems, the magnitudes over the frequency range are below the criterion for the closed-loop system robust performance.

File `ppf_mds.m`

```
%
% Performance of the perturbed closed-loop systems
%
% perturbed peformance
```

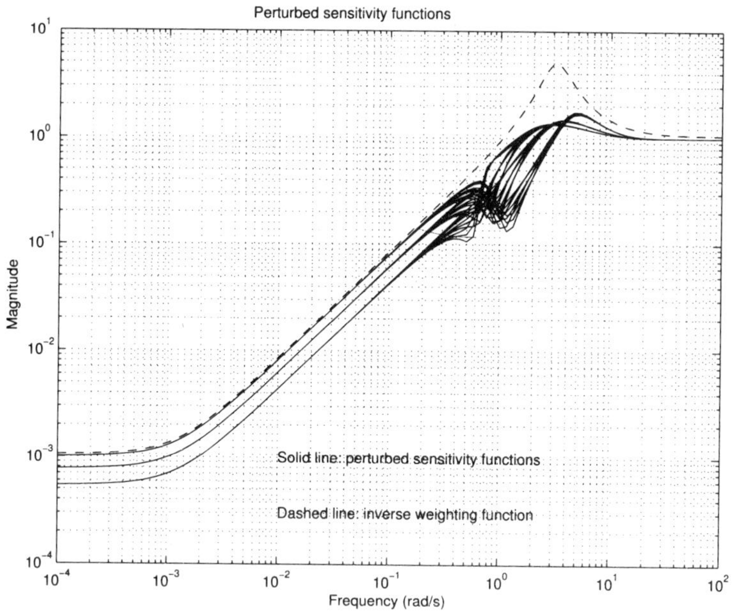

Fig. 8.35. Sensitivity functions of perturbed systems with $K_{\rm mu}$

```
omega = logspace(-2,2,100);
[delta1,delta2,delta3] = ndgrid([-1 0 1],[-1 0 1], ...
                                [-1 0 1]);
for j = 1:27
    delta = diag([delta1(j),delta2(j),delta3(j)]);
    clp = starp(delta,starp(sys_ic,K));
    clp_g = frsp(clp,omega);
    figure(1)
    vplot('liv,lm',vsvd(sel(clp_g,1:2,1)),'c-')
    hold on
end
%
% robust performance
clp_ic = starp(sys_ic,K);
clp_g = frsp(clp_ic,omega);
% real perturbations
blkrsR = [-1 1;-1 1;-1 1];
rob_perf = clp_g;
blkrp = [blkrsR;[1 2]];
bndsrp = mu(rob_perf,blkrp);
figure(1)
```

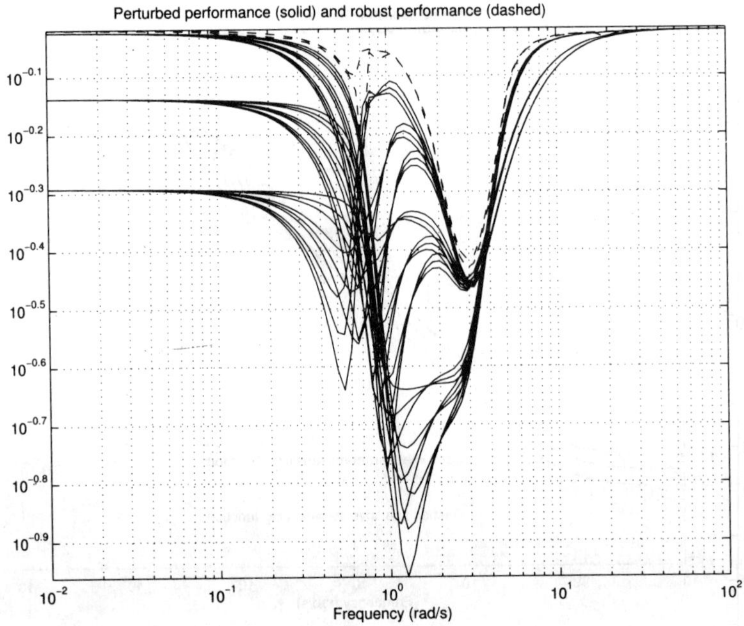

Fig. 8.36. Performance of perturbed systems with K_{mu}

```
vplot('liv,lm',sel(bndsrp,1,1),'y--',sel(bndsrp,1,2),'m--')
tmp1 = 'PERTURBED PERFORMANCE (solid) and';
tmp2 = ' ROBUST PERFORMANCE(dashed)';
title([tmp1 tmp2])
xlabel('Frequency (rad/s)')
hold off
```

In Figure 8.37 we show the frequency responses of the perturbed closed-loop systems. These responses are obtained by the commands included in the file pcf_mds.m.

File pcf_mds.m

```
%
% Frequency responses of the perturbed closed-loop
% systems
%
sim_mds
omega = logspace(-1,2,100);
[delta1,delta2,delta3] = ndgrid([-1 0 1],[-1 0 1], ...
                                [-1 0 1]);
```

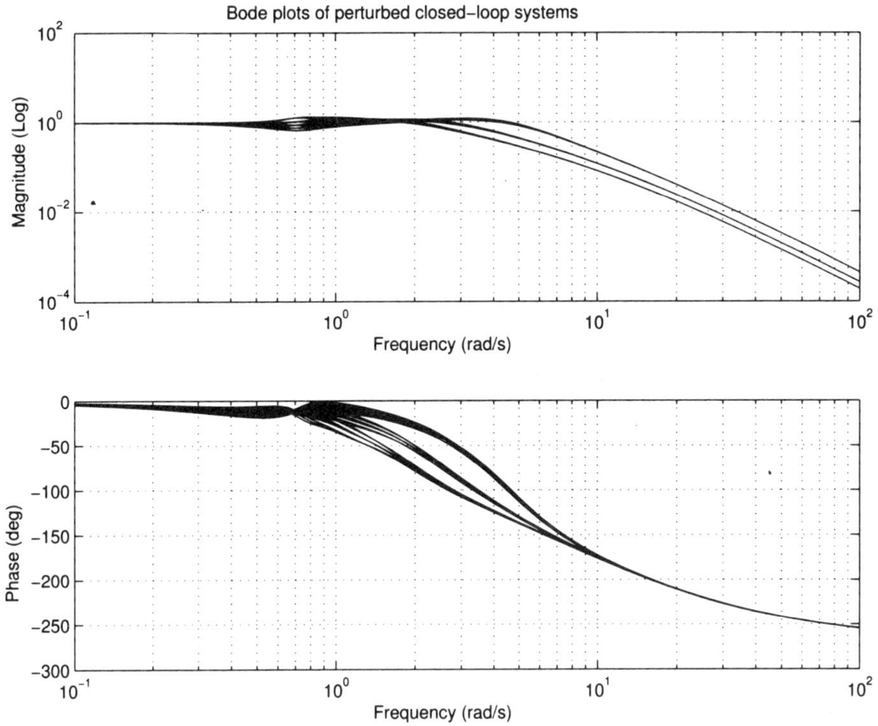

Fig. 8.37. Frequency responses of perturbed systems with K_{mu}

```
for j = 1:27
    delta = diag([delta1(j),delta2(j),delta3(j)]);
    clp = starp(delta,starp(sim_ic,K));
    clp_g = frsp(clp,omega);
    figure(1)
    vplot('bode',sel(clp_g,1,1),'c-')
    subplot(2,1,1)
    hold on
    subplot(2,1,2)
    hold on
end
subplot(2,1,1)
clp = starp(sim_ic,K);
clp_g = frsp(clp,omega);
vplot('bode',sel(clp_g,4,4),'r--')
subplot(2,1,1)
title('BODE PLOTS OF PERTURBED CLOSED-LOOP SYSTEMS')
hold off
```

```
subplot(2,1,2)
hold off
```

We see from Figure 8.37 that the frequency responses of the closed-loop, perturbed systems maintain their magnitude over a wider frequency band, in comparison to that of the open-loop system (Figure 8.7). Hence, faster responses would be expected with the designed closed-loop system.

In Figures 8.38 and 8.39 we show the transient responses of the system with μ-controller. Comparing with the responses in the case of the LSDP controller (Figures 8.25 and 8.26), we see that the μ-controller ensures smaller overshoot (10%) while maintaining the similar settling time.

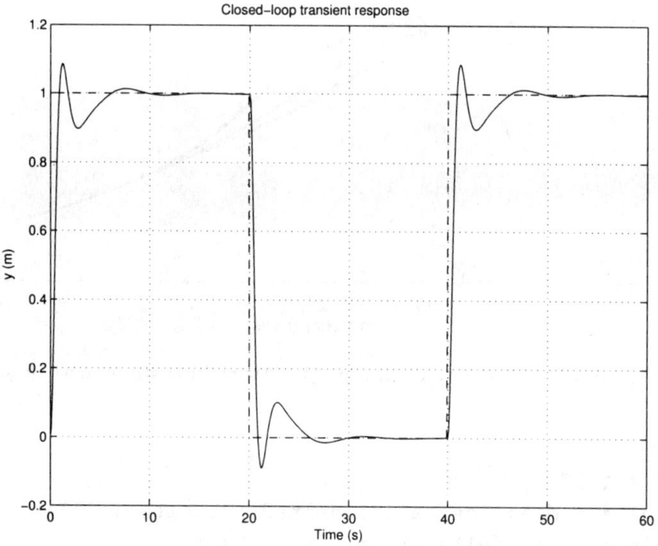

Fig. 8.38. Transient response to reference input ($K_{\rm mu}$)

In Figure 8.40 we show the transient responses (to the reference input) of a family of perturbed closed-loop systems with μ-controller. In all cases the overshoot does not exceed 20% that demonstrates satisfactory performance in the presence of parametric perturbations, *i.e.*, the closed-loop system achieves robust performance.

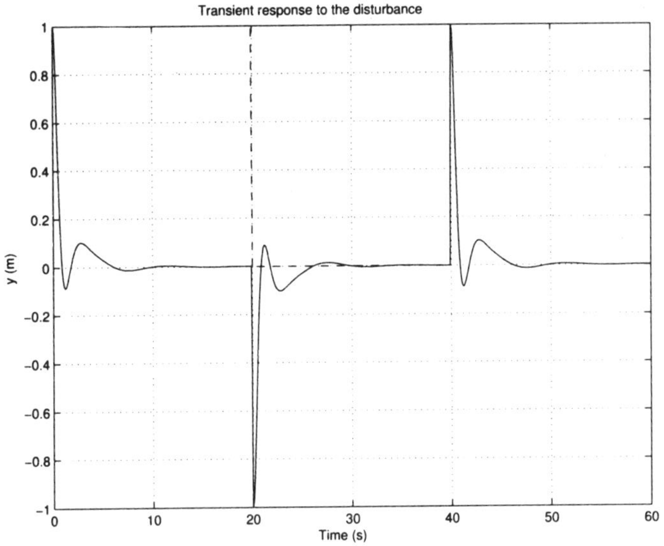

Fig. 8.39. Transient response to disturbance input (K_{mu})

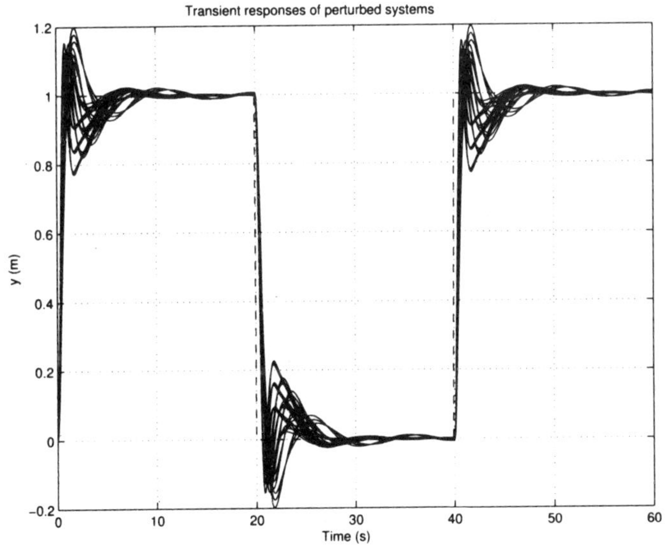

Fig. 8.40. Transient responses of perturbed systems (K_{mu})

8.11 Comparison of $\mathcal{H}_\infty$, $\mathcal{H}_\infty$ LSDP and μ-controllers

The comparison of the designed systems with $\mathcal{H}_\infty$, $\mathcal{H}_\infty$ loop-shaping and μ-controllers begins with the frequency responses of these three controllers. These responses are produced by using the file `kf_mds.m` below and are shown in Figure 8.41.

File `kf_mds.m`

```
omega = logspace(-2,3,100);
%
% H_infinity controller
K_hing = frsp(K_hin,omega);
%
% Loop Shaping controller
K_lshg = frsp(K_lsh,omega);
%
% mu-controller
K_mug = frsp(K_mu,omega);
%
figure(1)
vplot('bode',K_hing,'r-',K_lshg,'m--',K_mug,'c-.')
subplot(2,1,1)
title('BODE PLOTS OF ALL CONTROLLERS')
subplot(2,1,2)
```

It can be seen from Figure 8.41 that the $\mathcal{H}_\infty$ loop-shaping controller and μ-controller are characterised by larger gains compared with the $\mathcal{H}_\infty$ controller, in the frequency range above 5 rad/s. All the phase responses are close to each other up to about 3 rad/s and after that frequency the $\mathcal{H}_\infty$ controller continues to introduce a larger phase delay.

The comparison of frequency responses of the nominal closed-loop systems is conducted by the commands included in the file `clf_mds.m`, and the results are shown in Figure 8.42.

File `clf_mds.m`

```
sim_mds
omega = logspace(-2,2,100);
%
% H_infinity controller
clp_hin = starp(sim_ic,K_hin);
ref_hin = sel(clp_hin,4,4);
ref_hing = frsp(ref_hin,omega);
```

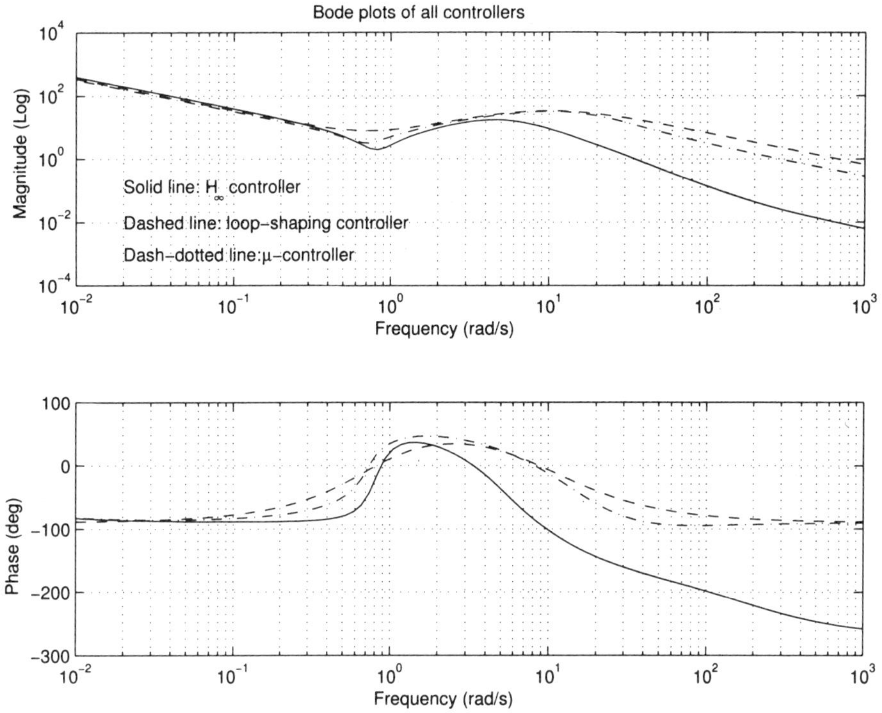

Fig. 8.41. Frequency responses of controllers

```
%
% Loop Shaping controller
clp_lsh = starp(sim_ic,K_lsh);
ref_lsh = sel(clp_lsh,4,4);
ref_lshg = frsp(ref_lsh,omega);
%
% mu-controller
clp_mu = starp(sim_ic,K_mu);
ref_mu = sel(clp_mu,4,4);
ref_mug = frsp(ref_mu,omega);
figure(1)
vplot('bode',ref_hing,'r-',ref_lshg,'m--',ref_mug,'c-.')
subplot(2,1,1)
title('BODE PLOTS OF CLOSED-LOOP SYSTEMS')
subplot(2,1,2)
```

Figure 8.42 shows that the systems with the $\mathcal{H}_\infty$ loop-shaping and μ-controllers are characterised by larger bandwidth that leads to faster dynamics of the corresponding closed-loop systems.

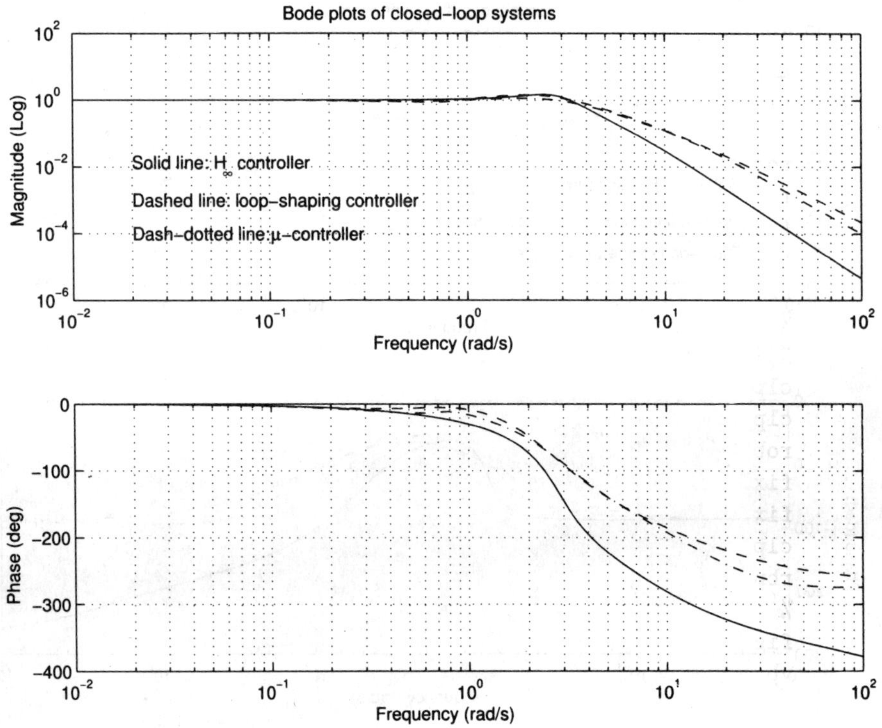

Fig. 8.42. Frequency responses of closed-loop systems

The comparison of the robust stability is conducted by the file **rbs_mds.m**.

File rbs_mds.m

```
omega = logspace(-1,2,100);
% Real perturbations
blkrs_R = [-1 1;-1 1;-1 1];
blkrs = [blkrsR; abs(blkrsR)];
pdim = 3;
%
% Hinf controller
clp_h = starp(sys_ic,K_hin);
clp_hg = frsp(clp_h,omega);
rob_stab = sel(clp_hg,[1:3],[1:3]);
fixl = [eye(pdim); 0.1*eye(pdim)]; % 1% Complex
fixr = fixl';
clp_mix = mmult(fixl,rob_stab,fixr);
```

```
rbnds_h = mu(clp_mix,blkrs);
%
% Loop Shaping controller
clp_lsh = starp(sys_ic,K_lsh);
clp_lshg = frsp(clp_lsh,omega);
rob_stab = sel(clp_lshg,[1:3],[1:3]);
fixl = [eye(pdim); 0.1*eye(pdim)]; % 1% Complex
fixr = fixl';
clp_mix = mmult(fixl,rob_stab,fixr);
rbnds_lsh = mu(clp_mix,blkrs);
%
% mu-controller
clp_mu = starp(sys_ic,K_mu);
clp_mug = frsp(clp_mu,omega);
rob_stab = sel(clp_mug,[1:3],[1:3]);
fixl = [eye(pdim); 0.1*eye(pdim)]; % 1% Complex
fixr = fixl';
clp_mix = mmult(fixl,rob_stab,fixr);
rbnds_mu = mu(clp_mix,blkrs);
%
figure(1)
vplot('liv,lm',sel(rbnds_h,1,1),'r-', ...
            sel(rbnds_lsh,1,1),'m--', ...
            sel(rbnds_mu,1,1),'c-.')
title('ROBUST STABILITY FOR ALL CONTROLLERS')
xlabel('Frequency (rad/s)')
ylabel('Upper bound of \mu')
```

The frequency responses of μ for three controllers are shown in Figure 8.43. The system with the $\mathcal{H}_\infty$ loop-shaping controller is characterized with best robust stability since in this case the destabilising perturbations have the largest norm (note that the norm of these perturbations is inversely proportional to the maximum value of μ).

The nominal performance of the closed-loop systems is obtained by the file `prf_mds.m`. The frequency responses of the weighted sensitivity functions of the nominal systems are shown in Figure 8.44. Again, the systems with the $\mathcal{H}_\infty$ loop-shaping and μ-controllers achieve better performance.

File `prf_mds.m`

```
omega = logspace(-2,2,100);
%
% H_infinity controller
clp_hin = starp(sys_ic,K_hin);
```

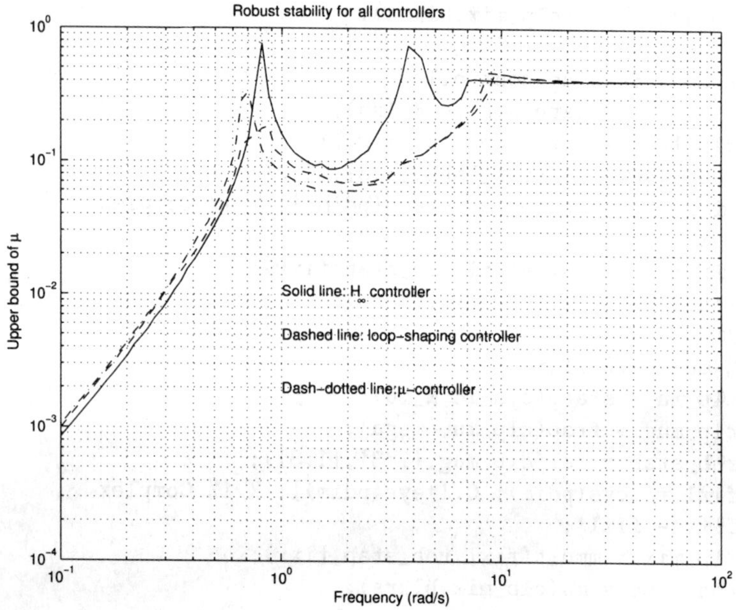

Fig. 8.43. Comparison of robust stability for 3 controllers

```
prf_hin = sel(clp_hin,4,4);
prf_hing = frsp(prf_hin,omega);
%
% Loop Shaping controller
clp_lsh = starp(sys_ic,K_lsh);
prf_lsh = sel(clp_lsh,4,4);
prf_lshg = frsp(prf_lsh,omega);
%
% mu-controller
clp_mu = starp(sys_ic,K_mu);
prf_mu = sel(clp_mu,4,4);
prf_mug = frsp(prf_mu,omega);
figure(1)
vplot('liv,m',vnorm(prf_hing),'r-',vnorm(prf_lshg), ...
      'm--',vnorm(prf_mug),'c-.')
title('NOMINAL PERFORMANCE: ALL CONTROLLERS')
xlabel('Frequency (rad/sec)')
```

The comparison of the robust performance is obtained by the file `rbp_mds.m`.

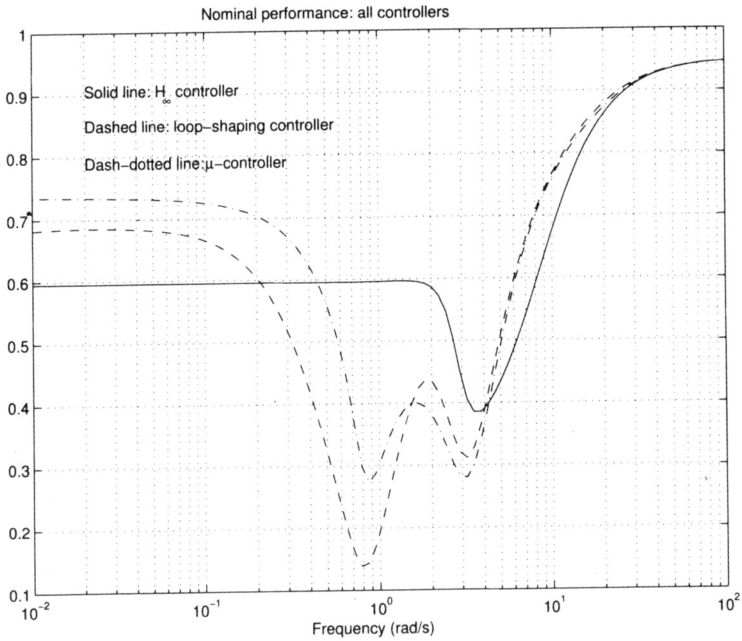

Fig. 8.44. Comparison of nominal performance for 3 controllers

File `rbp_mds.m`

```
blkrsR = [-1 1;-1 1;-1 1];
blkrp = [blkrsR;[1 2]];
omega = logspace(-2,2,100);
%
% H_infinity controller
clp_hin = starp(sys_ic,K_hin);
rbp_hin = frsp(clp_hin,omega);
bnd_hin = mu(rbp_hin,blkrp);
%
% Loop Shaping controller
clp_lsh = starp(sys_ic,K_lsh);
rbp_lsh = frsp(clp_lsh,omega);
bnd_lsh = mu(rbp_lsh,blkrp);
%
% mu-controller
clp_mu = starp(sys_ic,K_mu);
rbp_mu = frsp(clp_mu,omega);
bnd_mu = mu(rbp_mu,blkrp);
figure(1)
```

```
vplot('liv,m',bnd_hin,'r-',bnd_lsh,'m--',bnd_mu,'c-.')
title('ROBUST PERFORMANCE: ALL CONTROLLERS')
xlabel('Frequency (rad/sec)')
ylabel('\mu')
text(2.D+0,1.7D+0,'-   H_\infty controller')
text(2.D+0,1.6D+0,'--  Loop shaping controller')
text(2.D+0,1.5D+0,'-.  \mu-controller')
```

The μ values over the frequency range for all three designed systems are plotted in Figure 8.45. This confirms again that the system with the $\mathcal{H}_\infty$ controller does not achieve the robust performance criterion.

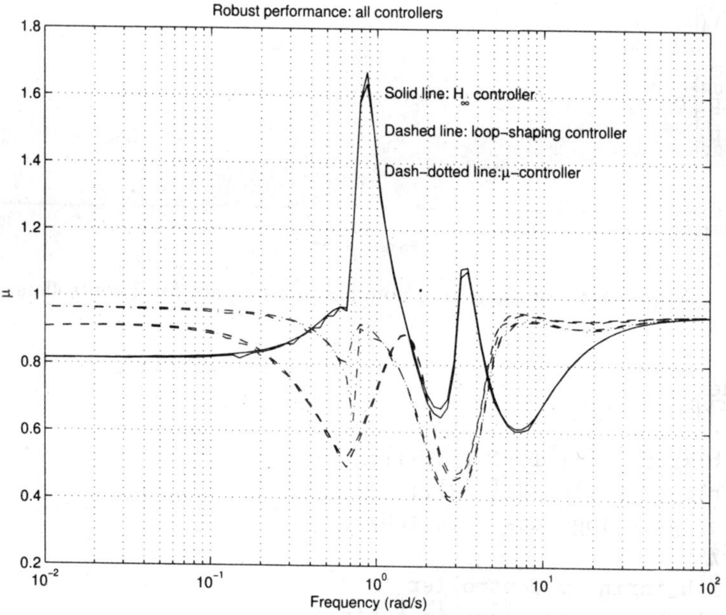

Fig. 8.45. Comparison of robust performance for 3 controllers

In summary, all three controllers ensure robust stability of the closed-loop system with respect to the parametric perturbations that are included in the 3×3 diagonal uncertainty matrix. However, the closed-loop system performance varies in a different way under the action of these diagonal perturbations. In the following file, `pdc_mds.m`, we use the function `wcperf`, which allows us to find the worst-case perturbation with respect to the performance, and to compare the performance of the three systems when the norm of perturbations increases. (The execution of this file may be accompanied by some warnings.)

File pdc_mds.m

```
omega = logspace(-2,3,100);
% Real perturbations
blks = [-1 1;-1 1;-1 1];
%
% H_infinity controller
clp_hin = starp(sys_ic,K_hin);
clp_hing = frsp(clp_hin,omega);
%
% Loop Shaping controller
clp_lsh = starp(sys_ic,K_lsh);
clp_lshg = frsp(clp_lsh,omega);
%
% mu-controller
clp_mu = starp(sys_ic,K_mu);
clp_mug = frsp(clp_mu,omega);
%
alpha = 0.1;
npts = 10;
[deltabadh,lowbndh,uppbndh] = wcperf(clp_hing,blks, ...
                                         alpha,npts);
[deltabadlsh,lowbndlsh,uppbndlsh] = wcperf(clp_lshg, ...
                                    blks,alpha,npts);
[deltabadmu,lowbndmu,uppbndmu] = wcperf(clp_mug,blks, ...
                                    alpha,npts);
figure(1)
vplot(lowbndh,'r-',uppbndh,'r-',lowbndlsh,'m--', ...
      uppbndlsh,'m--',lowbndmu,'c-.',uppbndmu,'c-.')
axis([0 1.8 0 2.0])
title('PERFORMANCE DEGRADATION CURVES FOR ALL CONTROLLERS')
xlabel('Size of Uncertainty')
text(1.0,0.6,'-H_\infty controller')
text(1.0,0.4,'-- Loop shaping controller')
text(1.0,0.2,'-. \mu-controller')
```

The results in Figure 8.46 show that the $\mathcal{H}_\infty$ loop-shaping controller and the μ-controller ensure robust performance for larger perturbations, with the former slightly outperforming the latter. The closed-loop performance deteriorates most rapidly with increasing perturbation magnitude in the case of the $\mathcal{H}_\infty$ controller.

The comparison results indicate that in the present exercise the $\mathcal{H}_\infty$ loop-shaping controller might be the best one to choose in terms of system robustness, though the μ-controller is not that far behind.

It is important to stress that in the above comparison the μ-controller is a reduced-order one of order 4. The order of the original μ-controller is actually

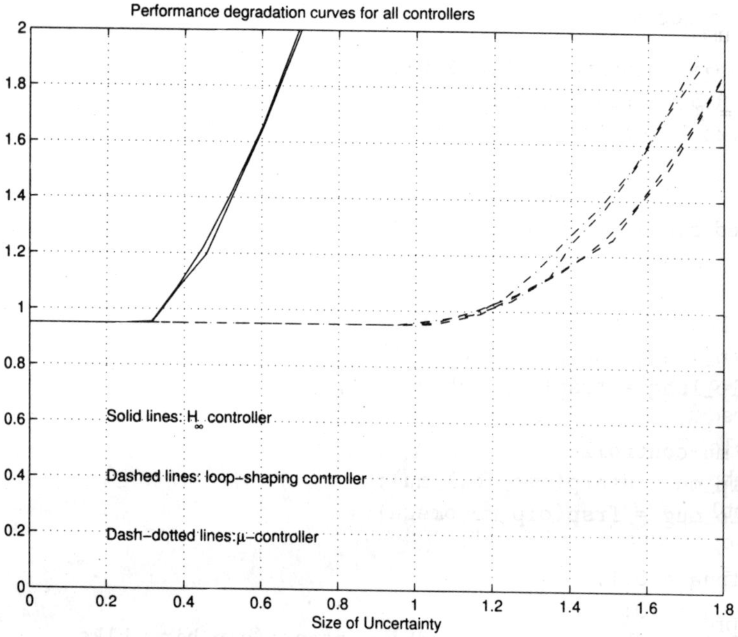

Fig. 8.46. Performance degradation for 3 controllers

20. However, the use of such a reduced-order μ-controller will be justified in the next section, which shows that the reduced-order controller does not seriously deteriorate the closed-loop system performance. It is also important to point out that for different design cases these three design approaches may well produce different results. That is, even the use of the $\mathcal{H}_\infty$ controller may lead to acceptable results in terms of performance and controller complexity. However, in general, in the case of structured uncertainty, the μ-synthesis will always produce more satisfactory and less conservative controllers.

8.12 Order Reduction of μ-controller

As shown in Section 8.9, the order of the μ-controller K_{mu} is 20, which makes it difficult in implementation. A reduced-order controller would be usually preferred. Various methods may be used to reduce the order, as discussed in Chapter 7. In this case study, the Hankel-norm approximation is used and implemented in the file `red_mds.m`. Two important commands in the file are **`sysbal`** and **`hankmr`**. The command **`sysbal`** generates the balanced realisation of a system (input argument) of the type `SYSTEM`. This command removes the unobservable and/or uncontrollable modes, if the system is not minimal. The command **`sysbal`** also returns the Hankel singular values of the system that can be used to choose the order to be reduced. To use the file **red_mds.m**, first define

```
K = K_mu;
```

File red_mds.m

```
omega = logspace(-2,4,100);
K_g = frsp(K,omega);
[Kb,hsig] = sysbal(K);
Kred = hankmr(Kb,hsig,4,'d');
Kred_g = frsp(Kred,omega);
vplot('bode',K_g,'y-',Kred_g,'m--')
subplot(2,1,1)
title('BODE PLOTS OF FULL AND REDUCED ORDER CONTROLLERS')
subplot(2,1,2)
K = Kred;
```

In the present case the balanced realization Kb turns out to be of 20th order as well, which means that the state-space model of the μ-controller is already minimal. The Hankel singular values are

```
hsig =

  1.0e+003 *

   1.28877293261018
   0.01757808424290
   0.01413607705650
   0.00185159191695
   0.00043945590100
   0.00027713314731
   0.00013630163054
   0.00006894336934
   0.00002465391647
   0.00000787583669
   0.00000779094168
   0.00000247258791
   0.00000053897770
   0.00000023805949
   0.00000016178585
   0.00000008448505
   0.00000001681482
   0.00000000066695
   0.00000000021363
   0.00000000000201
```

Some of these values are very small, which suggests that the controller order can be greatly reduced. In this exercise this order is set equal to 4 in the command **hankmr** and the test with this reduced-order controller shows that the closed-loop system obtained is stable. Actually, a 3rd-order controller has been tried as well, but noticeable differences exist in the frequency responses in the range of $10^{-1} - 10^{1}$ rad/s. The system matrices of this 4th order controller are

```
[Ak,Bk,Ck,Dk] = unpck(K)

Ak =

  -5.94029444678383 -12.71890180361086  11.09659323806946
  15.39552256762868 -29.53184484831400 -16.41287136235254
                  0                  0  -6.84730630402175
                  0                  0                  0

                                        -0.04081607183541
                                         0.00507186515891
                                         0.03584524485830
                                        -0.00125108980521
Bk =

 -15.96860717359783
   7.91326562204500
  13.97349755769254
  -1.79576743717989


Ck =

 -12.96086158489860 -14.24741790999237   9.28057043520263
                                        -1.79885756979541


Dk =

   0.54044998732827
```

In Figure 8.47 we display the frequency responses of the full-order and reduced-order controllers. The responses plots practically coincide with each other, for frequencies up to 100 rad/s (up to about 1000 rad/s in magnitude) that ensures almost the same closed-loop performance for both controllers. In particular, the transient responses of the closed-loop systems with full-order and reduced-order controllers are practically indistinguishable (compare Figure 8.48 with Figure 8.38). Clearly, the 4th-order controller is implemented much easier compared to the 20th-order controller.

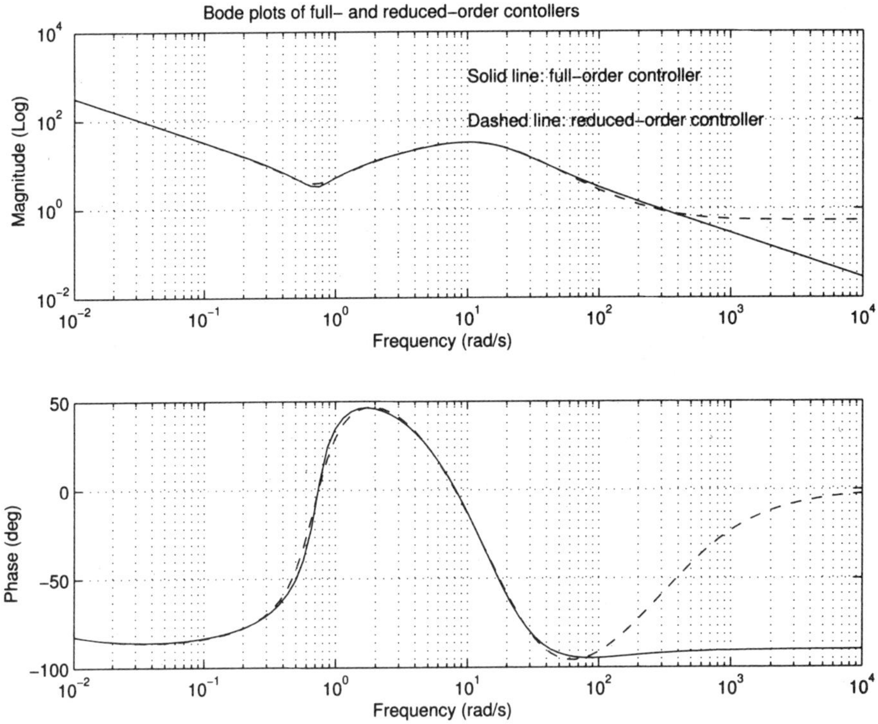

Fig. 8.47. Frequency responses of full- and reduced-order controllers

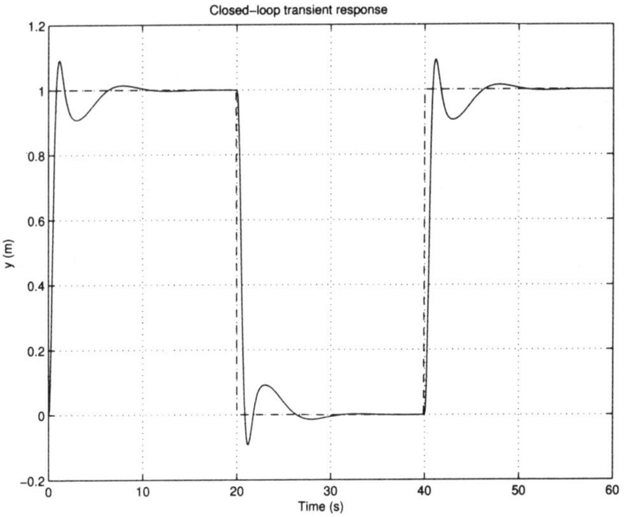

Fig. 8.48. Transient response for reduced-order controller

8.13 Conclusions

The design exercise of the mass-damper-spring system leads to the following conclusions.

- Even for a low-order plant such as the mass-damper-spring system the derivation of the uncertainty model may not be straightforward. For a system with parametric perturbations, it has to form a standard configuration in order to use the analysis and design commands from the Robust Control Toolbox. In this procedure, it may be convenient to consider separately the uncertain parameters, describe these perturbation influences by simple LFTs and construct the whole uncertainty model of the plant by using the function `sysic`.
- Finding appropriate weighting functions is a crucial step in robust control designs. It usually involves trial and error. Design experience and knowledge of the plants will help in choosing weighting functions.
- Best design results for this particular system are obtained by using the $\mathcal{H}_\infty$ loop-shaping design procedure and μ-synthesis. With nearly identical robustness properties, the μ-controller gives better transient responses with smaller overshoot than the $\mathcal{H}_\infty$ LSDP controller.
- Even for a second-order plant like this, the order of the μ-controller may be very high (20 in the given design). This makes it necessary to apply model reductions. Usually, the order may be reduced significantly without serious performance degradation.

Notes and References

The derivation of the uncertainty model of mass-damper-spring system is shown in [175, Chapter 10]. A model, obtained by using the function `sysic` and implemented by the file `sys_mds.m`, is presented in [9, Chapter 4].

9

A Triple Inverted Pendulum Control-system Design

Robust design of a triple inverted pendulum control system is discussed in this chapter.

The triple inverted pendulum is an interesting control system that resembles many features found in, for instance, walking robots and flexible space structures, and other industrial applications. This kind of pendulum system is difficult to control due to the inherent instability and nonlinear behaviour. Some of the pendulum parameters may not be known exactly in practice, which influences significantly the system dynamics.

In the design of a robust control system for triple inverted pendulums it is conventionally assumed that the system is affected by unstructured uncertainties and thus the robust properties of the closed-loop system could be achieved by using an $\mathcal{H}_\infty$ controller. In real cases, however, the uncertainties of such a pendulum system would be more reasonably considered to have some structures. For instance, because the moments of inertia and the friction coefficients are difficult to estimate precisely, it makes sense to assume unknown deviations in those parameters. Also, we would like to design the closed-loop control system to be more "robust"against those parameters that have a "larger"or more serious influence upon the system behaviour. For instance, the viscous friction in the joints may destroy the controllability of the linearised model. Hence, it would be important to treat uncertainties in such parameters individually rather than congregate them in an overall, unstructured uncertainty of the system dynamics. Consequently, it may be more suitable to apply the μ-synthesis technique that may lead to a less-conservative design to meet tighter design specifications.

In the pendulum control-system design we first model the uncertainties as a mixed type that consists of complex uncertainties in the actuators, real uncertainties in the moments of inertia and in the viscous friction coefficients. A two-degree-of-freedom (2DOF) design framework is adopted. Both $\mathcal{H}_\infty$ suboptimal and μ-controllers are designed. The $\mathcal{H}_\infty$ controller shows better transient and disturbance responses but does not ensure robust stability nor robust performance. The μ-controller achieves both robust stability and robust performance, however, at the price of poorer time responses. The μ-controller designed is initially of quite high order, which makes it unsuitable for implementation in practice. A model reduction is then conducted that leads to a reduced-order controller maintaining the required robust stability and robust performance of the closed-loop system.

9.1 System Description

The triple inverted pendulum considered is the experimental setup realized by Furuta *et al.* [41] (Figure 9.1). The pendulum consists of three arms that are hinged by ball bearings and can rotate in the vertical plane. The torques of the two upper hinges are controlled by motors, with the lowest hinge made free for rotation. By controlling the angles of the two upper arms around specified values, the pendulum can be stabilised inversely with the desired angle attitudes. A horizontal bar is fixed to each of the arms to ease the control by increasing the moment of inertia. Two dc motors, M_1 and M_2, are mounted on the first and third arm, respectively, acting as actuators that provide torques to the two upper hinges through timing belts. The potentiometers P_1, P_2 and P_3 are fixed to the hinges to measure the corresponding angles.

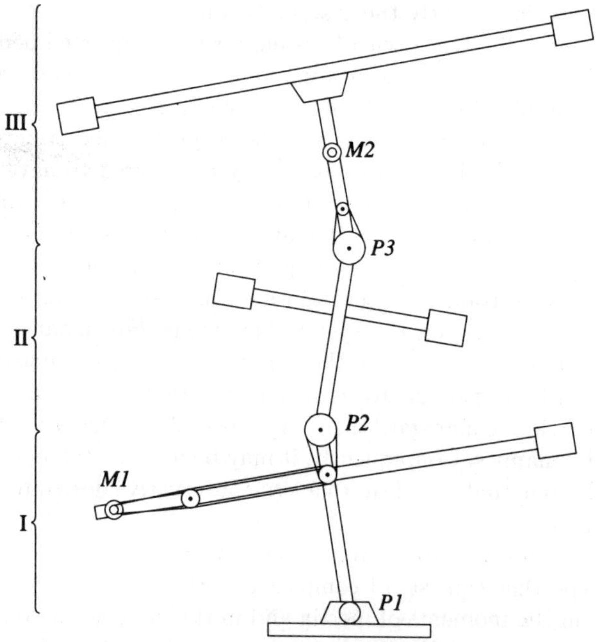

Fig. 9.1. Triple inverted pendulum

Let Θ_i denote the angle of the ith arm. The first potentiometer measures the angle Θ_1, and the second and third potentiometers measure the angles $\Theta_2 - \Theta_1$ and $\Theta_3 - \Theta_2$, respectively (Figure 9.2).

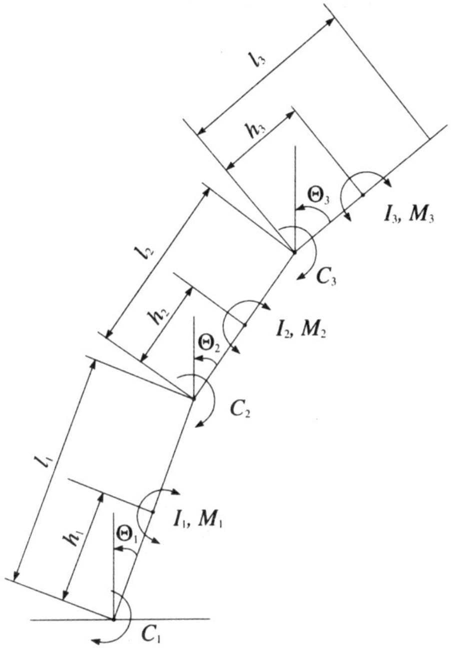

Fig. 9.2. Geometric relationship of potentiometers

The mathematical description of the triple inverted pendulum is derived under the following assumptions:

(a) each arm is a rigid body;
(b) the lengths of the belts remain constant during the operation of the system;
(c) the friction force in the bottom hinge is proportional to the velocity of the bottom arm and the friction forces in the upper hinges are proportional to the differences of the respective velocities of two neighbouring arms.

We shall first consider the mathematical model of the pendulum itself, without the actuators. The pendulum model is constructed using the Lagrange differential equations [41] that yield the following nonlinear vector-matrix differential equation

$$M(\Theta)\begin{bmatrix}\ddot{\Theta}_1\\ \ddot{\Theta}_2\\ \ddot{\Theta}_3\end{bmatrix} + N\begin{bmatrix}\dot{\Theta}_1\\ \dot{\Theta}_2\\ \dot{\Theta}_3\end{bmatrix} + \begin{bmatrix}q_1\\ q_2\\ q_3\end{bmatrix} + G\begin{bmatrix}t_{m_1}\\ t_{m_2}\end{bmatrix} = T\begin{bmatrix}\tau_1\\ \tau_2\\ \tau_3\end{bmatrix} \tag{9.1}$$

where, $\Theta = \begin{bmatrix} \Theta_1 \\ \Theta_2 \\ \Theta_3 \end{bmatrix}$

$$M(\Theta) =$$

$$\begin{bmatrix} J_1 + I_{p_1} & l_1M_2\cos(\Theta_1-\Theta_2) - I_{p_1} & l_1M_3\cos(\Theta_1-\Theta_3) \\ l_1M_2\cos(\Theta_1-\Theta_2) - I_{p_1} & J_2 + I_{p_1} + I_{p_2} & l_2M_3\cos(\Theta_2-\Theta_3) - I_{p_2} \\ l_1M_3\cos(\Theta_1-\Theta_3) & l_2M_3\cos(\Theta_2-\Theta_3) - I_{p_2} & J_3 + I_{p_2} \end{bmatrix}$$

$$N = \begin{bmatrix} C_1 + C_2 + C_{p_1} & -C_2 - C_{p_1} & 0 \\ -C_2 - C_{p_1} & C_{p_1} + C_{p_2} + C_2 + C_3 & -C_3 - C_{p_2} \\ 0 & -C_3 - C_{p_2} & C_3 + C_{p_2} \end{bmatrix}$$

$$q_1 = l_1M_2\sin(\Theta_1-\Theta_2)\dot{\Theta}_2^2 + l_1M_3\sin(\Theta_1-\Theta_3)\dot{\Theta}_3^2 - M_1g\sin(\Theta_1)$$
$$q_2 = l_1M_2\sin(\Theta_1-\Theta_2)\dot{\Theta}_1^2 + l_2M_3\sin(\Theta_2-\Theta_3)\dot{\Theta}_3^2 - M_2g\sin(\Theta_2)$$
$$q_3 = l_1M_3\sin(\Theta_1-\Theta_3)(\dot{\Theta}_1^2 - 2\dot{\Theta}_1\dot{\Theta}_3) + l_2M_3\sin(\Theta_2-\Theta_3)(\dot{\Theta}_2^2 - 2\dot{\Theta}_2\dot{\Theta}_3)$$
$$-M_3g\sin(\Theta_3),$$

$$G = \begin{bmatrix} K_1 & 0 \\ -K_1 & K_2 \\ 0 & -K_2 \end{bmatrix},\ T = \begin{bmatrix} 1 & -1 & 0 \\ 0 & 1 & -1 \\ 0 & 0 & 1 \end{bmatrix}$$

$$C_{p_i} = C_{p_i'} + K_i^2C_{m_i},\ I_{p_i} = I_{p_i'} + K_i^2I_{m_i},$$

$$M_1 = m_1h_1 + m_2l_1 + m_3l_1,\ M_2 = m_2h_2 + m_3l_2,\ M_3 = m_3h_3$$

$$J_1 = I_1 + m_1h_1^2 + m_2l_1^2 + m_3l_1^2,\ J_2 = I_2 + m_2h_2^2 + m_3l_2^2$$

$$J_3 = I_3 + m_3h_3^2$$

and all other parameters and variables are defined in Table 9.1.

After linearisation of (9.1) under the assumptions of small deviations of the pendulum from the vertical position and of small velocities, one obtains the following equation

$$M\begin{bmatrix} \ddot{\Theta}_1 \\ \ddot{\Theta}_2 \\ \ddot{\Theta}_3 \end{bmatrix} + N\begin{bmatrix} \dot{\Theta}_1 \\ \dot{\Theta}_2 \\ \dot{\Theta}_3 \end{bmatrix} + P\begin{bmatrix} \Theta_1 \\ \Theta_2 \\ \Theta_3 \end{bmatrix} + G\begin{bmatrix} t_{m_1} \\ t_{m_2} \end{bmatrix} = T\begin{bmatrix} \tau_1 \\ \tau_2 \\ \tau_3 \end{bmatrix} \tag{9.2}$$

where

$$M = \begin{bmatrix} J_1 + I_{p_1} & l_1M_2 - I_{p_1} & l_1M_3 \\ l_1M_2 - I_{p_1} & J_2 + I_{p_1} + I_{p_2} & l_2M_3 - I_{p_2} \\ l_1M_3 & l_2M_3 - I_{p_2} & J_3 + I_{p_2} \end{bmatrix}$$

$$P = \begin{bmatrix} M_1g & 0 & 0 \\ 0 & -M_2g & 0 \\ 0 & 0 & -M_3g \end{bmatrix}$$

By introducing the control torques vector $t_m = [t_{m_1}\ t_{m_2}]^T$ and the vector of disturbance torques $d = [\tau_1,\ \tau_2,\ \tau_3]^T$, (9.2) can be rewritten in the form

Table 9.1. System nomenclature

Symbol	Description
u_j	input voltage to the jth motor
t_{m_j}	control torque of the jth motor
τ_i	disturbance torque to the ith arm
l_i	length of the ith arm
h_i	the distance from the bottom to the centre of gravity of the ith arm
m_i	mass of the ith arm
I_i	moment of inertia of the i-th arm around the centre of gravity
C_i	coefficient of viscous friction of the ith hinge
Θ_i	angle of the ith arm from vertical line
C_{m_i}	viscous friction coefficient of the ith motor
I_{m_i}	moment of inertia of the ith motor
K_i	ratio of teeth of belt-pulley system of the ith hinge
$C_{p_i'}$	viscous friction coefficient of the belt-pulley system of the ith hinge
$I_{p_i'}$	moment of inertia of the belt-pulley system of the ith hinge
α_i	gain of the ith potentiometer
g	acceleration of gravity

$$M\ddot{\Theta} + N\dot{\Theta} + P\Theta + Gt_m = Td$$

i.e.,

$$\ddot{\Theta} = -M^{-1}N\dot{\Theta} - M^{-1}P\Theta - M^{-1}Gt_m + M^{-1}Td.$$

The block-diagram of the pendulum system is shown in Figure 9.3 and the nominal values of the parameters are given in Table 9.2.

9.2 Modelling of Uncertainties

Based on practical considerations, we in particular consider the variations of the moments of inertia I_1, I_2 and I_3 of the three arms and the variations of the viscous friction coefficients C_1, C_2, C_3 and C_{p_1}, C_{p_2}. It is assumed that the moments of inertia are constants but with possible relative error of 10% around the nominal values; similarly, the friction coefficients may have with 15% relative errors.

Notice that in the rest of the chapter, a parameter with a bar above denotes its nominal value.

Therefore, the actual moments of inertia are presented as

$$I_i = \overline{I}_i(1 + p_i\delta_{I_i}),\ i = 1,\ 2,\ 3$$

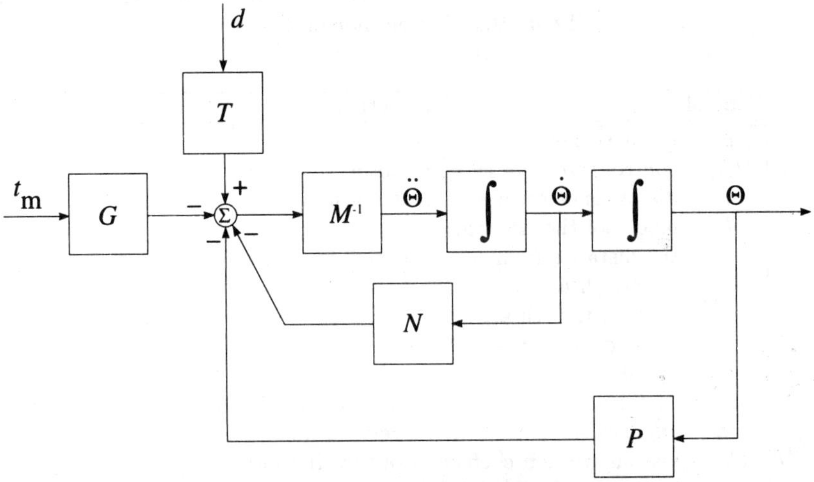

Fig. 9.3. Block diagram of the triple inverted pendulum system

Table 9.2. Nominal values of the parameters

Symbol (unit)	value	Symbol (unit)	value
l_1 (m)	0.5	α_1 (V/rad)	1.146
l_2 (m)	0.4	α_2 (V/rad)	1.146
h_1 (m)	0.35	α_3 (V/rad)	0.9964
h_2 (m)	0.181	C_{m_1} (N m s)	2.19×10^{-3}
h_3 (m)	0.245	C_{m_2} (N m s)	7.17×10^{-4}
m_1 (kg)	3.25	I_{m_1} (kg m^2)	2.40×10^{-5}
m_2 (kg)	1.90	I_{m_2} (kg m^2)	4.90×10^{-6}
m_3 (kg)	2.23	$C_{p'_1}$ (N m s)	0
I_1 (kg m^2)	0.654	$C_{p'_2}$ (N m s)	0
I_2 (kg m^2)	0.117	$I_{p'_1}$ (kg m^2)	7.95×10^{-3}
I_3 (kg m^2)	0.535	$I_{p'_2}$ (kg m^2)	3.97×10^{-3}
C_1 (N m s)	6.54×10^{-2}	K_1	30.72
C_2 (N m s)	2.32×10^{-2}	K_2	27.00
C_3 (N m s)	8.80×10^{-3}		

where $\overline{I}_i$ is the nominal value of the corresponding moment of inertia, $p_i = 0.10$ is the maximum relative uncertainty in each of these moments and $-1 \leq \delta_{I_i} \leq 1$, $i = 1, 2, 3$. As a result, the matrix M is obtained in the form

$$M = \overline{M} + M_p \Delta_I$$

where the elements of $\overline{M}$ are determined by the nominal values of the moments of inertia,

$$\overline{M} = \begin{bmatrix} \overline{J}_1 + \overline{I}_{p_1} & \overline{l}_1 \overline{M}_2 - \overline{I}_{p_1} & \overline{l}_1 \overline{M}_3 \\ \overline{l}_1 \overline{M}_2 - \overline{I}_{p_1} & \overline{J}_2 + \overline{I}_{p_1} + \overline{I}_{p_2} & \overline{l}_2 \overline{M}_3 - \overline{I}_{p_2} \\ \overline{l}_1 \overline{M}_3 & \overline{l}_2 \overline{M}_3 - \overline{I}_{p_2} & \overline{J}_3 + \overline{I}_{p_2} \end{bmatrix}$$

and

$$M_p = \begin{bmatrix} \overline{I}_1 p_1 & 0 & 0 \\ 0 & \overline{I}_2 p_2 & 0 \\ 0 & 0 & \overline{I}_3 p_3 \end{bmatrix}, \quad \Delta_I = \begin{bmatrix} \delta_{I_1} & 0 & 0 \\ 0 & \delta_{I_2} & 0 \\ 0 & 0 & \delta_{I_3} \end{bmatrix}$$

Next we calculate the matrix M^{-1}. We have that

$$M^{-1} = (M_p^{-1} \overline{M} + \Delta_I)^{-1} M_p^{-1}$$

Using the Matrix Inversion Lemma, we obtain

$$(M_p^{-1} \overline{M} + \Delta_I)^{-1} = \overline{M}^{-1} M_p - \overline{M}^{-1} M_p \Delta_I (\overline{M}^{-1} M_p \Delta_I + I_3)^{-1} \overline{M}^{-1} M_p$$

where I_3 is the 3×3 unit matrix. Therefore, we have

$$M^{-1} = \overline{M}^{-1} - \overline{M}^{-1} M_p \Delta_I (I_3 + \overline{M}^{-1} M_p \Delta_I)^{-1} \overline{M}^{-1}$$

The matrix M^{-1} can be represented as an upper linear fractional transformation (LFT)

$$M^{-1} = F_U(Q_I, \Delta_I) = Q_{I_{22}} + Q_{I_{21}} \Delta_I (I_3 - Q_{I_{11}} \Delta_I)^{-1} Q_{I_{12}}$$

where

$$Q_{I_{11}} = -\overline{M}^{-1} M_p, \; Q_{I_{12}} = \overline{M}^{-1}$$
$$Q_{I_{21}} = -\overline{M}^{-1} M_p, \; Q_{I_{22}} = \overline{M}^{-1}$$

such that

$$Q_I = \begin{bmatrix} Q_{I_{11}} & Q_{I_{12}} \\ Q_{I_{21}} & Q_{I_{22}} \end{bmatrix} = \begin{bmatrix} -\overline{M}^{-1} M_p & \overline{M}^{-1} \\ -\overline{M}^{-1} M_p & \overline{M}^{-1} \end{bmatrix}$$

Let us now consider the uncertainties in the friction coefficients. It can be seen that C_{p_1} and C_2 always appear together and so do C_{p_2} and C_3, in the matrix N. Further, the magnitude of C_{p_1} and C_{p_2} is much smaller in comparison to that of C_2 and C_3, Hence, the related uncertainties in C_{p_1} and C_{p_2} have much less influence with regard to the system dynamics. We may therefore assume

$$C_2 + C_{p_1} = (\overline{C}_2 + \overline{C}_{p_1})(1 + s_2 \delta_{C_2})$$

$$C_3 + C_{p_2} = (\overline{C}_3 + \overline{C}_{p_2})(1 + s_3 \delta_{C_3})$$

where $s_2 = s_3 = 0.15$ and $-1 \leq \delta_{C_2}, \delta_{C_3} \leq 1$. Similarly, we set

$$C_1 = \overline{C}_1 (1 + s_1 \delta_{C_1})$$

where $s_1 = 0.15$ and $-1 \leq \delta_{C_1} \leq 1$.

Taking into account the uncertainties in the friction coefficients, we obtain

$$N = \overline{N} + \Delta N$$

where

$$\overline{N} = \begin{bmatrix} \overline{C}_1 + \overline{C}_2 + \overline{C}_{p_1} & -\overline{C}_2 - \overline{C}_{p_1} & 0 \\ -\overline{C}_2 - \overline{C}_{p_1} & \overline{C}_2 + \overline{C}_{p_1} + \overline{C}_3 + \overline{C}_{p_2} & -\overline{C}_3 - \overline{C}_{p_2} \\ 0 & -\overline{C}_3 - \overline{C}_{p_2} & \overline{C}_3 + \overline{C}_{p_2} \end{bmatrix}$$

and

$$\Delta N =$$

$$\begin{bmatrix} \overline{C}_1 s_1 \delta_{C_1} + (\overline{C}_2 + \overline{C}_{p_1}) s_2 \delta_{C_2} & -(\overline{C}_2 + \overline{C}_{p_1}) s_2 \delta_{C_2} & 0 \\ -(\overline{C}_2 + \overline{C}_{p_1}) s_2 \delta_{C_2} & (\overline{C}_2 + \overline{C}_{p_1}) s_2 \delta_{C_2} & -(\overline{C}_3 + \overline{C}_{p_2}) s_3 \delta_{C_3} \\ & +(\overline{C}_3 + \overline{C}_{p_2}) s_3 \delta_{C_3} & \\ 0 & -(\overline{C}_3 + \overline{C}_{p_2}) s_3 \delta_{C_3} & (\overline{C}_3 + \overline{C}_{p_2}) s_3 \delta_{C_3} \end{bmatrix}$$

The matrix ΔN may be represented as

$$\Delta N = N_1 \Delta_C N_2$$

where

$$N_1 = \begin{bmatrix} \overline{C}_1 s_1 & -(\overline{C}_2 + \overline{C}_{p_1}) s_2 & 0 \\ 0 & (\overline{C}_2 + \overline{C}_{p_1}) s_2 & -(\overline{C}_3 + \overline{C}_{p_2}) s_3 \\ 0 & 0 & (\overline{C}_3 + \overline{C}_{p_2}) s_3 \end{bmatrix}$$

$$N_2 = \begin{bmatrix} 1 & 0 & 0 \\ -1 & 1 & 0 \\ 0 & -1 & 1 \end{bmatrix}, \quad \Delta_C = \begin{bmatrix} \delta_{C_1} & 0 & 0 \\ 0 & \delta_{C_2} & 0 \\ 0 & 0 & \delta_{C_3} \end{bmatrix}$$

In turn, the matrix $N = \overline{N} + N_1 \Delta_C N_2$ can be rewritten as an upper LFT

$$N = F_U(Q_C, \Delta_C) = Q_{C_{22}} + Q_{C_{21}} \Delta_C (I_3 - Q_{C_{11}} \Delta_C)^{-1} Q_{C_{12}}$$

where

$$Q_{C_{11}} = 0_{3\times3},\ Q_{C_{12}} = N_2,\ Q_{C_{21}} = N_1,\ Q_{C_{22}} = \overline{N}$$

such that

$$Q_C = \begin{bmatrix} Q_{C_{11}} & Q_{C_{12}} \\ Q_{C_{21}} & Q_{C_{22}} \end{bmatrix} = \begin{bmatrix} 0_{3\times3} & N_2 \\ N_1 & \overline{N} \end{bmatrix}$$

To represent the pendulum model as an LFT of the real uncertain parameters δ_{I_1}, δ_{I_2}, δ_{I_3}, and δ_{C_1}, δ_{C_2}, δ_{C_3}, we first extract out the uncertain parameters and then denote the inputs and outputs of Δ_I and Δ_C as y_I, y_C and u_I, u_C, respectively, as shown in Figure 9.4.

The dynamic equation of the pendulum system is then rearranged as

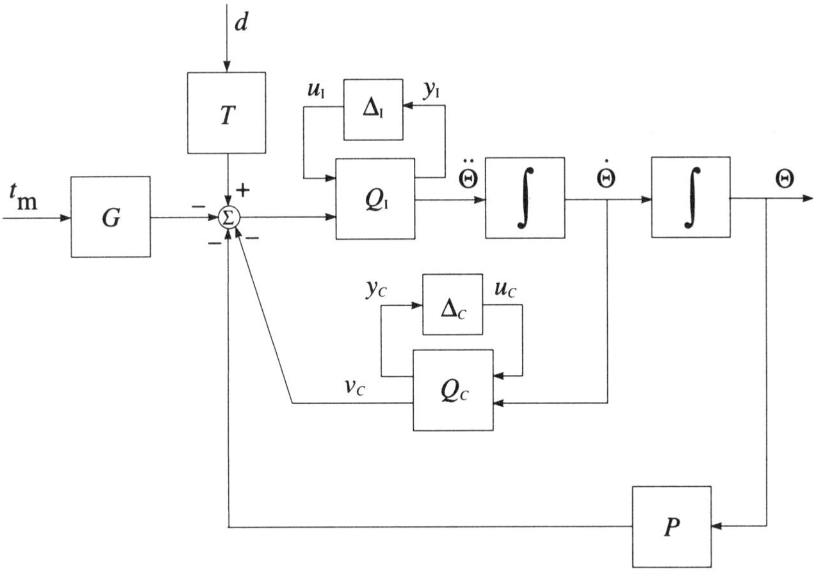

Fig. 9.4. System block diagram with uncertain parameters

$$\begin{bmatrix} y_I \\ \ddot{\Theta} \end{bmatrix} = \begin{bmatrix} -\overline{M}^{-1} M_p \, \overline{M}^{-1} \\ -\overline{M}^{-1} M_p \, \overline{M}^{-1} \end{bmatrix} \begin{bmatrix} u_I \\ Td - (Gt_m + v_C + P\Theta) \end{bmatrix}$$

$$\begin{bmatrix} y_c \\ v_c \end{bmatrix} = \begin{bmatrix} 0_{3\times 3} & N_2 \\ N_1 & \overline{N} \end{bmatrix} \begin{bmatrix} u_c \\ \dot{\Theta} \end{bmatrix}$$

$$u_I = \Delta_I y_I$$

$$u_C = \Delta_C y_C$$

The pendulum state vector $x = [x_1 \; x_2 \; x_3 \; x_4 \; x_5 \; x_6]^T$ is defined by

$$x_1 = \Theta_1, \; x_2 = \Theta_2, \; x_3 = \Theta_3, \; x_4 = \dot{\Theta}_1, \; x_5 = \dot{\Theta}_2, \; x_6 = \dot{\Theta}_3$$

Hence

$$\left[\dot{x}_1 \; \dot{x}_2 \; \dot{x}_3\right]^T = \dot{\Theta}$$

and

$$\left[\dot{x}_4 \; \dot{x}_5 \; \dot{x}_6\right]^T = \ddot{\Theta}$$

As for the output vector, we define

$$y = [\Theta_1 \; \Theta_2 \; \Theta_3]^T$$

The outputs are measured by linear potentiometers, whose voltages are given by

$$y_{p_1} = \alpha_1 \Theta_1, \; y_{p_2} = \alpha_2(\Theta_2 - \Theta_1), \; y_{p_3} = \alpha_3(\Theta_3 - \Theta_2)$$

By introducing the vector of the measured outputs

$$y_p = [y_{p1}\ y_{p2}\ y_{p3}]^T$$

we obtain that

$$y_p = C_p\Theta, \quad C_p = \begin{bmatrix} \alpha_1 & 0 & 0 \\ -\alpha_2 & \alpha_2 & 0 \\ 0 & -\alpha_3 & \alpha_3 \end{bmatrix}$$

As a result, we obtain the equations of the system

$$\begin{bmatrix} \dot{x}_1 \\ \dot{x}_2 \\ \dot{x}_3 \end{bmatrix} = \dot{\Theta}$$

$$\begin{aligned} \begin{bmatrix} \dot{x}_4 \\ \dot{x}_5 \\ \dot{x}_6 \end{bmatrix} &= -\overline{M}^{-1}M_p u_I + \overline{M}^{-1}Td - \overline{M}^{-1}(Gt_m + v_C + P\Theta) \\ y_I &= -\overline{M}^{-1}M_p u_I + \overline{M}^{-1}Td - \overline{M}^{-1}(Gt_m + v_C + P\Theta) \\ y_C &= N_2\dot{\Theta} \\ v_C &= N_1 u_C + \overline{N}\dot{\Theta} \\ y &= \Theta \\ y_p &= C_p\Theta \\ u_I &= \Delta_I y_I \\ u_C &= \Delta_C y_C \end{aligned}$$

By substituting the variable v_C in the equations of $\ddot{\Theta}$ and y_I, respectively, the system equations are summarised as

$$\begin{bmatrix} \dot{x}_1 \\ \dot{x}_2 \\ \dot{x}_3 \\ \dot{x}_4 \\ \dot{x}_5 \\ \dot{x}_6 \\ -- \\ y_I \\ y_C \\ -- \\ y \\ y_p \end{bmatrix} = \Pi \begin{bmatrix} x_1 \\ x_2 \\ x_3 \\ x_4 \\ x_5 \\ x_6 \\ -- \\ u_I \\ u_C \\ -- \\ d \\ t_m \end{bmatrix} \tag{9.3}$$

$$\begin{bmatrix} u_I \\ u_C \end{bmatrix} = \begin{bmatrix} \Delta_I & 0 \\ 0 & \Delta_C \end{bmatrix} \begin{bmatrix} y_I \\ y_C \end{bmatrix} \tag{9.4}$$

where

$$\Pi =$$

$$\left[\begin{array}{cc|cc|cc} 0_{3\times 3} & I_{3\times 3} & 0_{3\times 3} & 0_{3\times 3} & 0_{3\times 3} & 0_{3\times 2} \\ -\overline{M}^{-1}P & -\overline{M}^{-1}\overline{N} & -\overline{M}^{-1}M_p & -\overline{M}^{-1}N_1 & \overline{M}^{-1}T & -\overline{M}^{-1}G \\ \hdashline -\overline{M}^{-1}P & -\overline{M}^{-1}\overline{N} & -\overline{M}^{-1}M_p & -\overline{M}^{-1}N_1 & \overline{M}^{-1}T & -\overline{M}^{-1}G \\ 0_{3\times 3} & N_2 & 0_{3\times 3} & 0_{3\times 3} & 0_{3\times 3} & 0_{3\times 2} \\ \hdashline I_{3\times 3} & 0_{3\times 3} & 0_{3\times 3} & 0_{3\times 3} & 0_{3\times 3} & 0_{3\times 2} \\ C_p & 0_{3\times 3} & 0_{3\times 3} & 0_{3\times 3} & 0_{3\times 3} & 0_{3\times 2} \end{array}\right]$$

Thus, the open-loop, pendulum model

$$\begin{bmatrix} y_I \\ y_C \\ y \\ y_p \end{bmatrix} = G_{\text{pend}} \begin{bmatrix} u_I \\ u_C \\ d \\ t_m \end{bmatrix}$$

is an eleven-input, twelve-output system with six states, where

$$G_{\text{pend}} = \left[\begin{array}{c|cc} A & B_1 & B_2 \\ \hline C_1 & D_{11} & D_{12} \\ C_2 & D_{21} & D_{22} \end{array}\right]$$

and

$$A = \begin{bmatrix} 0_{3\times 3} & I_{3\times 3} \\ -\overline{M}^{-1}P & -\overline{M}^{-1}\overline{N} \end{bmatrix}, \quad B_1 = \begin{bmatrix} 0_{3\times 3} & 0_{3\times 3} \\ -\overline{M}^{-1}M_p & -\overline{M}^{-1}N_1 \end{bmatrix}$$

$$B_2 = \begin{bmatrix} 0_{3\times 3} & 0_{3\times 2} \\ \overline{M}^{-1}T & -\overline{M}^{-1}G \end{bmatrix}$$

$$C_1 = \begin{bmatrix} -\overline{M}^{-1}P & -\overline{M}^{-1}\overline{N} \\ 0_{3\times 3} & N_2 \end{bmatrix}, \quad D_{11} = \begin{bmatrix} -\overline{M}^{-1}M_p & -\overline{M}^{-1}N_1 \\ 0_{3\times 3} & 0_{3\times 3} \end{bmatrix}$$

$$D_{12} = \begin{bmatrix} \overline{M}^{-1}T & -\overline{M}^{-1}G \\ 0_{3\times 3} & 0_{3\times 2} \end{bmatrix}$$

$$C_2 = \begin{bmatrix} I_{3\times 3} & 0_{3\times 3} \\ C_p & 0_{3\times 3} \end{bmatrix}, \quad D_{21} = 0_{6\times 6}, \quad D_{22} = 0_{6\times 5}$$

This pendulum model is implemented by the M-file `mod_pend.m`. (Note that the system matrix G_{pend} may be obtained by using the function `sysic`, too.)

The input/output relation of the perturbed, pendulum system is described by the upper LFT

$$\begin{bmatrix} y \\ y_p \end{bmatrix} = F_U(G_{\text{pend}}, \Delta_{\text{pend}}) \begin{bmatrix} d \\ t_m \end{bmatrix}$$

with the diagonal, uncertain matrix

$$\Delta_{pend} = \begin{bmatrix} \Delta_I & 0 \\ 0 & \Delta_C \end{bmatrix}$$

as depicted in Figure 9.5.

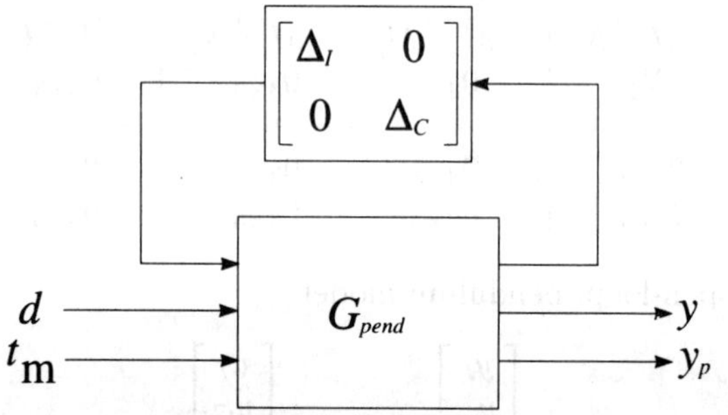

Fig. 9.5. The LFT representation of the perturbed triple inverted pendulum

The singular value plot of the triple, inverted pendulum system (the nominal system) is shown in Figure 9.6.

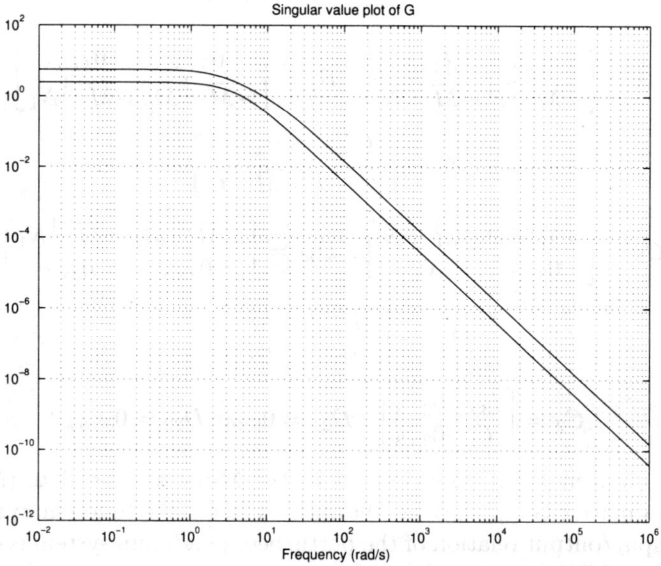

Fig. 9.6. Singular values of the pendulum system

We now consider the models of the actuators. The nominal transfer functions of the actuators are taken as first-order, phase-lag models of

$$\overline{G}_{m_1} = \frac{\overline{K}_{m_1}}{\overline{T}_{m_1}s+1}, \quad \overline{G}_{m_2} = \frac{\overline{K}_{m_2}}{\overline{T}_{m_2}s+1}$$

with parameters

$$\overline{K}_{m_1} = 1.08, \ \overline{T}_{m_1} = 0.005, \ \overline{K}_{m_2} = 0.335, \ \overline{T}_{m_2} = 0.002$$

It is assumed that the actual gain coefficients (K_{m_1}, K_{m_2}) are constants with relative error 10% around their nominal values and the time constants (T_{m_1}, T_{m_2}) with relative error 20%. The uncertain frequency responses of the actuators are shown in Figure 9.7.

In order to account for unmodelled dynamics and nonlinear effects, the uncertainties in the actuator models are approximated by input multiplicative uncertainties that give rise to the perturbed transfer functions

$$G_{m_1} = (1 + W_{m_1}\delta_{m_1})\overline{G}_{m_1}, \quad G_{m_2} = (1 + W_{m_2}\delta_{m_2})\overline{G}_{m_2}$$

where

$$|\delta_{m_1}| \leq 1, \quad |\delta_{m_2}| \leq 1$$

and the uncertainty weights W_{m_1}, W_{m_2} are so chosen that

$$\frac{\left|G_{m_1}(j\omega) - \overline{G}_{m_1}(j\omega)\right|}{\left|\overline{G}_{m_1}(j\omega)\right|} < |W_{m_1}(j\omega)|, \quad \frac{\left|G_{m_2}(j\omega) - \overline{G}_{m_2}(j\omega)\right|}{\left|\overline{G}_{m_2}(j\omega)\right|} < |W_{m_2}(j\omega)|.$$

The frequency responses of W_{m_1}, W_{m_2} are found graphically as shown in Figure 9.8 and then approximated by first-order transfer functions using the file `wfit.m`. As a result, we obtain

$$W_{m_1} = \frac{0.3877s + 25.6011}{1.0000s + 246.3606}, \quad W_{m_2} = \frac{0.3803s + 60.8973}{1.0000s + 599.5829}$$

By introducing the vector $u = [u_1, \ u_2]^T$, the equations of the actuators are rewritten as

$$t_m = \overline{G}_m(I_2 + W_m\Delta_m)u$$

where

$$\overline{G}_m = \begin{bmatrix} \overline{G}_{m_1} & 0 \\ 0 & \overline{G}_{m_2} \end{bmatrix}, \quad W_m = \begin{bmatrix} W_{m_1} & 0 \\ 0 & W_{m_2} \end{bmatrix}, \quad \Delta_m = \begin{bmatrix} \delta_{m_1} & 0 \\ 0 & \delta_{m_2} \end{bmatrix}$$

The block diagram of the actuators with the input multiplicative uncertainty is shown in Figure 9.9.

Let the inputs and outputs of the uncertain blocks (δ_{m_1}, δ_{m_2}) be denoted by y_{m_1}, y_{m_2} and u_{m_1}, u_{m_2}, respectively. Defining the vectors $u_m = [u_{m_1} \ u_{m_2}]^T$ and $y_m = [y_{m_1} \ y_{m_2}]^T$, we may rearrange the actuator models as

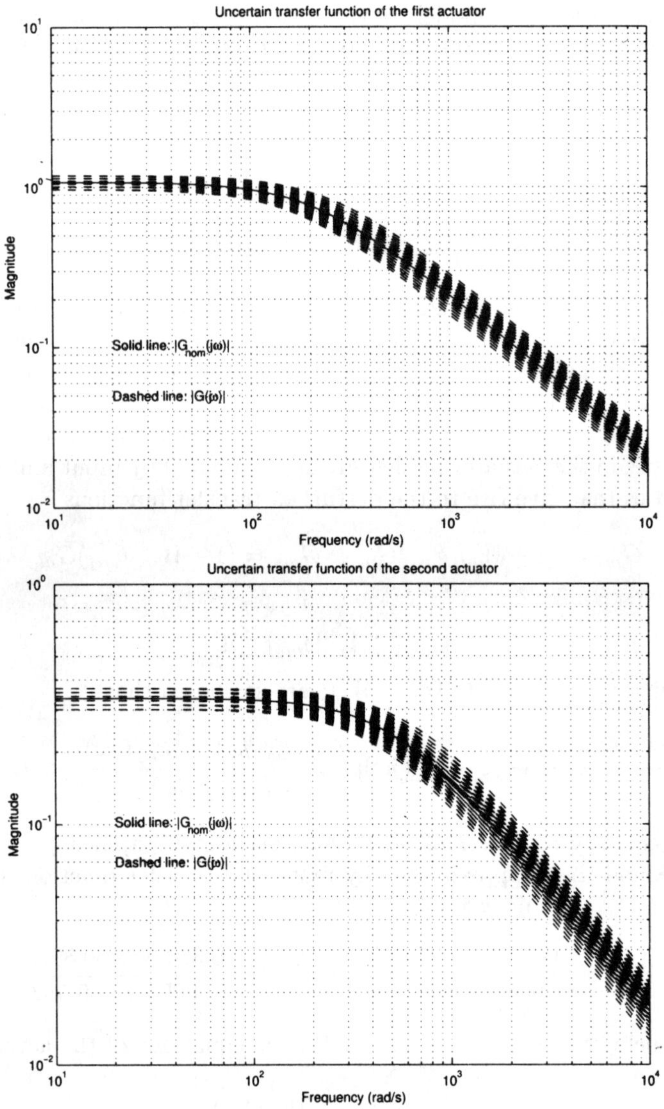

Fig. 9.7. Uncertain frequency responses of the actuators

$$\begin{bmatrix} y_m \\ t_m \end{bmatrix} = G_{\text{act}} \begin{bmatrix} u_m \\ u \end{bmatrix} \tag{9.5}$$

$$u_m = \Delta_m y_m \tag{9.6}$$

where

$$G_{\text{act}} = \begin{bmatrix} 0_{2\times 2} & W_m \\ \overline{G}_m & \overline{G}_m \end{bmatrix}$$

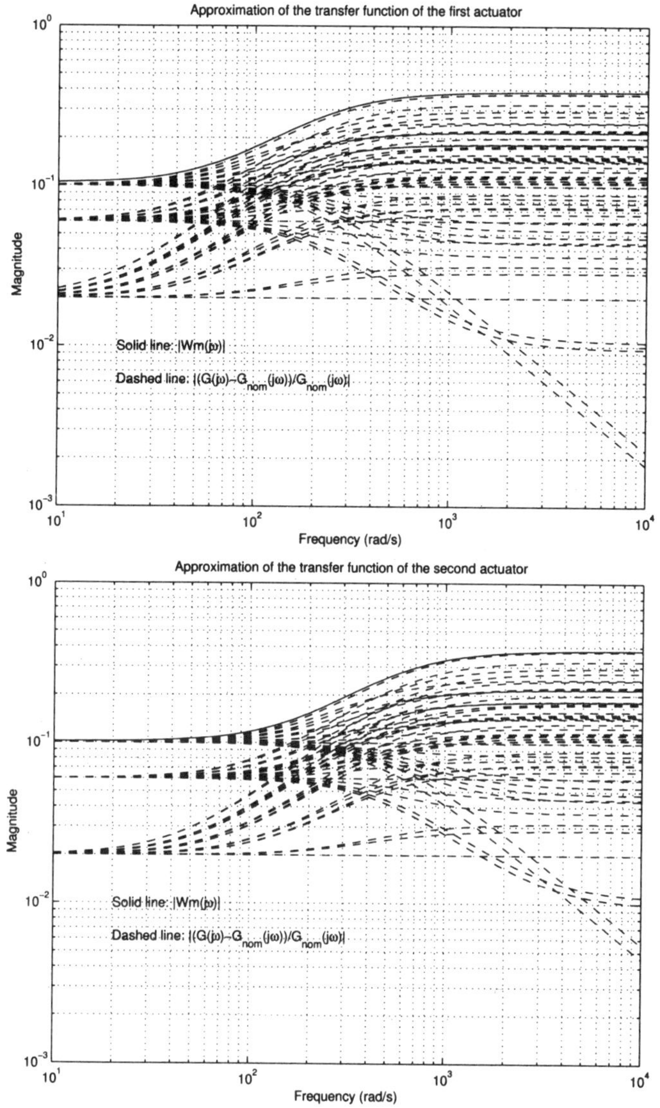

Fig. 9.8. Actuators uncertainty approximations

The perturbed actuators can be described by the following upper LFT and shown in Figure 9.10,

$$t_m = F_U(G_{act}, \Delta_m)u$$

Having modelled both the pendulum and the actuators with consideration of possible perturbations, the block-diagram of the whole system is seen as in Figure 9.11. Note that Δ_m is a complex uncertainty, while Δ_I and Δ_C are real

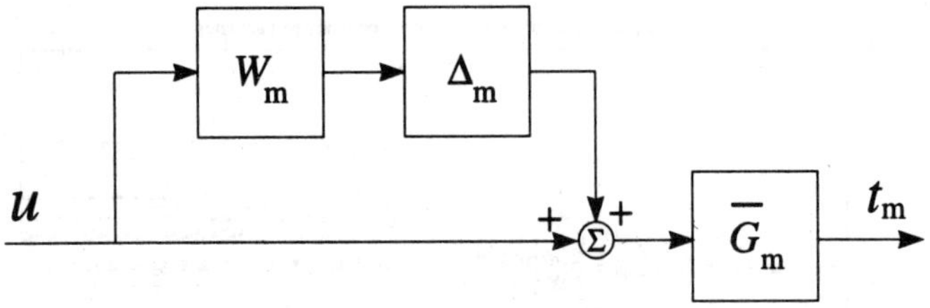

Fig. 9.9. Block diagram of the perturbed actuators

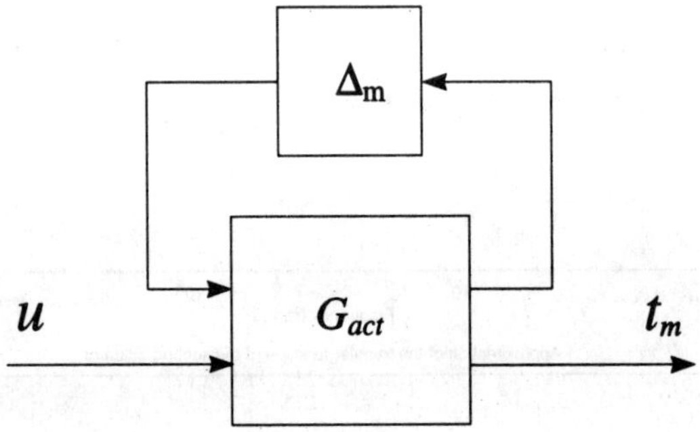

Fig. 9.10. LFT representation of the perturbed actuators

uncertainties. We thus have a case of a mixed, real and complex, uncertainty configuration.

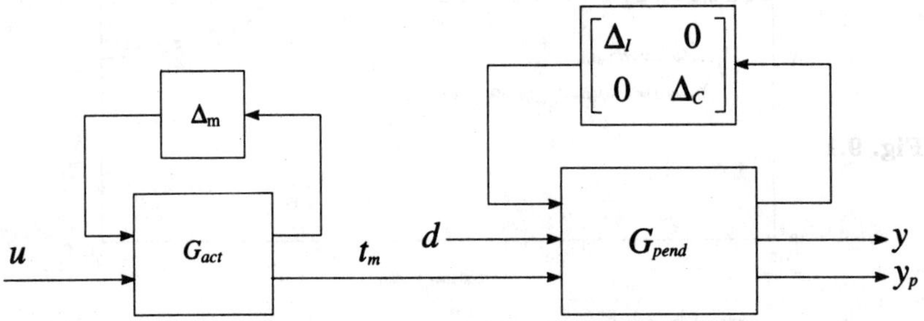

Fig. 9.11. Triple inverted pendulum system with uncertainties

The model of the whole triple inverted pendulum system is given by the equations

$$\begin{bmatrix} y_m \\ y_I \\ y_C \\ y \\ y_p \end{bmatrix} = G_{\text{sys}} \begin{bmatrix} u_m \\ u_I \\ u_C \\ d \\ u \end{bmatrix} \tag{9.7}$$

where the transfer matrix G_{sys} is determined by the matrices G_{act} and G_{pend}, and can be easily obtained by using the function `sysic`.

The whole, perturbed, triple inverted pendulum system can be described by the following upper LFT, as shown in Figure 9.12,

$$\begin{bmatrix} y \\ y_p \end{bmatrix} = F_U(G_{\text{sys}}, \Delta) \begin{bmatrix} d \\ u \end{bmatrix}$$

with the diagonal, uncertainty matrix

$$\Delta = \begin{bmatrix} \Delta_m & 0 & 0 \\ 0 & \Delta_I & 0 \\ 0 & 0 & \Delta_C \end{bmatrix} \tag{9.8}$$

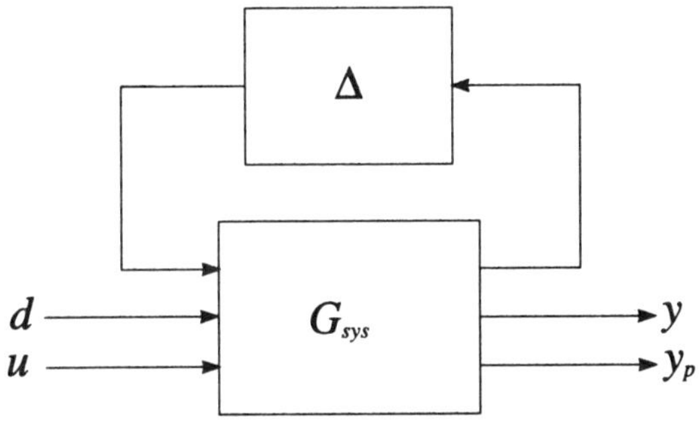

Fig. 9.12. LFT representation of triple inverted pendulum system with uncertainty

9.3 Design Specifications

The block diagram of the closed-loop system, which includes the triple inverted pendulum model, the feedback structure and the controller, as well as the elements reflecting the model uncertainty and the performance objectives, is depicted in Figure 9.13. In order to achieve better performance, we shall make use of the configuration of a two-degree-of-freedom (2DOF) controller.

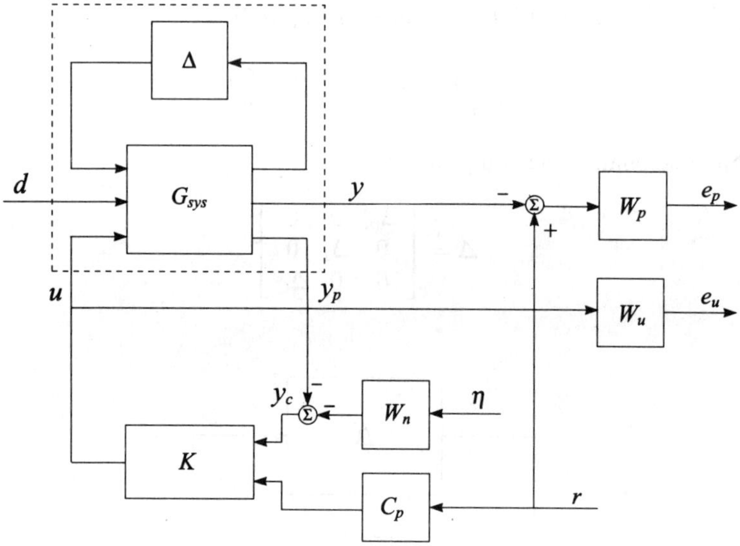

Fig. 9.13. Interconnection structure of the closed-loop system

The model of the triple inverted pendulum system represents an upper LFT in the form $G = F_U(G_{sys}, \Delta)$. The rectangle, shown with dashed lines, represents the transfer function matrix G. Inside the rectangle is the nominal model $G_{\rm sys}$ and the block Δ, which parametrises the model uncertainties. The matrix Δ is unknown, but with the structure defined in (9.8) and with the norm bound $\|\Delta\|_\infty < 1$.

The feedback signal is generated on the basis of the potentiometers output $y_p = C_p y$, which requires the reference r to be multiplied by the matrix C_p (the other way to deal with the situation is to multiply y_p by C_p^{-1}). The measurement of the arm angles is accompanied by introduction of frequency-dependent noises that are inevitable in practice and are thus added to the corresponding measurements. This is why at the outputs of the potentiometers one obtains the signal $y_c = -(y_p + W_n\eta)$, where W_n is a weighting function (shaping filter on the measurement noise) and $\eta = [\eta_1\ \eta_2\ \eta_3]^T$ is an arbitrary noise signal satisfying $\|\eta\|_2 \leq 1$. By choosing an appropriate W_n it is possible to form the desired spectral contents of the actual noise signal (in the form of $W_n\eta$).

In the given case the matrix W_n is chosen as

$$W_n(s) = \begin{bmatrix} w_n(s) & 0 & 0 \\ 0 & w_n(s) & 0 \\ 0 & 0 & w_n(s) \end{bmatrix}$$

where the weighting transfer function $w_n = 2 \times 10^{-5} \frac{10s+1}{0.1s+1}$ is a high pass filter that shapes the noise spectral density for the type of potentiometers under consideration. The magnitude plot of this filter is shown in Figure 9.14. This transfer function means that in the low-frequency range the magnitude of the measurement error is about 2×10^{-5} V, and in the high-frequency range, about 2×10^{-3} V.

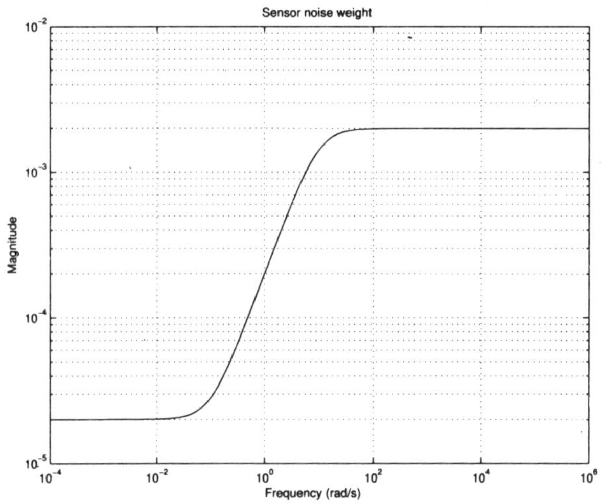

Fig. 9.14. Measurement noise weighting function

The closed-loop system error $e_y = r - y$ may be determined by

$$e_y = S \begin{bmatrix} r \\ d \\ \eta \end{bmatrix}$$

where the matrix $S = S(G_{\text{sys}})$ is considered as the nominal sensitivity function with respect to the inputs r, d and η. As for feedback signals, we shall use $y_c = -(y_p + W_n\eta)$ and $C_p r$.

The performance objective requires the transfer function matrices from r, d and η to e_y and e_u to be small in the sense of $\|\cdot\|_\infty$, for all possible (stable) uncertainty matrices Δ. The transfer function matrices W_p and W_u are used to reflect the relative significance over different frequency ranges for which the performance is required.

The design problem for the triple inverted pendulum system is to find a linear, output controller $K(s)$ to generate the output feedback

$$u(s) = K(s) \begin{bmatrix} y_c(s) \\ C_p r(s) \end{bmatrix}$$

to ensure the following properties of the closed-loop system.

Nominal performance:

The closed-loop system achieves nominal performance if the following performance objective is satisfied for the nominal plant model $G_{\rm sys}$,

$$\left\| \begin{bmatrix} W_p S(G_{\rm sys}) \\ W_u K S(G_{\rm sys}) \end{bmatrix} \right\|_\infty < 1 \tag{9.9}$$

where W_p and W_u are appropriately chosen weighting functions. This objective corresponds to the mixed S/KS sensitivity optimisation.

Robust stability:

The closed-loop system achieves robust stability if the closed-loop system is internally stable for each possible, perturbed plant dynamics $G = F_U(G_{\rm sys}, \Delta)$.

Robust performance:

The closed-loop system must maintain, for each $G = F_U(G_{\rm sys}, \Delta)$, the performance objective

$$\left\| \begin{bmatrix} W_p S(G) \\ W_u K S(G) \end{bmatrix} \right\|_\infty < 1 \tag{9.10}$$

where $S(G)$ is the perturbed sensitivity function of $G = F_U(G_{\rm sys}, \Delta)$.

In addition to the above requirements, it is desirable that the controller designed would have acceptable complexity, *i.e.* it is of reasonably low order.

9.4 System Interconnections

The internal structure of the nineteen-input, nineteen-output open-loop system, which is saved as the variable `pend_ic`, is shown in Figure 9.15. The inputs and outputs of the uncertainties are saved in the variables `pertin` and `pertout`, respectively, and the reference, the disturbance and the noise in the

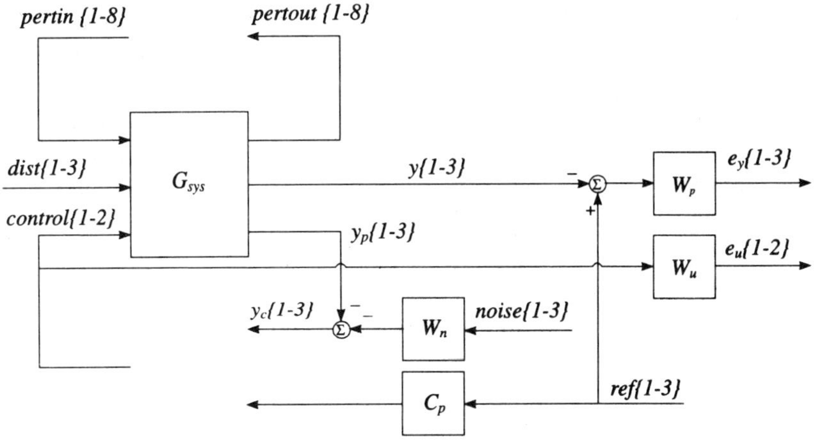

Fig. 9.15. Open-Loop interconnection of the triple inverted pendulum system

variables `ref`, `dist`, `noise`, respectively. The control signal is saved in the variable `control`.

The variables `pertin` and `pertout` each have eight elements. The variables `ref`, `dist`, `noise`, `y`, `y_p`, `y_c`, `e_y` have all three elements and the variables `control`, `e_u` have two elements. The open-loop connection is assigned by the M-file `olp_pend`. The schematic diagram showing the specific input/output ordering for the variable `pend_ic` is shown in Figure 9.16.

The block-diagram used in the closed-loop system simulation is shown in Figure 9.17. The corresponding closed-loop interconnection, which is saved in the variable `pend_sm`, is obtained by the M-file `sim_pend`.

Figure 9.18 shows the schematic diagram of the specific input/output ordering for the variable `pend_sm`.

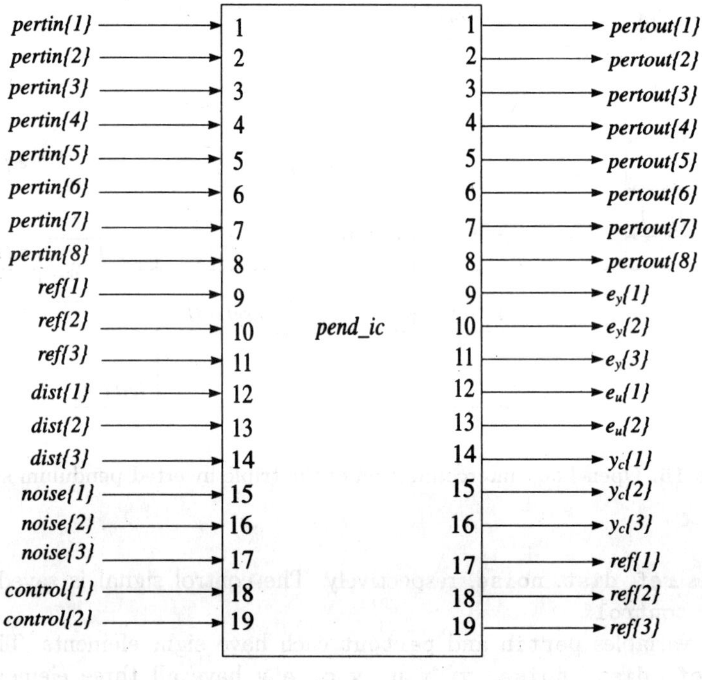

Fig. 9.16. Schematic diagram of the open-loop system

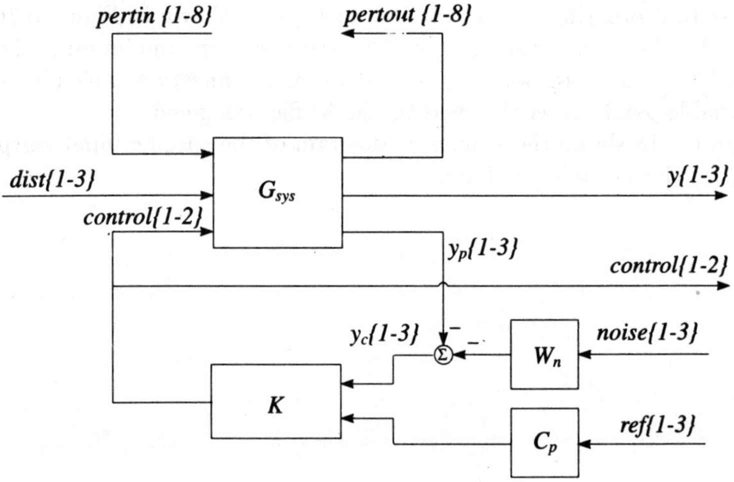

Fig. 9.17. Closed-loop interconnection structure of the pendulum system

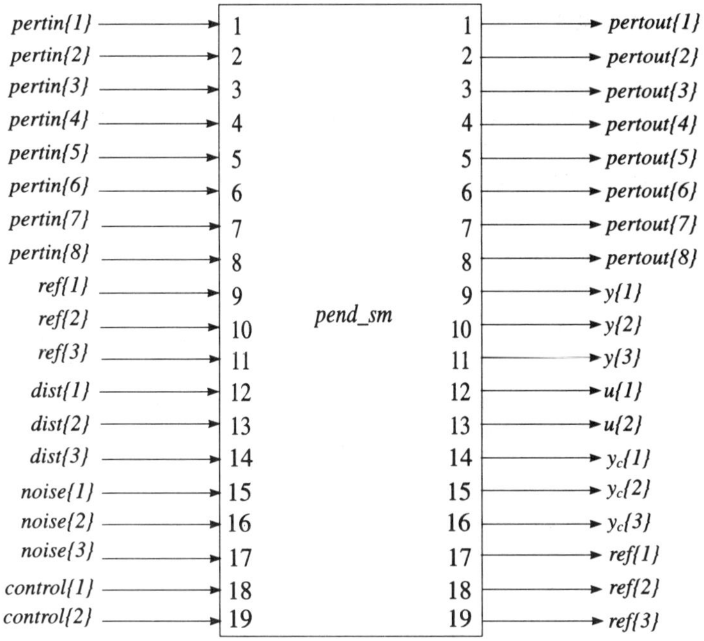

Fig. 9.18. Schematic diagram of the closed-loop system

9.5 $\mathcal{H}_\infty$ Design

The design goal in this case is to find an $\mathcal{H}_\infty$ (sub)optimal control law for the interconnection shown in Figure 9.19, in which we neglect the uncertainty inputs and outputs. The variable `hin_ic`, which corresponds to the transfer function matrix of the augmented system, may be obtained by the command line

```
hin_ic = sel(pend_ic,[9:19],[9:19])
```

The $\mathcal{H}_\infty$ optimal control minimises the $\|\cdot\|_\infty$ norm of $F_L(P, K)$ over the stabilising controller transfer matrix K, where P is the transfer function matrix of the augmented system. In the given case $F_L(P, K)$ is the nominal closed-loop system transfer matrix from the references, disturbances and noises (the signals r, d and η) to the weighted outputs e_y and e_u (Figure 9.19).

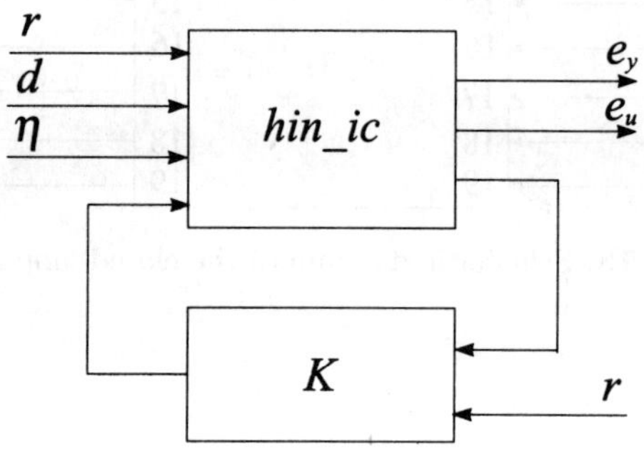

Fig. 9.19. Block diagram for $\mathcal{H}_\infty$ design

In the given case the weighting performance functions are chosen in the forms of

$$W_p(s) = \begin{bmatrix} w_{p1}(s) & 0 & 0 \\ 0 & w_{p2}(s) & 0 \\ 0 & 0 & w_{p3}(s) \end{bmatrix}, \quad W_u(s) = \begin{bmatrix} w_u(s) & 0 \\ 0 & w_u(s) \end{bmatrix}$$

where

$$w_{p1}(s) = 10^{-6} \frac{s^4 + 12s^3 + 80s^2 + 80s + 65}{s^4 + 1500s^3 + 13000s^2 + 12000s + 0.005}$$

$$w_{p2}(s) = \frac{s^4 + 12s^3 + 80s^2 + 80s + 65}{s^4 + 1500s^3 + 13000s^2 + 12000s + 0.005}$$

$$w_{p3}(s) = \frac{s^4 + 12s^3 + 80s^2 + 80s + 65}{s^4 + 1500s^3 + 13000s^2 + 12000s + 0.005}$$

and $w_u(s) = 10^{-6}$.

The small gain in the first weighting function (w_{p1}) indicates that the first error may be allowed to be large. This is due to the fact that the requirement on precision position of the first arm may be relaxed a little in this case.

The weighting functions are set in the M-file **wts_pend.m**.

The $\mathcal{H}_\infty$ design is conducted by using the M-file **hinf_pend.m**. It utilises the function **hinfsyn**, which determines a (sub)optimal $\mathcal{H}_\infty$ control law, based on the prescribed open-loop interconnection. The interval for γ iteration is chosen between 0 and 10 with a tolerance tol = 0.001. The controller obtained is of 28th order and for this controller the closed-loop system achieves $\mathcal{H}_\infty$ norm equal to 1.0005. An undesired property of the controller is that it has a pole at 33.6, *i.e.* this controller is unstable, which makes it less favourable in practice.

The singular value plot of the $\mathcal{H}_\infty$ controller transfer matrix is shown in Figure 9.20.

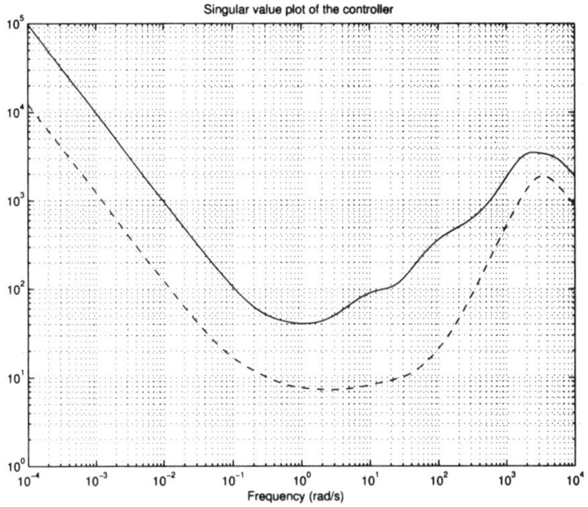

Fig. 9.20. Singular values of the $\mathcal{H}_\infty$ controller

The closed-loop, transient responses (reference tracking) are obtained by using the file clp_pend. The transient response of the closed-loop system with $\mathcal{H}_\infty$ controller for the reference vector

$$r = \begin{bmatrix} 0 \\ -0.1 \\ 0.2 \end{bmatrix}$$

(the references are measured in radians) is shown in Figure 9.21. The response is fast (settling time approximately equal to 6 s) with small overshoots of the output variables. The steady-state errors are small, except for a small error on the position of the first arm.

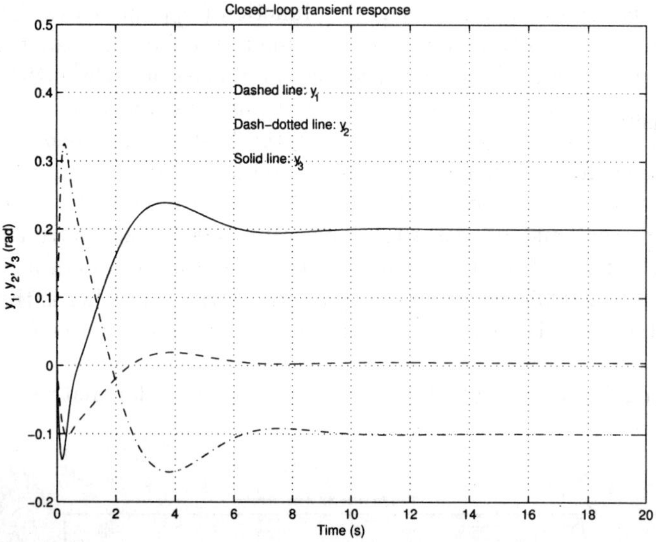

Fig. 9.21. Closed-loop transient response of $\mathcal{H}_\infty$ controller

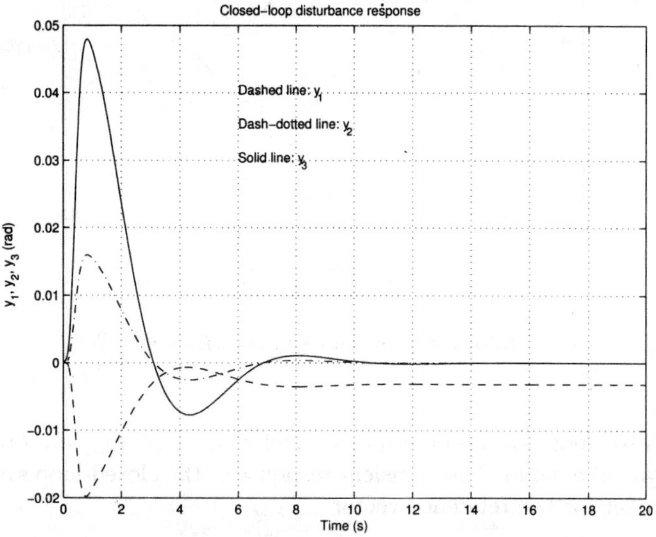

Fig. 9.22. Disturbance rejection of $\mathcal{H}_\infty$ controller

The disturbance rejection of the closed-loop system with the $\mathcal{H}_\infty$ controller is shown in Figure 9.22. The disturbance vector is set to

$$d = \begin{bmatrix} 0.1 \\ 0.1 \\ 0.1 \end{bmatrix}$$

and measured in N m.

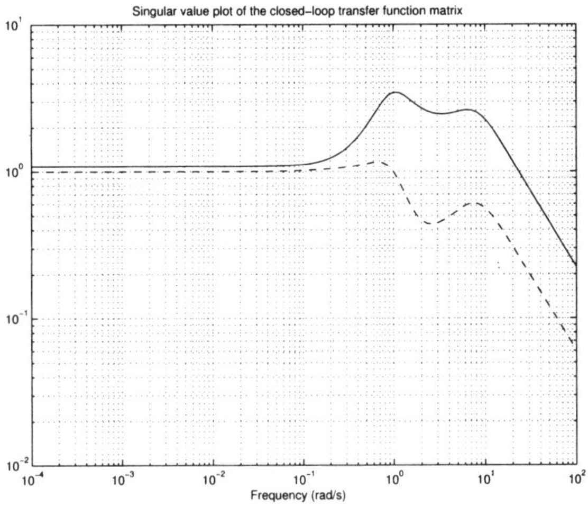

Fig. 9.23. Closed-loop singular values with $\mathcal{H}_\infty$ controller

In Figure 9.23 we show the singular values plot of the closed-loop transfer matrix with the $\mathcal{H}_\infty$ controller.

The test for robust stability is based on the μ values of the upper 8×8 block of the closed-loop transfer function matrix. The block structure of the uncertainty is set as

```
blks = [1 1;1 1;-1 1;-1 1;-1 1;-1 1;-1 1;-1 1]
```

The upper and lower bounds of the structured singular value μ (over the frequency range) are shown in Figure 9.24. It is clear that the closed-loop system with $\mathcal{H}_\infty$ controller does not achieve robust stability, since the maximum value of μ is equal to 1.666.

The robust performance of the closed-loop system with $\mathcal{H}_\infty$ controller is checked with the μ-analysis, too. The block-structure of the uncertainty Δ_P consists of an 8×8 diagonal uncertainty block and a 9×5 full (complex) uncertainty performance block, i.e.,

$$\Delta_P := \left\{ \begin{bmatrix} \Delta & 0 \\ 0 & \Delta_F \end{bmatrix} : \Delta \in \mathcal{C}^{8\times 8},\ \Delta_F \in \mathcal{C}^{9\times 5} \right\}$$

The μ values corresponding to the case of robust performance analysis are shown in Figure 9.25. Again, the closed-loop system does not achieve robust performance either, because the maximum value of μ is equal to 1.673.

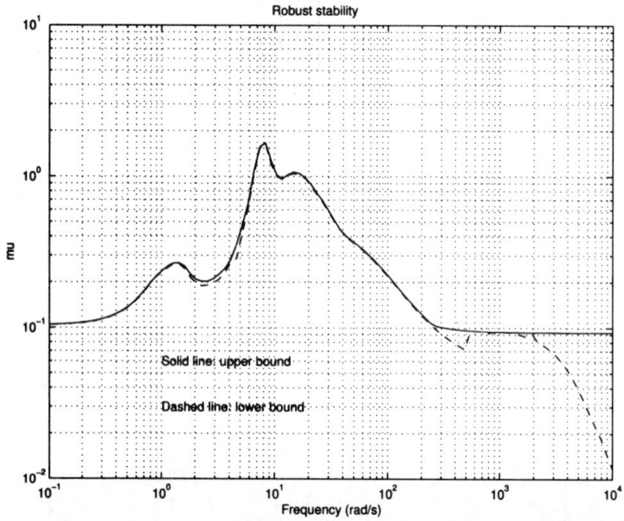

Fig. 9.24. Robust stability test of $\mathcal{H}_\infty$ controller

Hence, it is concluded that the designed $\mathcal{H}_\infty$ controller leads to good closed-loop transient responses, but does not ensure the necessary robustness of the closed-loop system as required.

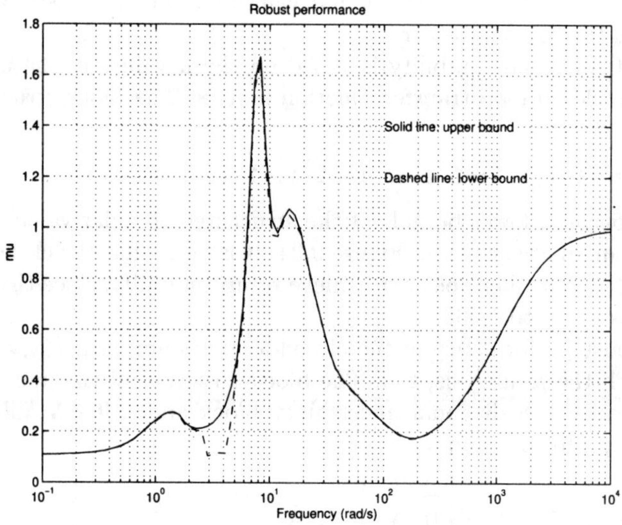

Fig. 9.25. Robust performance of $\mathcal{H}_\infty$ controller

9.6 μ-Synthesis

Let us denote by $P(s)$ the transfer function matrix of the nineteen-input, sixteen-output, open-loop system consisting of the pendulum system model plus the weighting functions, and let the block structure Δ_P of uncertainties be defined by

$$\Delta_P := \left\{ \begin{bmatrix} \Delta & 0 \\ 0 & \Delta_F \end{bmatrix} : \Delta \in \mathcal{R}^{8\times 8},\ \Delta_F \in \mathcal{C}^{9\times 5} \right\}$$

The first block of the matrix Δ_P, the uncertainty block Δ, corresponds to the parametric uncertainties modelled in the triple inverted pendulum system. The second block, Δ_F, is a fictitious uncertainty block, introduced to include the performance objectives in the framework of the μ-approach.

To meet the design objectives a stabilising controller K is to be found such that, at each frequency $\omega \in [0, \infty]$, the structured singular value μ satisfies the condition

$$\mu_{\Delta_P}[F_L(P, K)(j\omega)] < 1$$

The fulfillment of this condition guarantees robust performance of the closed-loop system, *i.e.*,

$$\left\| \begin{bmatrix} W_p S(G) \\ W_u K S(G) \end{bmatrix} \right\|_\infty < 1$$

where $G = F_U(G_{\rm sys}, \Delta)$, for all stable perturbations Δ with $\|\Delta\|_\infty < 1$.

The μ-synthesis is conducted by using the M-file `ms_pend.m`. The uncertainty structure and other parameters used in the D-K iteration are assigned in the auxiliary file `dk_pend.m`.

The experiments with μ-synthesis show that the use of the weighting functions, used in the $\mathcal{H}_\infty$ design, leads to very slow transient responses. Hence, these functions are appropriately modified to

$$w_{p1}(s) = 10^{-6}\frac{s^4 + 12s^3 + 80s^2 + 80s + 65}{15s^4 + 1500s^3 + 13000s^2 + 12000s + 0.005}$$

$$w_{p2}(s) = 2.5\frac{s^4 + 12s^3 + 80s^2 + 80s + 65}{15s^4 + 1500s^3 + 13000s^2 + 12000s + 0.005}$$

$$w_{p3}(s) = 10\frac{s^4 + 12s^3 + 80s^2 + 80s + 65}{15s^4 + 1500s^3 + 13000s^2 + 12000s + 0.005}$$

The control weighting functions are again taken as $w_u(s) = 10^{-6}$.

The progress of the D-K iteration is shown in Table 9.3.

It can be seen from the table that after the third iteration the maximum value of μ is equal to 0.960, which indicates that the robust performance has been achieved. The final controller obtained is of 84th order. Note that this controller is stable.

Table 9.3. D-K iterations results in μ-synthesis

Iteration	Controller order	Maximum value of μ
1	28	13.852
2	84	1.422
3	84	0.960

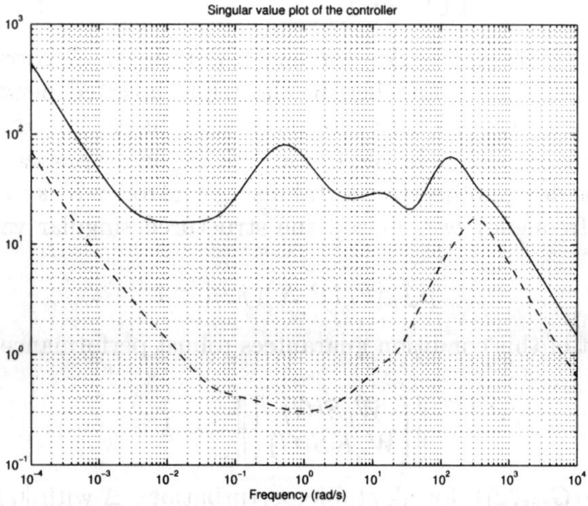

Fig. 9.26. Singular values of the μ-controller

The singular value plot of the μ-controller is shown in Figure 9.26. Similar to the $\mathcal{H}_\infty$ controller, the μ-controller has an integrating action in the low frequency range.

The structured singular values for the robust stability study (over the frequency range) are shown in Figure 9.27. The maximum value of μ is 0.626 which means that the stability of the system is preserved under perturbations that satisfy $\|\Delta\|_\infty < \frac{1}{0.626}$.

The μ values over the frequency range for the case of robust performance analysis are shown in Figure 9.28. The closed-loop system achieves robust performance, since the maximum value of μ is equal to 0.820.

Consider now the closed-loop transient responses obtained by using the M-file `clp_pend`. The same reference signal as used in the $\mathcal{H}_\infty$ simulation is used here. That is,

$$r = \begin{bmatrix} 0 \\ -0.1 \\ 0.2 \end{bmatrix}$$

The transient response is shown in Figure 9.29. To compare with those of the

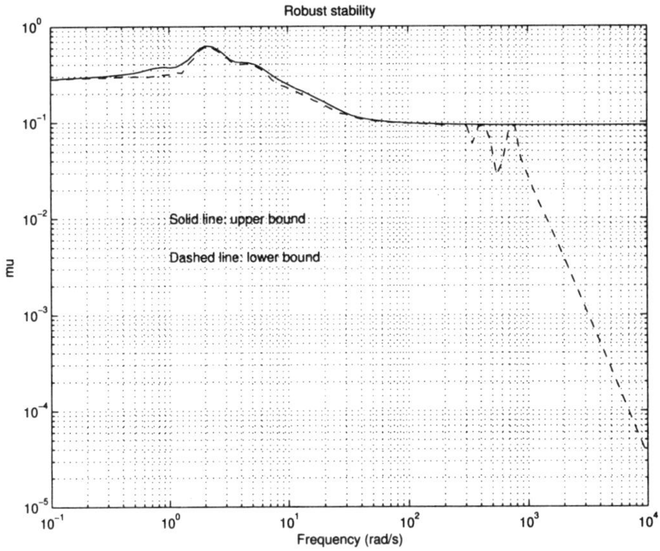

Fig. 9.27. Robust stability of μ controller

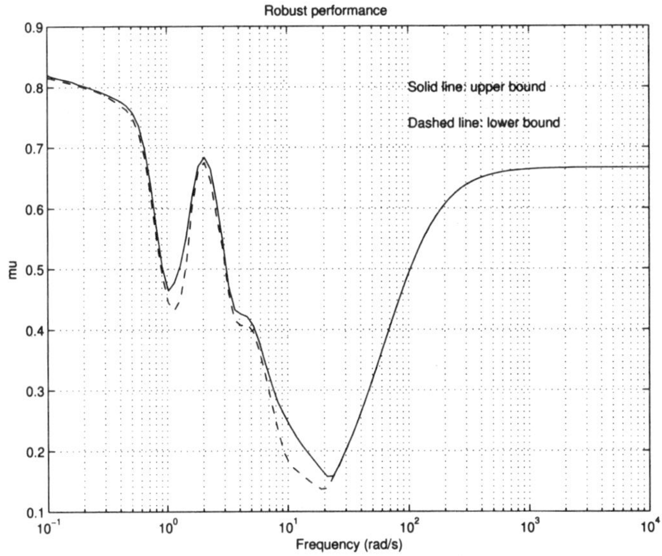

Fig. 9.28. Robust performance of μ controller

$\mathcal{H}_\infty$ controller, it can be seen that the transient response (reference tracking) of the μ-controller is slower and the steady-state errors are slightly larger, though with less overshoot (undershoot).

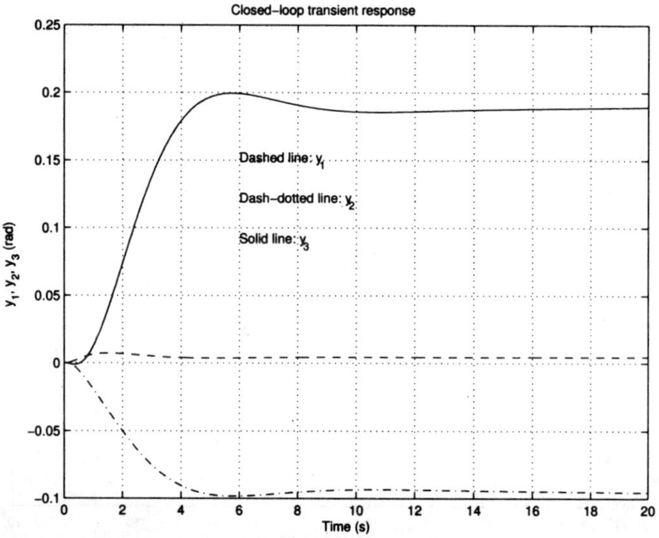

Fig. 9.29. Closed-loop transient response of μ-controller

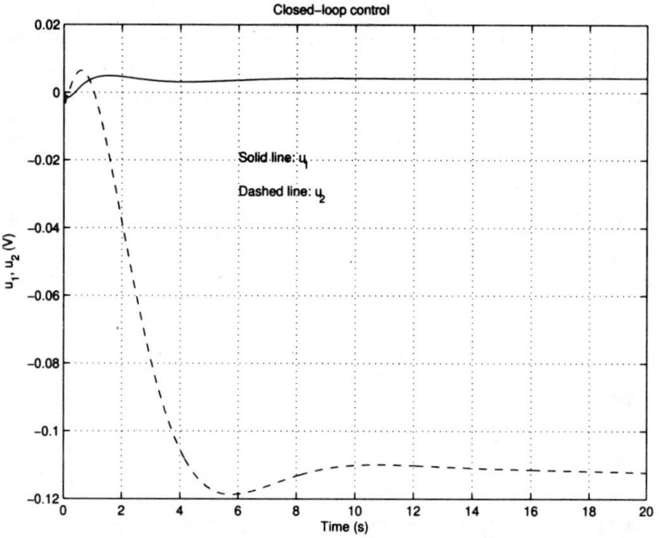

Fig. 9.30. Control action of μ-controller

The control action in the case of a μ-controller is shown in Figure 9.30. The motor voltages are kept within 0.12 V.

The disturbance rejection of the closed-loop system with the μ-controller is shown in Figure 9.31. We see that the overshoot of the third output is almost 7 times larger than in the case of the $\mathcal{H}_\infty$ controller.

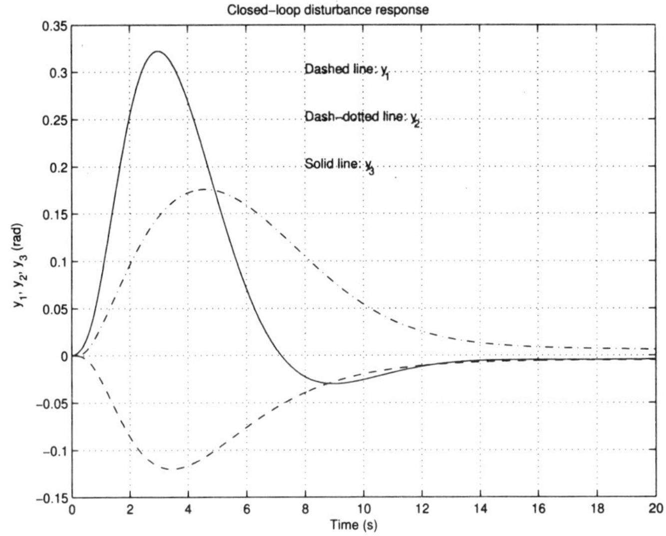

Fig. 9.31. Disturbance rejection of μ-controller

Hence, it is clear that the (nominal) transient responses in terms of reference tracking and disturbance attenuation are worse in the case of the μ-controller. This is the price that has to be paid to ensure the robust stability and robust performance of the closed-loop system.

In Figure 9.32, we show the noise $w_n\eta_3$ that acts at the output of the third potentiometer, and the corresponding response of the output y_3. From the magnitudes of $w_n\eta_3$ and y_3, one may conclude that the closed-loop system is susceptible to noises.

The closed-loop frequency responses are obtained by the M-file **frs_pend.m**. The singular values of the closed-loop transfer matrix with the μ-controller are plotted in Figure 9.33. The comparison with the plot shown in Figure 9.23 reveals that the closed-loop bandwidth in the case of a μ-controller is smaller, which leads to a slower response and worse disturbance attenuation.

The singular values of the transfer function matrix concerning disturbance rejection are shown in Figure 9.34. We see that the disturbance attenuation is worst for frequencies around 1 rad/s.

From the singular values plot of the transfer function matrix concerning noise attenuation, shown in Figure 9.35, we see that the influence of noises on the system output would be maximal for frequencies between 1 rad/s and 10 rad/s.

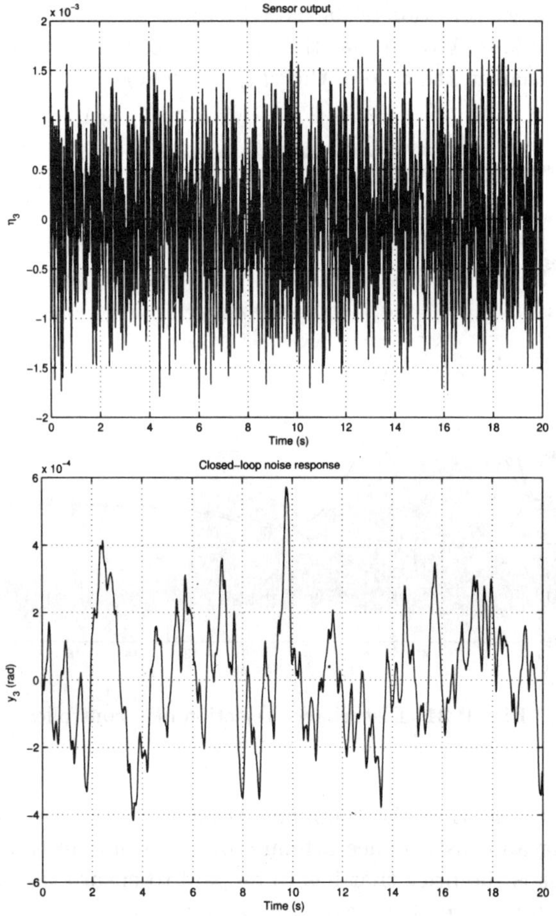

Fig. 9.32. Sensor noise and closed-loop noise response

As mentioned earlier, the controller obtained by μ-synthesis is initially of 80th order, which makes its implementation in practice very difficult. Therefore, it is necessary to reduce the controller order. For this purpose, we implements the M-file **red_pend.m**. We use the model reduction algorithm based on system balancing followed by optimal Hankel approximation. This algorithm allows us to reduce the controller order to 27. Further reduction of the controller order leads to deterioration of the closed-loop transient responses and would even cause the instability of the closed-loop system.

In Figure 9.36 we compare the frequency responses (the maximum singular values) of the full-order and reduced-order μ-controllers. Up to the frequency 10^4 rad/s the frequency plots of both controllers coincide with each other, which implies very similar performance of the closed-loop systems. In fact,

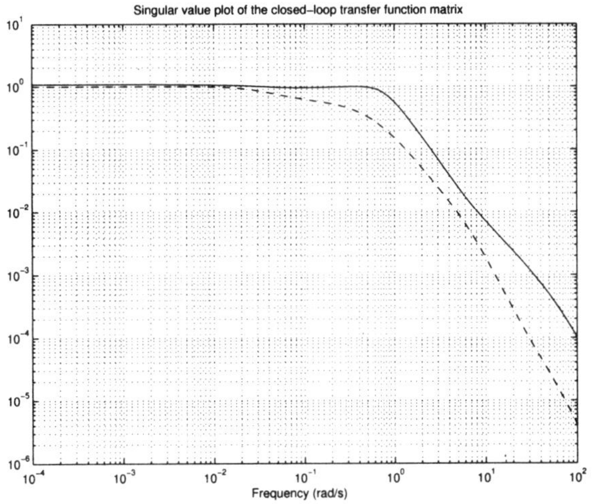

Fig. 9.33. Closed-loop singular value plot

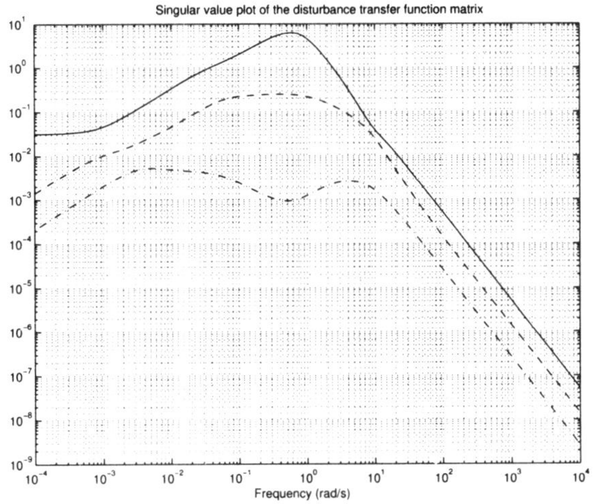

Fig. 9.34. Singular values of the disturbance rejection transfer function matrix

the closed-loop transient responses with the full-order controller and those with the reduced-order controller (not included in the book) are practically undistinguishable.

A discrete-time controller may be obtained by sampling the already designed continuous-time controller for a sampling frequency $f_s = 1/T_s$. This can be conducted by the M-file `dcl_pend.m` that utilises the Robust Control Toolbox function `samhld`. The resulting, sampled-data closed-loop system can be simulated by using the function `sdtrsp`. Since the inverted pendu-

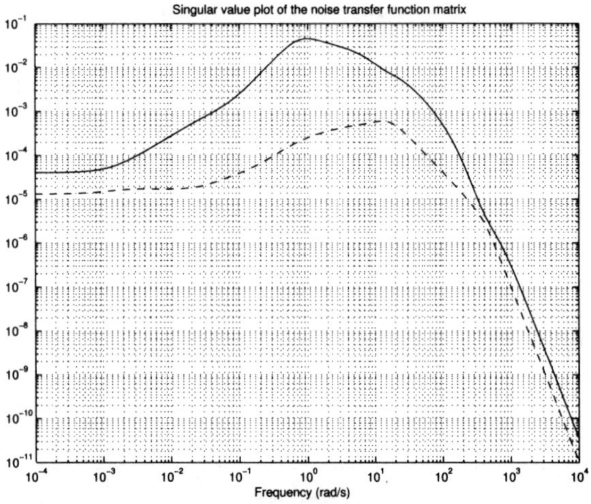

Fig. 9.35. Singular values of the noise-attenuation transfer function matrix

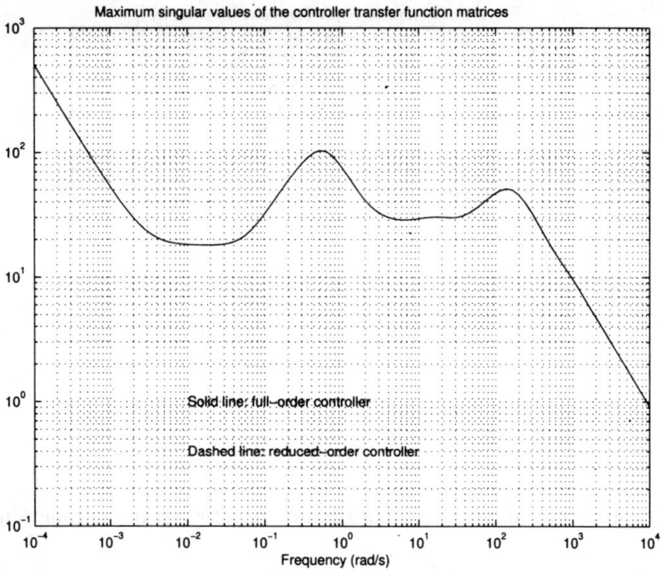

Fig. 9.36. Frequency responses of the full- and reduced-order controllers

lum is a relatively slow system, the sampling frequency may be chosen to be small. For instance, if this frequency is 100 Hz then the transient responses of the sampled-data system are close to the corresponding responses of the continuous-time system.

9.7 Nonlinear System Simulation

The nonlinear, closed-loop system of the triple inverted pendulum is simulated by using the Simulink® models `c_pend.mdl` (for the continuous-time system) and `d_pend.mdl` (for the sampled-data system). Both models allow us to simulate the closed-loop system for different references, disturbances and noise signals. Both models utilise the M-file S-function `s_pend.m` that includes the nonlinear differential equations (9.1) of the triple inverted pendulum. The initial conditions of the pendulum are assigned in the M-file `inc_pend.m`.

The sampled-data model consists of the discretised controller, a 14-bit analogue-to-digital converter with maximum input voltage 5 V, and a 14-bit digital-to-analogue converter with maximum output voltage 5 V. It is assumed that the calculation of the control action requires no longer than one sampling period T_s.

Before simulating the system it is necessary to set the model parameters by using the M-file `init_c_pend.m` (in the continuous-time case) or the M-file `init_d_pend.m` (in the discrete-time case).

The simulation of the nonlinear system for small references and disturbances produces results that are very close to the results obtained for the linearised model.

The Simulink® model `c_pend.mdl` of the continuous-time, nonlinear, pendulum system is shown in Figure 9.37.

In Figure 9.38 we show the transient response of the continuous-time, nonlinear system for a reference $r = [0\ \ -0.1\ 0.2]^T$. The transient response is close to that of the linear system (Figure 9.29).

Since the triple inverted pendulum is essentially nonlinear, the simulation results may differ in the case of large references and disturbances. In Figure 9.39 we show the transient response of the continuous-time, nonlinear system for a reference $r = [0\ 0\ 0.6]^T$. It is seen that for this relatively large reference the closed-loop system even becomes unstable.

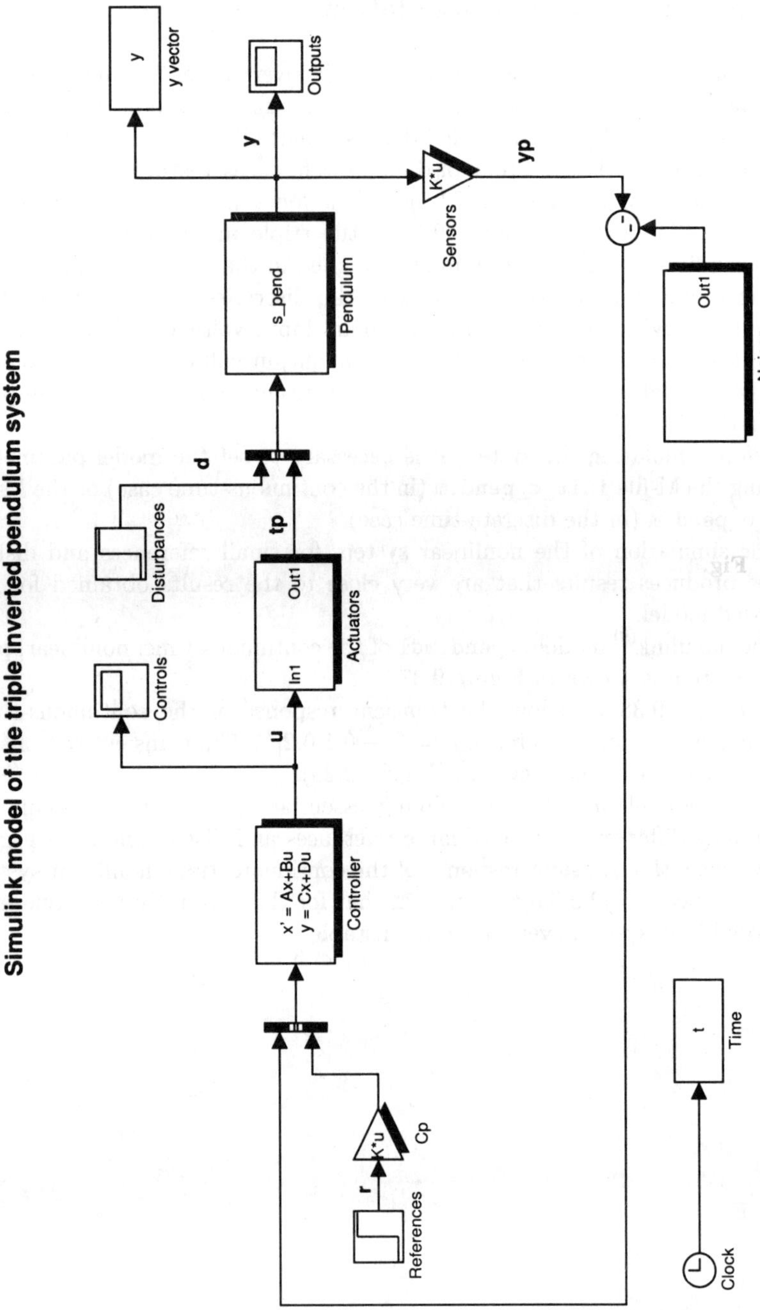

Fig. 9.37. Simulation model of the nonlinear system

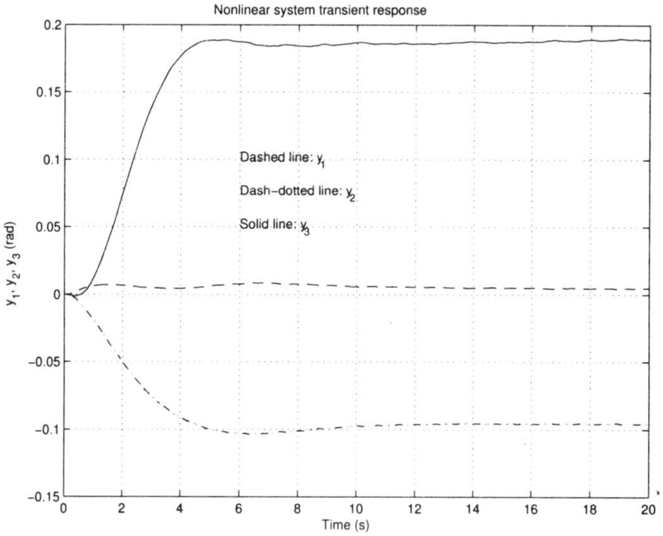

Fig. 9.38. Transient response of the nonlinear system - small reference

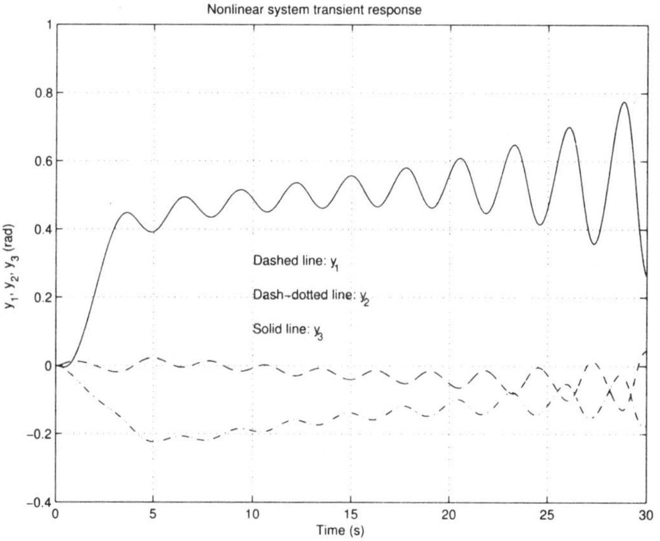

Fig. 9.39. Transient response of the nonlinear system - large reference

9.8 Conclusions

The experience gained in the design of the triple inverted pendulum control system makes it possible to make the following comments.

- The derivation of the uncertainty model of the triple inverted pendulum system is a difficult task. It is necessary to make several justifiable assumptions in order to obtain a relatively simple model.
- The use of a two-degree-of-freedom control structure allows us to obtain acceptable performance as well as robustness of the closed-loop system.
- The necessity to achieve robust stability and robust performance of the closed-loop system leads to decrease of the closed-loop bandwidth that in turn leads to slower response and worse disturbance attenuation. Thus, there is a trade-off between nominal performance and robustness of the closed-loop system. In the given case an acceptable trade-off is achieved by using a μ-controller.
- The order of the μ-controller is very high that thus requires controller order reduction to preserve the closed-loop performance and robustness while making implementation of the controller easier and more reliable.
- The triple inverted pendulum is an essentially nonlinear system and the results obtained by using the linearised model are valid only for sufficiently small arm angles and arm velocities.

Notes and References

As noted in [166] the inverted pendulum is frequently used as a good example to show the power of modern control theory. The design of inverted pendulum control systems is initially done by using state-space methods [110].

The single and double inverted pendulums are relatively easy to stabilize by using linear controllers [40, 80, 165]. The triple inverted pendulum is more difficult to control and it is very sensitive to torque disturbances, joint frictions and measurement noises [149]. The design of analogue and digital controllers for triple inverted pendulum systems has been considered in several papers (see, for instance, [41, 104]). References [28, 102] address the challenging problem of stabilising a triple inverted pendulum, a cart system that uses only one actuator.

The robust control of different types of inverted pendulums is considered in several publications, see, for instance, [72, 103, 149, 150].

10

Robust Control of a Hard Disk Drive

This chapter considers the design of a robust servo system of a *hard disk drive(HDD)*. Three robust design methods are applied, namely, the μ-synthesis, $\mathcal{H}_\infty$ optimal design and the $\mathcal{H}_\infty$ loop-shaping design procedure (LSDP). After a description of the HDD servo-system dynamics in the first section, it is shown in detail (in Section 10.2) how to derive the plant model that involves several uncertain parameters. Then we consider the synthesis of continuous-time controllers using the available methods for robust control design. These controllers are compared in aspects of robustness of closed-loop system stability and of performance in the frequency domain and in the time-domain. The design of discrete-time controller is also included, for two sampling frequencies. Finally, we present the simulation results of the nonlinear continuous-time and discrete-time closed-loop systems, followed by conclusions at the end of the chapter.

The prevalent trend in hard-disk design is towards smaller disks with increasingly larger capacities. This implies that the track width has to be smaller leading to lower error tolerance in the positioning of the read/write heads. The controller for track following has to achieve tighter regulation in the presence of parameter variations, nonlinearities and noises. Hence, it is appropriate to use advanced design methods like μ-synthesis and $\mathcal{H}_\infty$ optimisation in order to achieve robust stability and robust performance of the closed-loop servo system.

10.1 Hard Disk Drive Servo System

The schematic diagram of a typical hard disk drive is shown in Figure 10.1. The disk assembly consists of several flat disks called *platters* coated on both sides with very thin layers of magnetic material (thin-film media). The magnetic material is used to store the data in the form of magnetic patterns. The platters rotate at high speed, driven by a spindle motor. Recently, the spindle speed is 5400 r.p.m., 7200 r.p.m. or even 10,000 and 15,000 r.p.m. The

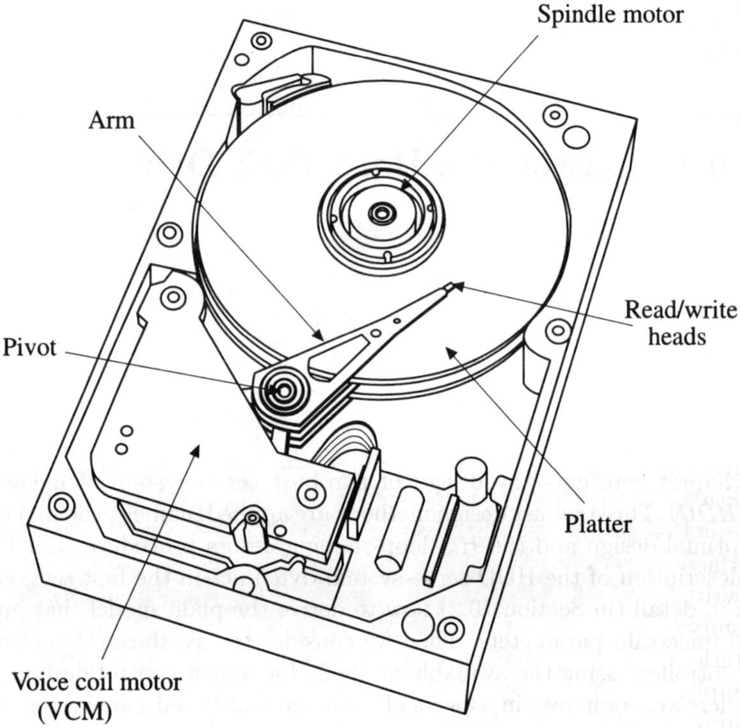

Fig. 10.1. Schematic diagram of a hard disk drive

data are retrieved from, or recorded onto, the platters by electromagnetic *read/write(R/W)* heads that are mounted at the bottom of *sliders*. Today's hard disks read data using *giant-magneto-resistive(GMR)* heads and write data with thin-film inductive heads. The sliders with read/write heads are mounted onto head arms. The arms are lightweight rigid constructions allowing them to be moved rapidly on the platter surface. There is one arm per read/write head and all arms are mounted in a stack assembly that moves over multiple disk surfaces simultaneously. The heads are suspended several microinches above the disk surface. The appropriate flying height of the heads is achieved thanks to the air flow generated by the spinning disk.

The data recorded on the platters are in concentric circles called *tracks*. Modern hard disks have tens of thousands of tracks resulting in track density as high as 30 000 *tracks per inch(TPI)*. Thus, the distance between adjacent tracks is of the order of a microinch. Each track is divided into smaller pieces called *sectors* that contain 512 bytes of information. There may be several hundreds or even thousands of sectors in a track. The drive density may reach more than 20 GB per platter.

The head arms are moved on the surface of the platter by a rotary voice coil actuator frequently called the *Voice Coil Motor(VCM)*. The VCM consists of a voice coil, mounted at the end of the head arm assembly, and permanent electromagnets. By controlling the current in the coil, the heads can move in one direction or the other in order to follow precisely the data track.

The goal of the hard disk drive servo control system is to achieve a precise positioning of the read/write heads on the desired track (track-following mode) while data are being written or read, and a quick transition from one track to another target track (seeking mode). As the drive initiates a seek command, it switches to a seek control algorithm that is a type of time-optimal (bang-bang) control algorithm. When the head is positioned over the desired track, the drive switches into track-following mode. In the present work we consider the servo controller of the track-following mode.

All modern hard disk drives read the relative position of the head to track directly from the disk media by using a method called *embedded servo* or *sector servo*. In this method the servo information is interleaved with user data across the entire surface of all platters. Position information is placed on the disk surfaces in servo frames during the manufacturing of the disk and cannot be rewritten. This is why the disk heads are locked out by the drive controller from writing to the areas where servo information is written. Because of the interleaving of servo information with data, the embedded servo system is a sampled-data system. Increasing the number of servo frames within a track improves the performance of the servo system due to the higher sample rate but limits the maximum data storage. The servo information is read by the same head that reads the data. The position information usually consists of two parts: coarse position giving track number and fine position information relative to each track. The position error, representing the difference between the reference track position and the head position, is measured by making use of *position bursts* that are part of the servo information. The position bursts are patterns of alternating magnetic polarity written on the disk surface with a particular frequency. The periodic signal, obtained from the read head passing over a burst pattern, has an amplitude that is proportional to how much the read head is directly over the burst pattern. The signals corresponding to 2 or 4 bursts of position information are demodulated by a servo demodulator in order to compose the *position error signal(PES)*. This signal is used by the servo controller to change the voice coil current appropriately and hence the read/write heads position. It should be pointed out that the absolute head position is not generally known from the servo information that is read off of the disk. Only a signal proportional to the position error is available, which means that for track following the reference signal is ideally equal to zero.

The failure of the head to follow the track on a platter surface faithfully as the disk spins is referred to as *runout*. The runout could have serious consequences, especially during writing where data in adjacent tracks might be overwritten. There are two types of runout, namely the *repeatable runout(RRO)* and *nonrepeatable runout(NRRO)*. Repeatable runout is caused by both imbalance in the spindle and imperfections in the servowriting process that result in noncircular position information. This information is encoded at the spindle frequency, yielding a repeatable noncircular track for the disk servo to follow. The nonrepeatable runout is caused by a windage induced actuator arm, slider motion and mechanical vibrations that can arise from various sources such as ball bearing defects, spindle motor vibrations and slider vibrations. The RRO may be reduced through feedforward compensation at the corresponding frequency by using *harmonic correctors.*

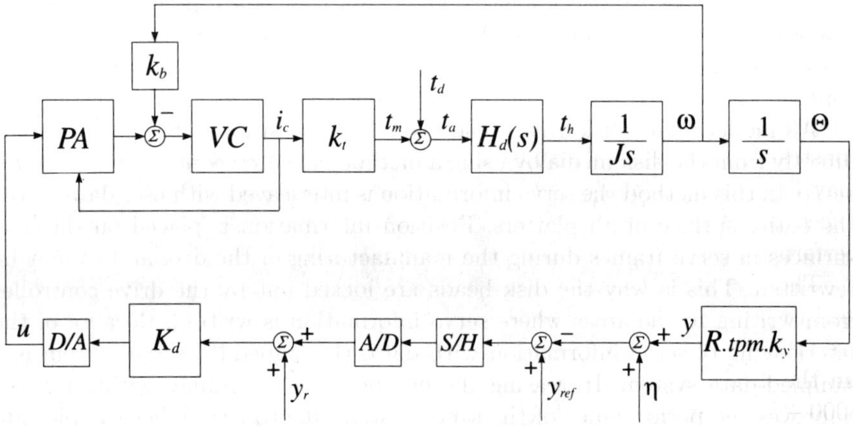

Fig. 10.2. Block diagram of the hard disk drive servo system

The block-diagram of the HDD servo system is shown in Figure 10.2. The R/W heads are moved by a VCM that is driven by the output current i_c of a *power amplifier(PA)*. The actual position signal y is compared with the signal y_{ref} that represents the desired head position. For the track following, the reference input y_{ref} is theoretically equal to zero and $y(t)$ appears as an error signal. In practice, y_{ref} must be set equal to the signal representing both RRO and NRRO. The digital signal y_r is a reference for the desired track and is used during the seeking mode. The error signal is sampled by an *analogue-to-digital(A/D)* converter and serves as part of an input to the digital servo controller K_d that is typically implemented on a DSP chip. The output of the controller is converted to analogue form by a *digital-to-analog(D/A)* converter and amplified by the PA. Since the motor torque is proportional to the voice coil current, the amplifier is configured as a current source. The exogenous signal t_d is the torque disturbance due to external shock and vibrations, power amplifier noise, digital-to-analogue converter noise, pivot bearing friction and flex cable bias. The increase of the spindle speed increases the air flow inside the disk (windage) that in turn increases the disturbance torque on the actuator. The disturbance is a low-frequency signal with spectral content usually below 500 Hz. The position noise signal η includes quantisation errors due to servo demodulator noise, finite resolution of the analogue-to-digital converter, media noise and preamplifier noise. The position noise is a high-frequency signal with spectral content usually above 1 kHz. Since the measured PES is contaminated with noise, the true PES, $y_{\text{ref}} - y$, is not available.

One of the limitations inherent in the design of servo controllers for high track density HDD is the influence of actuator mechanical resonant modes on the head-positioning servo. If the actuator input contains a periodic component with frequency equal to a resonance frequency, this component may be amplified greatly that results in large off-track deviation of the read/write heads. Usually, the actuator is mechanically designed in such a way that the resonant modes occur at frequencies that will be attenuated by the servo system. However, as servo bandwidth increases to meet higher performance requirements, this attenuation may not be achieved due to mechanical design constraints. It is also important to note that the presence of resonant modes may limit the servo bandwidth via stability margin constraints. With a reduced bandwidth, the servo system may not be able to achieve the desired performance.

The rotary actuators of hard disks may have tens of resonances that may lead to a high order model. However, in practice only 3 to 4 main resonances are taken into account. Usually, these are the first and second torsion modes (in the range 1500 – 2500 Hz) as well as the first sway mode (in the range 8000 – 12000 Hz).

A common approach to reduce the effect of resonance modes is to put notch filters in the servoloop that attenuates or filters out vibrations at selected major resonant frequencies. However, each notch filter introduces phase margin loss at low frequencies thus reducing the system robustness. Also, the presence of uncertainty in the resonant modes may decrease significantly the efficiency of those filters.

Our goal in this chapter is to design a robust track-following servo control system for a 3.5-inch HDD with track density 25 400 TPI. The desired settling time is about 1 ms in the presence of four resonances, several uncertain parameters, position sensing noise and disturbances. The parameters of the rigid-body model and their tolerances are given in Table 10.1. For dimensional compatibility, the track density is given in tracks per metre, instead of in TPI.

We now first consider the derivation of a HDD servo system model.

The dynamics of the rotary arm is described by the equation

$$J\frac{\mathrm{d}^2\Theta}{\mathrm{d}t^2} = t_m + t_d$$

where J is the arm moment of inertia, Θ is the angle of arm rotation, t_m is the VCM torque and t_d is the disturbance torque.

The VCM torque is given by

$$t_m = k_t i_c$$

Table 10.1. Rigid body model parameters and tolerances

Parameter	Description	Value	Units	Tolerance
J	arm moment of inertia	6.3857×10^{-6}	kg m^2	$\pm 10.0\%$
R	arm length	5.08×10^{-2}	m	$\pm 0.1\%$
k_{PA}	amplifier gain	10.0	V/V	$\pm 0.0\%$
k_t	VCM torque constant	9.183×10^{-2}	N m/A	$\pm 10.0\%$
k_b	back e.m.f. constant	9.183×10^{-2}	N m/A	$\pm 10.0\%$
tpm	tracks per meter	10^6	tracks	–
k_y	position measurement gain	1.2	V/track	$\pm 5.0\%$
R_{coil}	coil resistance	8.00	Ω	$\pm 20.0\%$
R_s	sense resistance in the power amplifier feedback	0.2	Ω	$\pm 1.0\%$
L_{coil}	coil inductance	0.001	H	$+0, -15\%$
$e_{\max}$	saturated power amplifier voltage	12.0	V	$-0, +5\%$
RPM	disk rotation rate	7200	rev/min	$\pm 1.0\%$
t_w	track width	1	μ m	$\pm 1.0\%$

where k_t is the motor torque constant and i_c is the current through the VCM coil.

The voice coil has a resistance R_{coil} and an inductance L_{coil}. An additional current sense resistance R_s is connected in serial with the voice coil to implement a feedback from the power amplifier output. Hence the voice coil admittance is described by the transfer function

$$G_{\text{vca}}(s) = \frac{i_c(s)}{e_c(s)} = \frac{1/R_c}{\tau s + 1}$$

where e_c is the input voltage to the voice coil, $\tau = L_{coil}/R_c$ and $R_c = R_{coil} + R_s$.

The block diagram of the power amplifier with voice coil is shown in Figure 10.3. The input of the voice coil is the difference $e_c = e_p - e_b$, where e_p is the output voltage of the amplifier and $e_b = k_b\omega$ is the back electromotive force (e.m.f.) that is generated during the moving of the coil in the magnetic field. Since the saturation voltage of the amplifier is $e_{\max}$, in the absence of back e.m.f., the amplifier will saturate for an input voltage greater than $e_{\max}/k_{PA}$. In the study limited to linear systems, the amplifier saturation is neglected.

The length of the arc, corresponding to the arm rotation angle Θ, is equal to $R\Theta$. For small values of Θ the number of tracks contained in the arc is $R \cdot \Theta \cdot tpm$. This gives an output signal $y = R \cdot tpm \cdot k_y \cdot \Theta$.

If we neglect the dynamics of the voice coil, the transfer function of the servoactuator considered as a rigid body is obtained in the form of a dou-

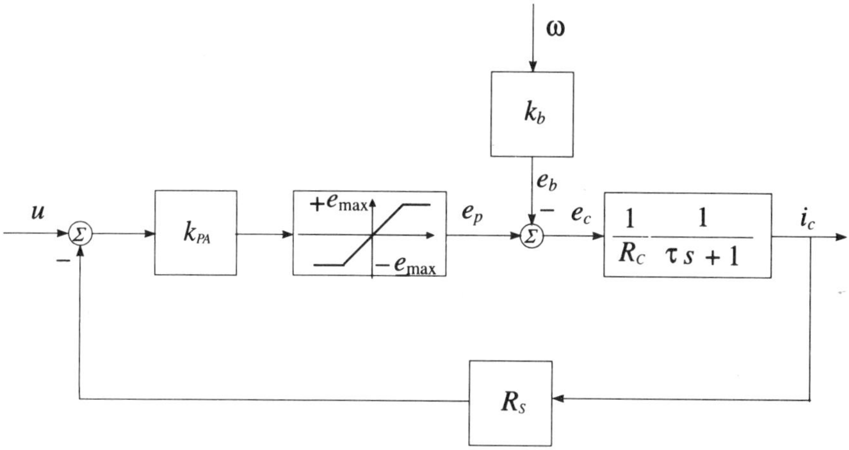

Fig. 10.3. Block diagram of the power amplifier with voice coil

ble integrator. Such a model oversimplifies the system dynamics and cannot produce reliable results if used in the design.

The next step is to take into account the high frequency resonant modes of the head disk assembly represented by the transfer function $H_d(s)$. In the given case $H_d(s)$ consists of four resonant modes and is obtained as

$$H_d(s) = \sum_{j=1}^{4} \frac{b_{2j}\omega_j s + b_{2j-1}\omega_j^2}{s^2 + 2\xi_j\omega_j s + \omega_j^2}$$

(see Figure 10.4).

Here ω_j, ξ_j and b_{2j}, b_{2j-1} are respectively the resonance frequency, the damping coefficient and the coupling coefficients of the jth mode, for $j = 1, ..., 4$. The resonance parameters are usually determined experimentally and for the servo system under consideration their values are shown in Table 10.2.

It is important to note that all model parameters are known with some tolerances and may vary with the changing of working conditions as well as with time. Also, the closed-loop system can be very sensitive to the external disturbance t_d and the position-sensing noise η. Both factors would lead to an actual system dynamics far away from the dynamics of the nominal closed-loop system. Thus, it is necessary to use control-system design methods that ensure the desired closed-loop stability and performance in the presence of uncertain parameters, noises and disturbances.

10.2 Derivation of Uncertainty Model

In order to implement robust control design methods we must have a plant model that incorporates uncertain parameters. As is seen from Tables 10.1

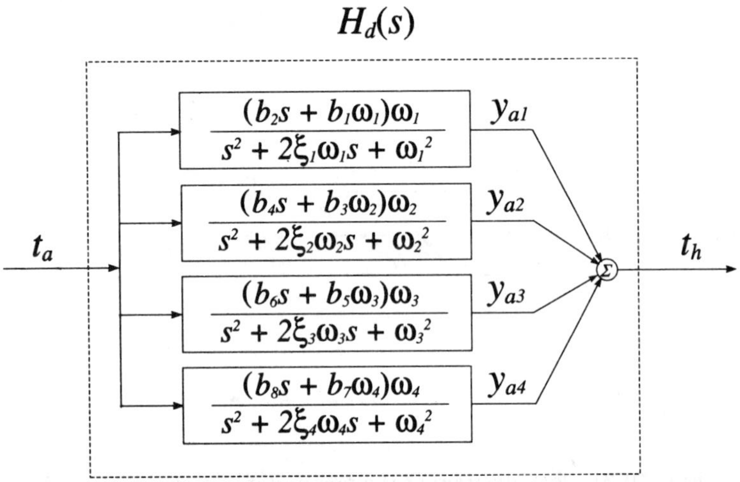

Fig. 10.4. Transfer functions of resonant modes

Table 10.2. Resonance parameters and tolerances

Parameter	Description	Value	Units	Tolerance
ω_1	pivot bearing resonance	$2\pi 50$	rad/s	±5.0%
ω_2	first torsional resonance	$2\pi 2200$	rad/s	±12.0%
ω_3	second torsional resonance	$2\pi 6400$	rad/s	±8.0%
ω_4	first sway resonance	$2\pi 8800$	rad/s	±15.0%
b_1	first resonance coupling	0.006	–	±7.0%
b_2	first resonance coupling	0	1/s	±7.0%
b_3	second resonance coupling	0.013	–	±10.0%
b_4	second resonance coupling	−0.0018	1/s	±7.0%
b_5	third resonance coupling	0.723	–	±5.0%
b_6	third resonance coupling	−0.0015	1/s	±10.0%
b_7	fourth resonance coupling	0.235	–	±5.0%
b_8	fourth resonance coupling	−0.0263	1/s	±10.0%
ξ_1	first resonance damping	0.05	–	±5.0%
ξ_2	second resonance damping	0.024	–	±8.0%
ξ_3	third resonance damping	0.129	–	±10.0%
ξ_4	fourth resonance damping	0.173	–	±10.0%

and 10.2, the total number of uncertain parameters is greater than 25, which complicates the analysis and design of a HDD servo system. In this study, we shall concentrate on those uncertainty parameters that influence the closed-loop system behaviour most.

The block diagram of the plant is shown in Figure 10.5.

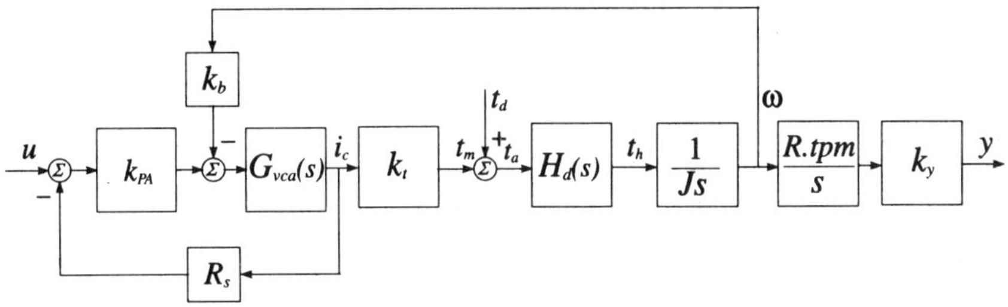

Fig. 10.5. Block diagram of the plant

Consider first the derivation of the uncertainty model for the resonant modes. All four modes have similar transfer functions. A state-space model is

$$\begin{aligned}\dot{x}_1 &= \omega x_2,\\ \dot{x}_2 &= \omega(-x_1 - 2\xi x_2 + t_a)\\ y_a &= b_1 x_1 + b_2 x_2\end{aligned}$$

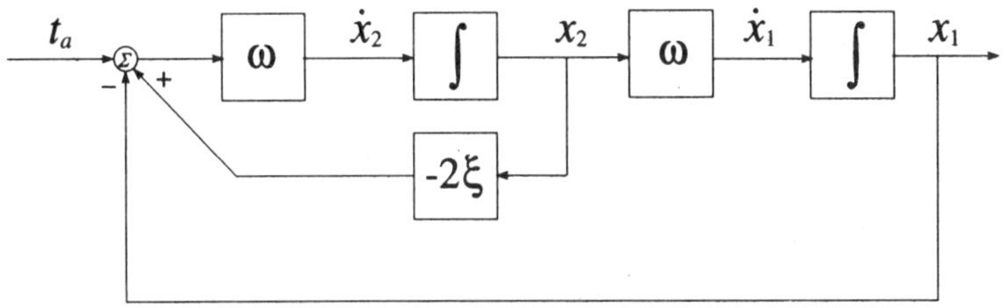

Fig. 10.6. Block diagram of a resonant mode

The block diagram corresponding to the state equations of a resonant mode is shown in Figure 10.6.

The variations in the frequency ω and the damping coefficient ξ are represented, respectively, by

$$\omega = \overline{\omega}(1 + p_\omega \delta_\omega)$$

and

$$\xi = \overline{\xi}(1 + p_\xi \delta_\xi)$$

where $\overline{\omega}$ and $\overline{\xi}$ are the nominal values, p_ω and p_ξ are the maximum relative uncertainties with

$$-1 \leq \delta_\omega \leq 1$$

$$-1 \leq \delta_\xi \leq 1$$

being the relative variations in these parameters.

The parameter ω may be represented as an upper linear fractional transformation (LFT) in δ_ω,

$$\omega = F_U(M_\omega, \delta_\omega)$$

with

$$M_\omega = \begin{bmatrix} 0 & \overline{\omega} \\ p_\omega & \overline{\omega} \end{bmatrix}$$

and the parameter ξ may be represented similarly as

$$\xi = F_U(M_\xi, \delta_\xi)$$

with

$$M_\xi = \begin{bmatrix} 0 & \overline{\xi} \\ p_\xi & \overline{\xi} \end{bmatrix}$$

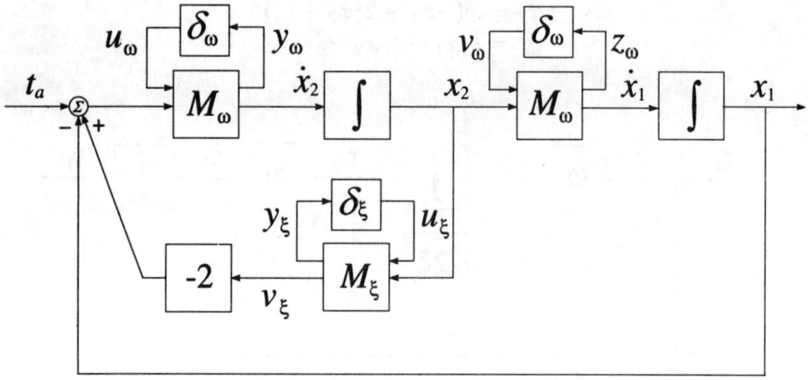

Fig. 10.7. Resonant mode with uncertain parameters

The block diagram of a resonant mode with uncertain parameters is given in Figure 10.7. Based on this diagram we derive the following equations

$$\begin{bmatrix} y_\omega \\ \dot{x}_2 \end{bmatrix} = \begin{bmatrix} 0 & \overline{\omega} \\ p_\omega & \overline{\omega} \end{bmatrix} \begin{bmatrix} u_\omega \\ t_a - 2v_\xi - x_1 \end{bmatrix}$$

$$\begin{bmatrix} z_\omega \\ \dot{x}_1 \end{bmatrix} = \begin{bmatrix} 0 & \overline{\omega} \\ p_\omega & \overline{\omega} \end{bmatrix} \begin{bmatrix} v_\omega \\ x_2 \end{bmatrix}$$

$$\begin{bmatrix} y_\xi \\ v_\xi \end{bmatrix} = \begin{bmatrix} 0 & \overline{\xi} \\ p_\omega & \overline{\xi} \end{bmatrix} \begin{bmatrix} u_\xi \\ x_2 \end{bmatrix},$$

and

$$y_a = \begin{bmatrix} b_1 & b_2 \end{bmatrix} \begin{bmatrix} x_1 \\ x_2 \end{bmatrix}$$

From these equations we further obtain the perturbed model of a resonant mode in the form of

$$\begin{bmatrix} \dot{x}_1 \\ \dot{x}_2 \\ -- \\ y_\omega \\ z_\omega \\ y_\xi \\ -- \\ y_a \end{bmatrix} = \Pi \begin{bmatrix} x_1 \\ x_2 \\ -- \\ u_\omega \\ v_\omega \\ u_\xi \\ -- \\ t_a \end{bmatrix}$$

where

$$\Pi = \left[\begin{array}{cc|ccc|c} 0 & \overline{\omega} & 0 & p_\omega & 0 & 0 \\ -\overline{\omega} & -2\overline{\omega}\xi & p_\omega & 0 & -2\overline{\omega}p_\xi & \overline{\omega} \\ \hline -\overline{\omega} & -2\overline{\omega}\xi & 0 & 0 & -2\overline{\omega}p_\xi & \overline{\omega} \\ 0 & \overline{\omega} & 0 & 0 & 0 & 0 \\ 0 & \overline{\xi} & 0 & 0 & 0 & 0 \\ \hline b_1 & b_2 & 0 & 0 & 0 & 0 \end{array}\right]$$

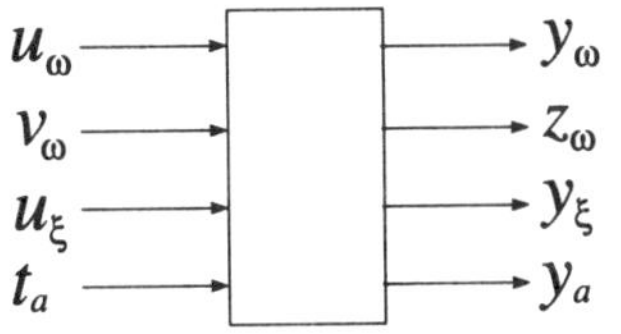

Fig. 10.8. Block representation of a resonant mode

A block with four inputs and four outputs can be used to represent a resonant mode (Figure 10.8).

We now consider other uncertain parameters of the plant.

The uncertain parameters of the rigid-body model are taken as

$$\begin{aligned} k_t &= \overline{k}_t(1 + p_{k_t}\delta_{k_t}) \\ J &= \overline{J}(1 + p_J\delta_J) \\ k_y &= \overline{k}_y(1 + p_{k_y}\delta_{k_y}) \end{aligned}$$

where $\overline{k}_t$, $\overline{J}$, $\overline{k}_y$ are the corresponding nominal parameters and

$$-1 \le \delta_{k_t} \le 1$$

$$-1 \le \delta_J \le 1$$

$$-1 \le \delta_{k_y} \le 1$$

are variations. These parameters are represented as LFTs in the related uncertainties δ_{k_t}, δ_J, δ_{k_y}:

$$\begin{aligned} k_t &= F_U(M_{k_t}, \delta_{k_t}) \\ J &= F_U(M_{Ji}, \delta_J) \\ k_y &= F_U(M_{k_y}, \delta_{k_y}) \end{aligned}$$

where

$$M_{k_t} = \begin{bmatrix} 0 & \overline{k}_t \\ p_{k_t} & \overline{k}_t \end{bmatrix}, \quad M_{Ji} = \begin{bmatrix} -p_J & \frac{1}{\overline{J}} \\ -p_J & \frac{1}{\overline{J}} \end{bmatrix}, \quad M_{k_y} = \begin{bmatrix} 0 & \overline{k}_y \\ p_{k_y} & \overline{k}_y \end{bmatrix}$$

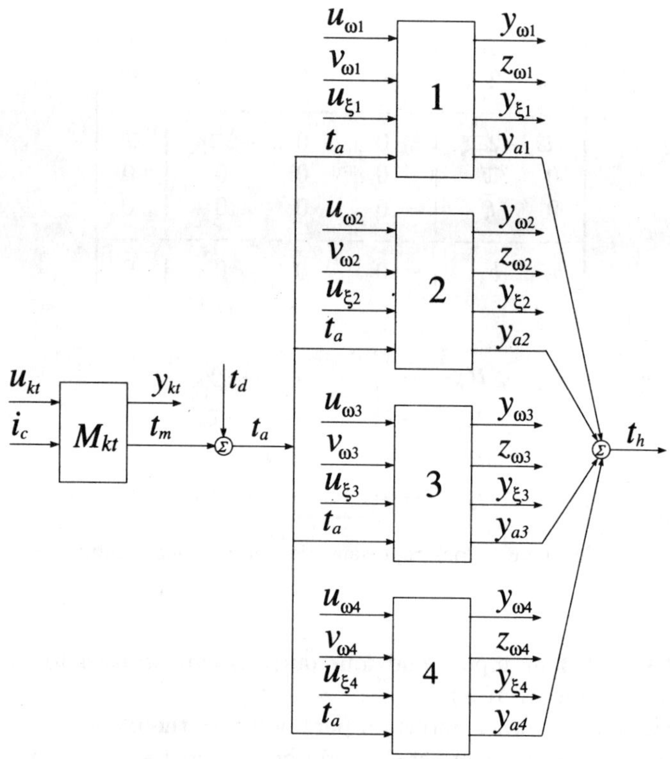

Fig. 10.9. Representations of uncertain motor torque constant and resonant modes

In Figure 10.9 we show the part of the uncertainty model involving the motor torque constant and the four resonant modes, and in Figure 10.10 we show the part of the model involving the arm moment of inertia and position measurement gain.

Overall, we have twelve uncertain parameters but four of them (ω_1, ω_2, ω_3 and ω_4) are repeated twice. By "puling out" the uncertain parameters from

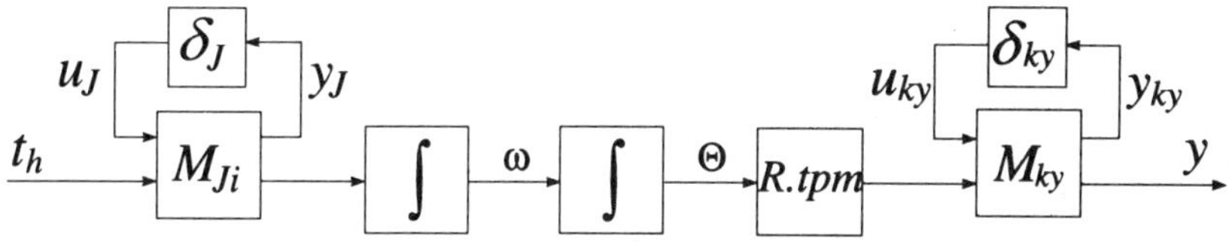

Fig. 10.10. Block diagram of uncertain arm moment of inertia and position gain

the nominal part of the model, we obtain a perturbed plant model in the form of an upper LFT $F_U(G_{nom}, \Delta)$ as shown in Figure 10.11 with a 15×15 matrix Δ,

$$\Delta = \text{diag}(\delta_{\omega_1}, \delta_{\omega_1}, \delta_{\xi_1}, \delta_{\omega_2}, \delta_{\omega_2}, \delta_{\xi_2}, \delta_{\omega_3}, \delta_{\omega_3}, \delta_{\xi_3}, \delta_{\omega_4}, \delta_{\omega_4}, \delta_{\xi_4}, \delta_{k_t}, \delta_J, \delta_{k_y})$$

that contains all the uncertain parameters. In the given case the matrix G_{nom} is obtained by the function `sysic` using the interconnections shown in Figures 10.5, 10.7, 10.9 and 10.10. The system model is of order 11. The uncertainty model of the HDD servo system is implemented by the M-file **mod_hdd.m**.

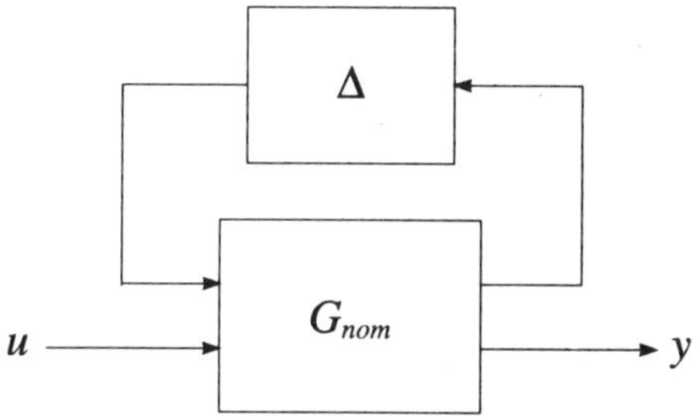

Fig. 10.11. Plant model in the form of an upper LFT

10.3 Closed-loop System-design Specifications

The design of the HDD servo controller will be first conducted in the continuous-time case. In general, it may obtain best possible performance in the continuous-time case that can then be considered as a limit for discrete-time designs. Also, in the continuous-time case it is easier to find appropriate performance weighting functions that again may be implemented in the discrete-time design.

The block diagram of the closed-loop system, which includes the feedback structure and the controller as well as the elements representing the model uncertainty and the performance objectives, is shown in Figure 10.12.

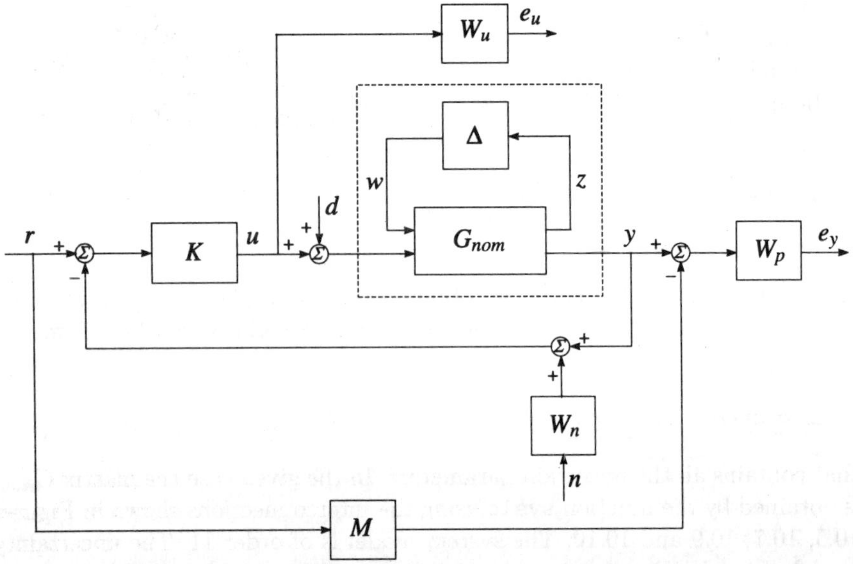

Fig. 10.12. Block diagram of the closed-loop system with performance specifications

The system has a reference input (r), an input disturbance (d), a noise (n) and two output costs (e_y and e_u). The system M is an ideal model of performance, to which the designed closed-loop system tries to match. The rectangle, shown with dashed lines, represents the plant transfer function matrix G. Inside the rectangle is the nominal model G_{nom} of the Hard Disk Drive plant and the block Δ that parameterises the model uncertainties. The matrix Δ is unknown, but has a diagonal structure and norm bound $\|\Delta\|_\infty < 1$. The performance objective requires the transfer matrices from r, d and n to e_y and e_u to be small in the sense of $\|\cdot\|_\infty$, for all uncertain matrices Δ under consideration. The position noise signal is obtained by passing the unit-bounded signal n through the weighting transfer matrix W_n. The transfer matrices W_p and W_u are used to reflect the relative significance of the different frequency ranges for which the performance is required. Hence, the performance objective can be recast, with possible slight conservativeness, as that the $\|\cdot\|_\infty$ of the transfer function matrix from r, d and n to e_y and e_u is less than 1.

It is straightforward to show that

$$\begin{bmatrix} e_y \\ e_u \end{bmatrix} = \begin{bmatrix} W_p(S_oGK - M) & W_pS_oG & -W_pS_oGKW_n \\ W_uS_iK & -W_uKS_oG & -W_uKS_oW_n \end{bmatrix} \begin{bmatrix} r \\ d \\ n \end{bmatrix} \qquad (10.1)$$

where $S_i = (I + KG)^{-1}$, $S_o = (I + GK)^{-1}$ are the input and output sensitivities, respectively. Note that S_oG is the transfer function between d and y.

This objective is similar to the usual mixed S/KS sensitivity optimization and it would meet both robust stability and performance criteria by incorporating performance specifications in the matching model M. The six functions to be minimised are described in Table 10.3.

Table 10.3. $\mathcal{H}_\infty$ functions to be minimised

Function	Description
$W_p(S_oGK - M)$	Weighted difference between the ideal and actual closed-loop systems
W_pS_oG	Weighted disturbance sensitivity
$W_pS_oGKW_n$	Weighted noise sensitivity
W_uS_iK	Weighted control effort due to reference
W_uKS_oG	Weighted control effort due to disturbance
$W_uKS_oW_n$	Weighted control effort due to noise

The controller synthesis problem of the Hard Disk Drive Servo System is to find a linear, output feedback controller $K(s)$ that has to ensure the following properties of the closed-loop system.

Nominal performance:

The closed-loop system achieves nominal performance if the performance objective is satisfied for the nominal plant model.

The nominal performance objective is to satisfy the inequality

$$\left\| \begin{bmatrix} W_p(S_oG_{nom}K - M) & W_pS_oG_{nom} & -W_pS_oG_{nom}KW_n \\ W_uS_iK & -W_uKS_oG_{nom} & -W_uKS_oW_n \end{bmatrix} \right\|_\infty < 1 \quad (10.2)$$

Robust stability:

The closed-loop system achieves robust stability if the closed-loop system is internally stable for each possible plant dynamics $G = F_U(G_{nom}, \Delta)$.

Robust performance:

The closed-loop system must remain internally stable for each $G = F_U(G_{nom}, \Delta)$ and in addition the performance criterion

$$\left\| \begin{bmatrix} W_p(S_oGK - M) & W_pS_oG & -W_pS_oGKW_n \\ W_uS_iK & -W_uKS_oG & -W_uKS_oW_n \end{bmatrix} \right\|_\infty < 1 \quad (10.3)$$

should be satisfied for each $G = F_U(G_{nom}, \Delta)$. This means that the structured singular value, corresponding to the transfer function matrix from $\begin{bmatrix} w \\ r \\ d \\ n \end{bmatrix}$ to $\begin{bmatrix} z \\ e_y \\ e_u \end{bmatrix}$ (in Figure 10.12) should be less than 1, with regard to $\begin{bmatrix} \Delta & 0 \\ 0 & \Delta_F \end{bmatrix}$ where Δ_F is a fictitious 3×2 complex uncertainty block.

In addition to these requirements it is desirable that the controller designed would have acceptable complexity, *i.e.* it is of reasonably low order.

According to the above considerations, the aim of the design is to determine a controller K, such that for all stable perturbations Δ with $\|\Delta\|_\infty < 1$, the perturbed closed-loop system remains stable and the performance objective is satisfied for all such perturbations.

10.4 System Interconnections

The internal structure of the eighteen-input, seventeen-output system, which is saved in the variable `sys_ic`, is shown in Figure 10.13. The inputs and outputs of the uncertainties are saved in the variables `pertin` and `pertout`, the reference, the disturbance and the noise in the variables `ref`, `dist`, `noise`, respectively, and the control signal in the variable `control`.

Both variables `pertin` and `pertout` have fifteen elements and `ref, dist, noise, y, y_c, e_y` and `e_u` are scalar variables.

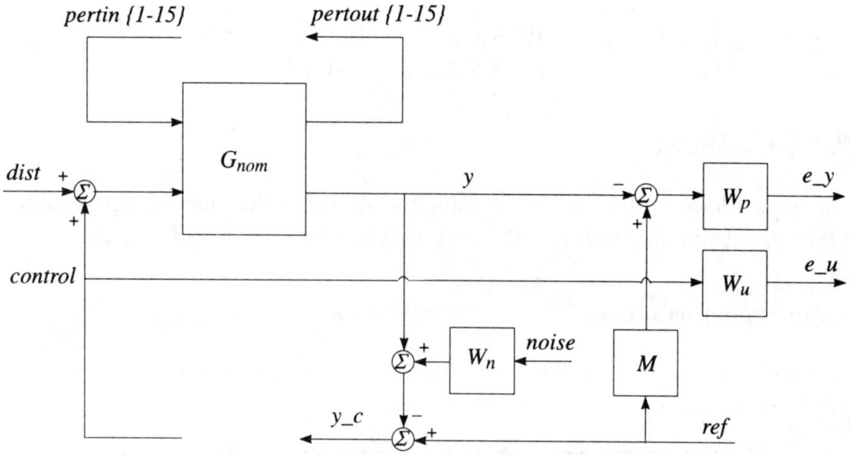

Fig. 10.13. Block diagram of the open-loop system with performance specifications

The open-loop connection is obtained by the M-file `olp_hdd`. The schematic diagram showing the specific input/output ordering for the variable `sys_ic` is shown in Figure 10.14.

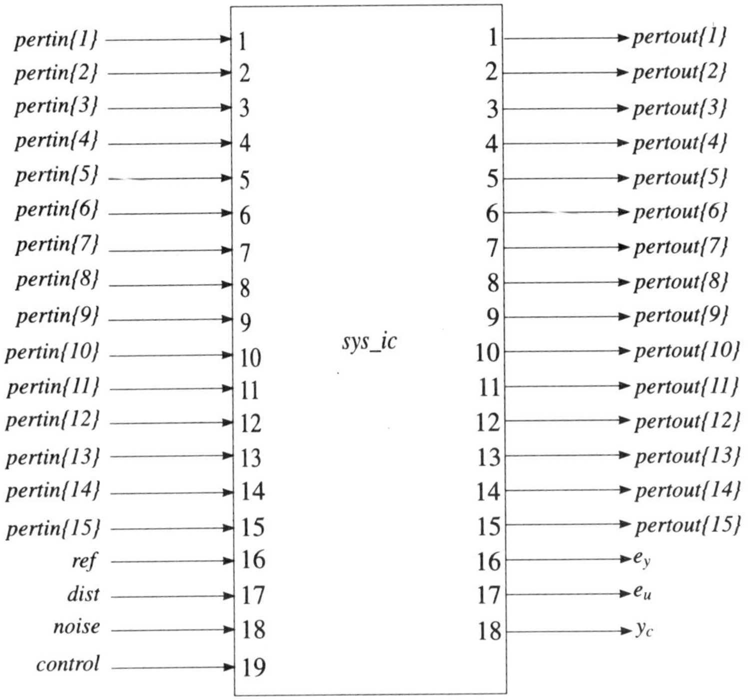

Fig. 10.14. Schematic diagram of the open-loop connection

The block-diagram used in the simulation of the closed-loop system is shown in Figure 10.15. The corresponding closed-loop interconnection, which is saved in the variable `sim_ic`, is obtained by the M-file `sim_hdd`.

The schematic diagram showing the specific input/output ordering for the variable `sim_ic` is shown in Figure 10.16.

10.5 Controller Design in Continuous-time

There are a few further "hard" constraints for the controller design, as listed below.

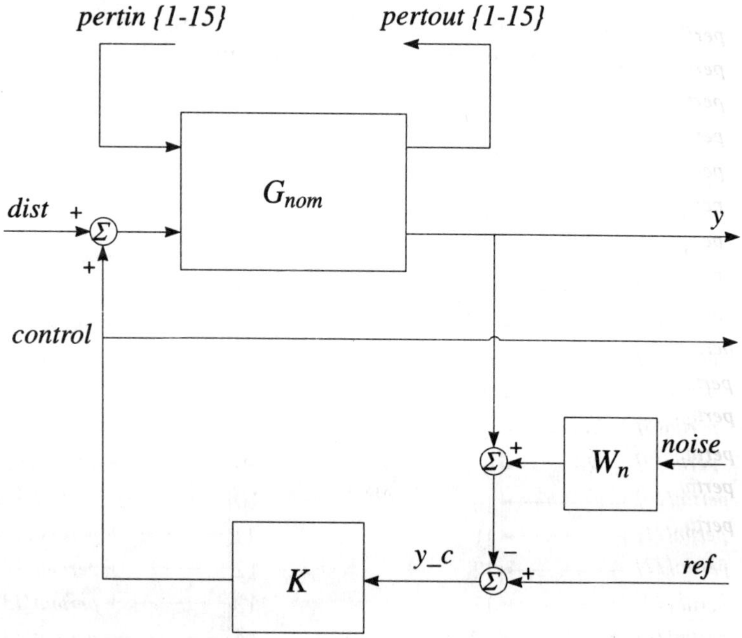

Fig. 10.15. Block diagram of the closed-loop system

Peak closed-loop gain	< 4 dB
Open-loop gain	> 20 dB at 100 Hz
Steady-state error	$< 0.1\ \mu$ m
Settling time	< 1.5 ms
Closed-loop bandwidth	> 1000 Hz
Gain margin	> 5 dB
Phase margin	> 40 deg

The designed control system must achieve good disturbance rejection and noise attenuation. In addition, it is necessary to have control action smaller than 1.2 V in order to avoid power-amplifier saturation.

To design the controller we shall use μ-synthesis, $\mathcal{H}_\infty$ optimisation and the $\mathcal{H}_\infty$ loop shaping design procedure (LSDP) in this exercise.

In the case of μ-synthesis and $\mathcal{H}_\infty$ optimisation design, we have to specify the model transfer function M and the weighting transfer functions W_n, W_p and W_u.

The model transfer function is chosen so that the time response to the reference signal would have an overshoot less than 20 % and a settling time less than 1 ms. A possible model satisfying the requirements is

$$M = \frac{1}{3.75 \times 10^{-9} s^2 + 1.2 \times 10^{-4} s + 1}$$

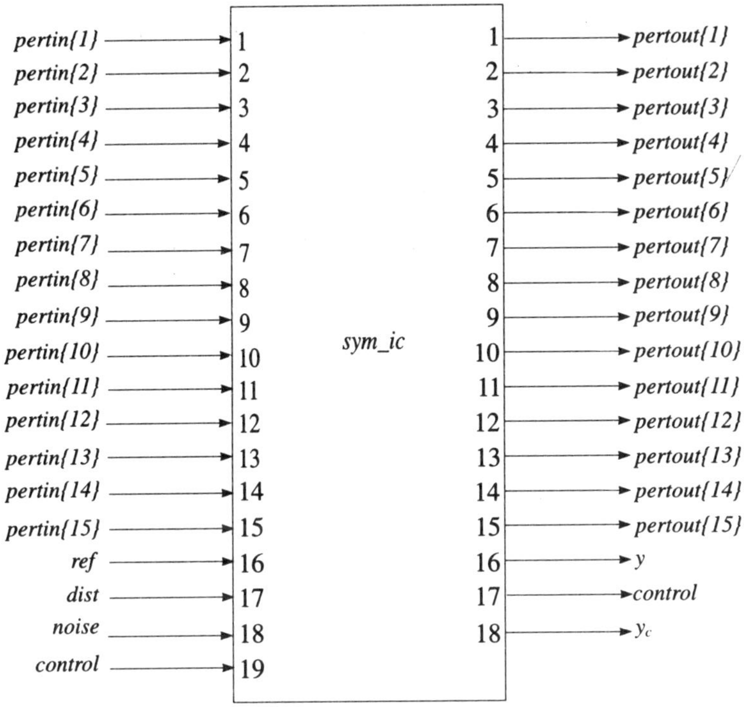

Fig. 10.16. Schematic diagram of the closed-loop connection

The noise shaping function W_n is determined on the basis of the spectral density of the position noise signal. In the given case it is taken as the high-pass filter

$$W_n = 6 \times 10^{-4} \frac{0.1s + 1}{0.001s + 1}$$

whose output has a significant spectral content above 500 Hz. For this shaping filter, the position noise signal is only 0.6 mV in the low-frequency range but it is 60 mV in the high-frequency range that corresponds to a position error of 5% of the track width.

The weighting functions W_p and W_u have to be chosen so as to ensure an acceptable trade-off between the nominal performance and the robust performance of the closed-loop system. They are selected in the course of the μ-synthesis, since this particular design method allows us to achieve maximum performance of the perturbed, closed-loop system.

10.5.1 μ-Design

In the μ-synthesis it is necessary to specify individually the inputs and outputs of the uncertainty blocks. In this exercise, only the inputs and outputs of the uncertainty in the rigid-body model (*i.e.* the parameters k_t, J and k_y) will be considered. The inclusion of the uncertainties of resonant modes would make the $D-K$ iterations difficult to converge. This confirms that the resonant modes may create difficulties in the controller design. However, these resonant modes (with nominal values) are included in the plant dynamics and will be considered, with parametric variations, in the assessment of designed controllers in Section 10.6.

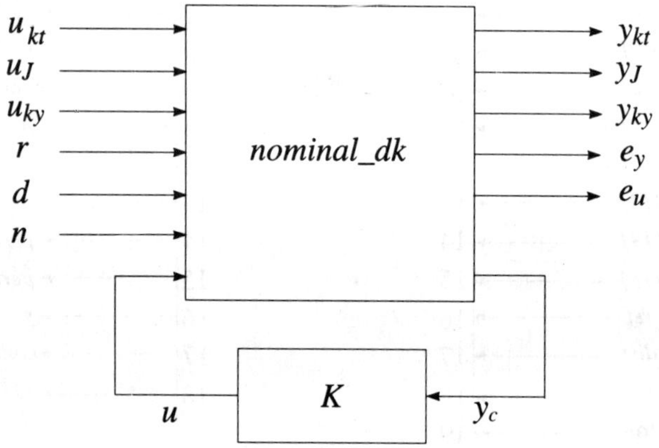

Fig. 10.17. Block diagram of the closed-loop system with μ-controller

The block diagram of the closed-loop system used in the μ-synthesis is shown in Figure 10.17. Denote by $P(s)$ the transfer matrix of the seven-input, six-output, open-loop system `nominal_dk` and let the block structure of the uncertainty Δ_P be defined by

$$\Delta_P := \left\{ \begin{bmatrix} \Delta_r & 0 \\ 0 & \Delta_F \end{bmatrix} : \Delta_r \in \mathcal{R}^{3\times 3},\ \Delta_F \in \mathcal{C}^{3\times 2} \right\}$$

The first block of this uncertainty matrix corresponds to the block Δ_r containing the uncertainties in the rigid-body model. The second block Δ_F is a fictitious uncertainty block that is used for the performance requirements in the framework of the μ approach. In order to satisfy the robust performance requirements it is necessary to find a stabilising controller $K(s)$, such that for each frequency $\omega \in [0, \infty]$ the structured singular value satisfies the condition

$$\mu_{\Delta_P}[F_L(P, K)(j\omega)] < 1$$

The fulfillment of this condition guarantees robust performance of the closed-loop system, *i.e.*

$$\left\| \begin{bmatrix} W_p(S_oGK - M) & W_pS_oG & -W_pS_oGKW_n \\ W_uS_iK & -W_uKS_oG & -W_uKS_oW_n \end{bmatrix} \right\|_\infty < 1 \tag{10.4}$$

The μ-synthesis is conducted by using the M-file `ms_hdd` along with the file `dk_hdd` describing the interconnection of the design. It should be noted that the robust performance achieved during the $D-K$ iteration is only with respect to the uncertainties in the rigid-body model, since only these uncertainties are taken into account in the design. Hence it is necessary to make additional robust stability and robust performance analysis that takes into account the other uncertainties.

The closed-loop system performance specifications are reflected by the weighting performance function $W_p(s)$. Three performance weighting functions are considered in the design. They are

$$W_{p1}(s) = 10^{-4}\frac{s^2 + 8 \times 10^4 s + 10^8}{s^2 + 7 \times 10^4 s + 2.5 \times 10^4}$$

$$W_{p2}(s) = 10^{-4}\frac{s^2 + 4 \times 10^5 s + 2.5 \times 10^9}{s^2 + 3.9 \times 10^5 s + 6.25 \times 10^5}$$

and

$$W_{p3}(s) = 10^{-4}\frac{s^2 + 1.15 \times 10^6 s + 3.6 \times 10^{10}}{s^2 + 1.05 \times 10^6 s + 9 \times 10^6}$$

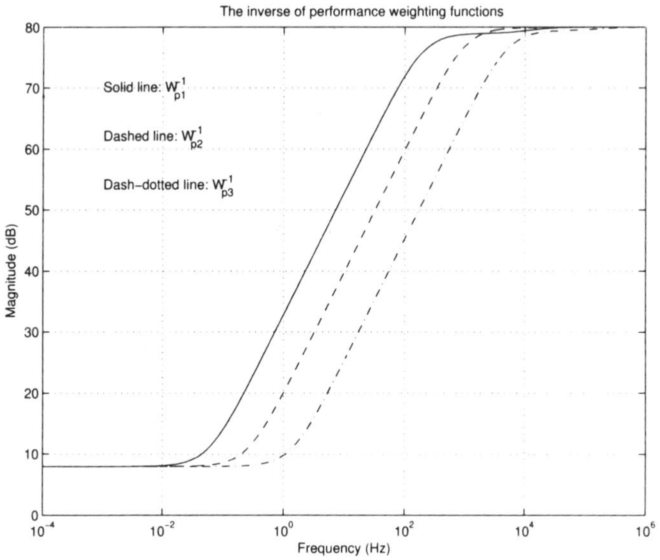

Fig. 10.18. Frequency response of the inverse of W_p

In Figure 10.18 we show the frequency responses of the inverses of these three weighting functions, *i.e.* W_{p1}^{-1}, W_{p2}^{-1} and W_{p3}^{-1}. It is seen that in all three selections, the aim is to achieve a small difference between the system and model outputs, and a small effect of the disturbance on the system outputs. This will ensure good tracking of the reference input and small error due to low-frequency disturbances. Changing the performance weighting function from W_{p1} to W_{p3} moves the inverse weighting frequency response to the right (to higher frequencies) which forces the system to match the model in over a larger frequency range.

The control weighting function is usually chosen as high-pass filters in order to ensure that the control action will not exceed 1.2 V. Again, three such weighting functions are considered in the design and are listed below:

$$W_{u1}(s) = 10^{-6}\frac{0.385s^2 + s + 1}{10^{-4}s^2 + 2 \times 10^{-3}s + 1}$$

$$W_{u2}(s) = 10^{-6}\frac{0.55s^2 + s + 1}{10^{-4}s^2 + 2.1 \times 10^{-3}s + 1}$$

and

$$W_{u3}(s) = 3 \times 10^{-6}\frac{4.05s^2 + s + 1}{10^{-4}s^2 + 2 \times 10^{-3}s + 1}$$

These three control weighting functions are paired with the three performance weighting functions, in the given order, in the μ-synthesis.

The final choice of the appropriate performance and control weighting functions is obtained by comparing the results from the corresponding μ-designs. (Six $D-K$ iterations are used in each case.) The robust stability and robust performance analysis conducted by the file `mu_hdd` gives the results shown in Table 10.4.

Table 10.4. Robust stability and robust performance for three controllers

Controller	Order	Robust stability $\mu_{\max}$	Robust performance $\mu_{\max}$
1	38	0.3730	0.436
2	38	0.412	0.549
3	30	0.577	0.904

In the first and third design cases, corresponding to the weighting functions (W_{p1}, W_{u1}) and (W_{p3}, W_{u3}), we take the controllers from the 6th $D-K$ iteration and in the second case we take the controller obtained at the 5th $D-K$ iteration. The closed-loop system achieves robust stability and robust performance for all three controllers; however, the best results, in terms of gain and phase margins, come from the first controller. This can also be seen from the corresponding μ-values. The gain and phase margins for all three design cases are listed in Table 10.5.

Table 10.5. Gain and phase margins for three controllers

Controller	Gain margin dB	Phase margin deg
1	10.9	59.0
2	8.9	50.5
3	8.9	36.9

The closed-loop transient responses are shown in Figure 10.19, for a simulated runout of 1 track width (1μ m) and a torque disturbance $t_d = 0.0005$ N m.

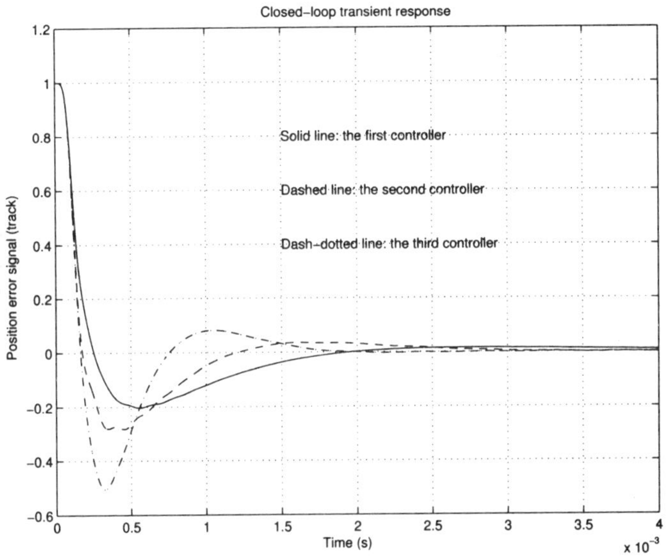

Fig. 10.19. Transient responses of three μ-controllers

It is seen from Figure 10.19 that the first controller gives a response with the smallest undershoot (about 20%), but this response is the slowest one. The third controller gives the fastest response but the undershoot in this case is the largest one (about 50%).

Figure 10.20 shows that as a result of the appropriate tuning of the control weighting functions all three controllers produce a control action whose amplitude is slightly less than 1.2 V.

The comparison of the transient responses to disturbance, shown in Figure 10.21, reveals that the worst disturbance rejection is found in the first

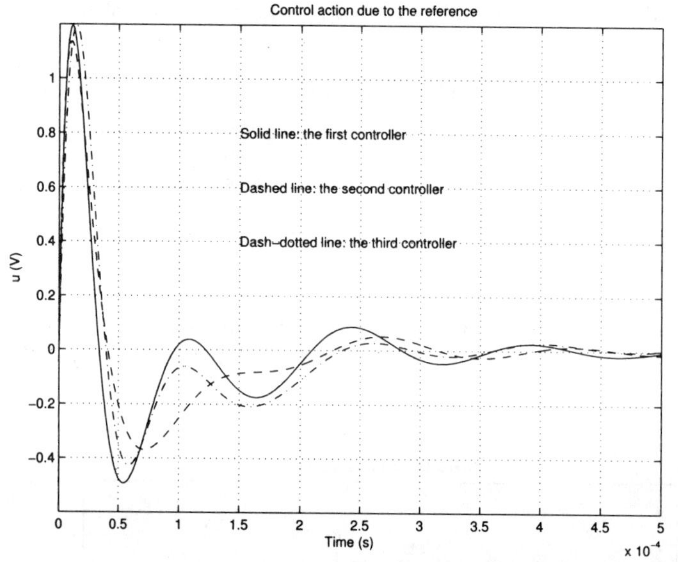

Fig. 10.20. Control actions of three μ-controllers

controller case and the best in the third controller case. This is a result of the tightest closed-loop bandwidth of 2 kHz (measured at -3 dB) in the first controller case compared with the bandwidth of 3.2 kHz in the third controller case (see Figure 10.22). Note that the largest peak of the magnitude is found in the third controller case, which results in the largest undershoot of the transient response.

The results obtained from different weighting functions show that moving the frequency response of the inverse performance weighting function to the right would lead to larger closed-loop system bandwidth, and consequently, faster time-responses of the closed-loop system, though may introduce larger over(under)shoot. However, at the same time, this may reduce the robustness of the closed-loop system. Hence, one has to compromise between the different objectives in the design. In the present design case, it seems that the second controller leads to a good trade-off between the requirements in terms of transient response, disturbance rejection and robustness. Hence, we will use the weighting functions W_{p2} and W_{u2} in both the μ-synthesis and $\mathcal{H}_\infty$ design.

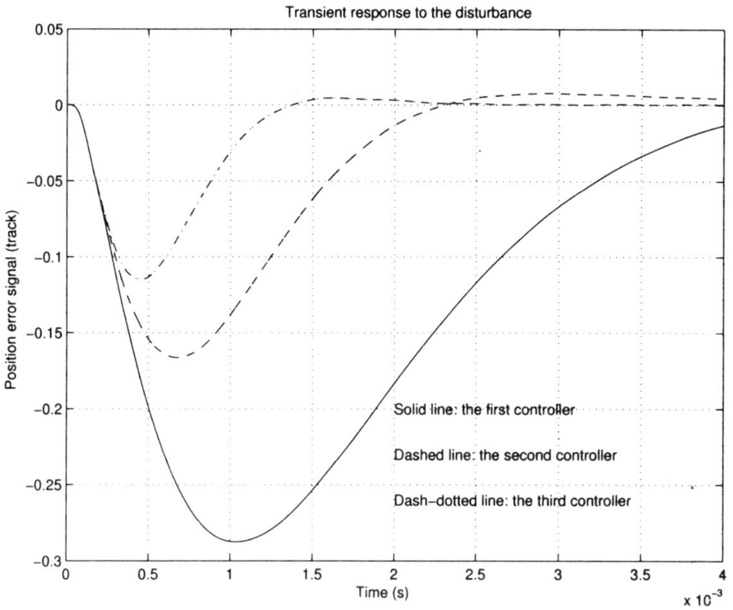

Fig. 10.21. Transient responses to disturbance of three μ-controllers

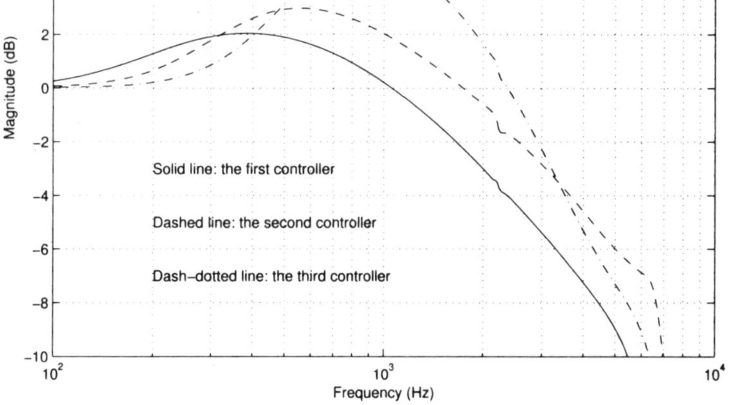

Fig. 10.22. Closed-loop magnitude plots of three μ-controllers

10.5.2 $\mathcal{H}_\infty$ Design

The aim of the design in this case is to find an $\mathcal{H}_\infty$ (sub)optimal, output controller for the interconnection shown in Figure 10.23 in which we exclude the inputs and outputs of the uncertainty block.

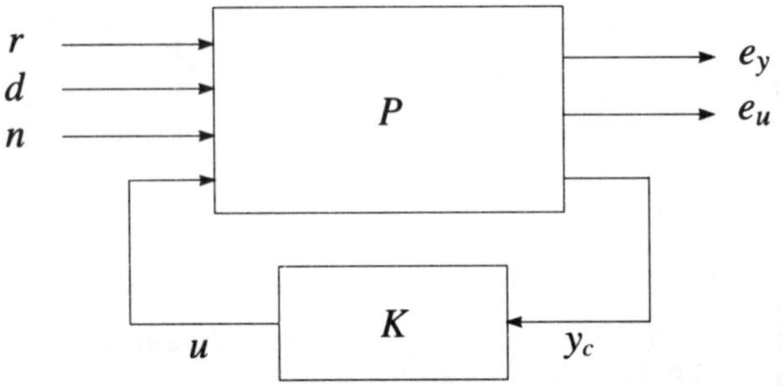

Fig. 10.23. Closed-loop system with $\mathcal{H}_\infty$ controller

The variable `hin_ic` that corresponds to the transfer function P of the open-loop system is obtained by the command line

```
hin_ic = sel(sys_ic,[16:18],[16:19])
```

The $\mathcal{H}_\infty$ optimal control minimises the ∞-norm of $F_L(P,K)$ with respect to the transfer function K of the controller. In the given case $F_L(P,K)$ (as the transfer function matrix in (10.2)) is the nominal closed-loop transfer function matrix from the reference, disturbance and noise signals (the variables **ref**, **dist** and **noise**) to the weighted outputs **e_y** and **e_u**. The design is conducted using the M-file **hinf_hdd.m**, which computes a (sub)optimal $\mathcal{H}_\infty$ control law for a given open-loop system. The value of γ is chosen 10% higher than the minimum possible value. The controller such obtained is of 18th order.

10.5.3 $\mathcal{H}_\infty$ Loop-shaping Design

The design of a robust control for the Disk Drive System can be successfully accomplished using the $\mathcal{H}_\infty$ loop-shaping design procedure (LSDP) as well, as described in this subsection. Note that in the case of $\mathcal{H}_\infty$ LSDP, we do not use the performance weighting function implemented in the cases of μ and $\mathcal{H}_\infty$ designs. Instead, we use a prefilter W_1 and a postfilter W_2 in order to shape appropriately the frequency response of the augmented, open-loop transfer function W_2GW_1 (see Chapter 5).

In the present case we choose a prefilter with transfer function

$$W_1 = 4\frac{0.05s+1}{s}$$

The gain of 4 is chosen to ensure a steady-state error, due to the disturbance, of less than 10% of the track width. Larger gain leads to smaller steady-state errors but possibly worse transient response. The postfilter is taken simply as $W_2 = 1$. The magnitude plots of the original and shaped systems are shown in Figure 10.24. The design of the $\mathcal{H}_\infty$ LSDP controller uses the M-file **lsh_hdd** that implements the function **ncfsyn**. The controller obtained is of order 13.

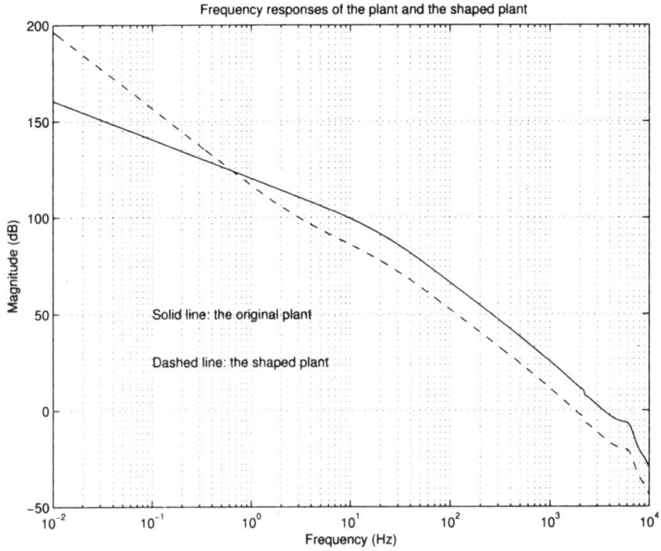

Fig. 10.24. Magnitudes of the original and shaped systems

10.6 Comparison of Designed Controllers

The comparison of the closed-loop system with μ, $\mathcal{H}_\infty$ and $\mathcal{H}_\infty$ LSDP controllers begins with the robust stability and performance analysis.

The robust stability is tested on the upper 15×15 block of the closed-loop transfer function matrix. To achieve robust stability it is necessary that the μ-values are less than 1 over the frequency range. In Figure 10.25 we compare the structured singular values, for the robust stability analysis, of the closed-loop systems with the three controllers (μ, $\mathcal{H}_\infty$, and $\mathcal{H}_\infty$ LSDP).

This shows that all three closed-loop systems achieve the robust stability. The best robustness is obtained by the μ-controller.

The nominal performance is tested on the bottom 3×2 block of the closed-loop transfer matrix. The comparison of the nominal performance for the three controllers, in Figure 10.26, shows that the performance in the $\mathcal{H}_\infty$

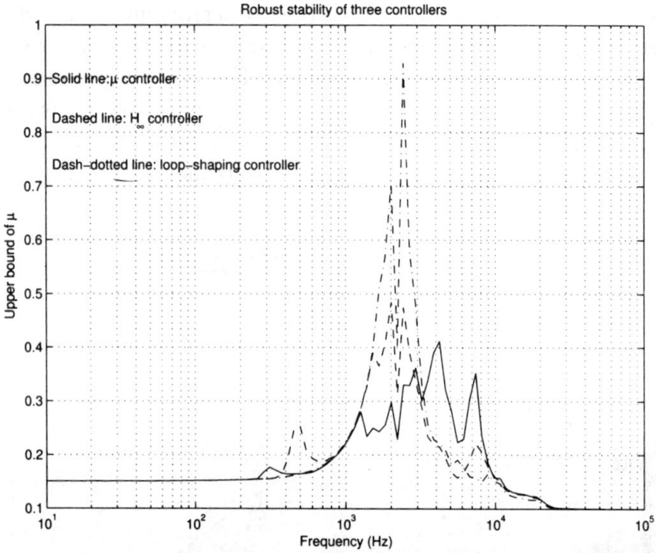

Fig. 10.25. Robust stability of the closed-loop systems

LSDP controller case over the low-frequency range is much worse than that in the other two cases. This is a consequence of the fact that the performance specifications used in the μ-design and $\mathcal{H}_\infty$ design are not explicitly adopted in the design of the $\mathcal{H}_\infty$ LSDP controller. The larger magnitude over the low frequencies leads to an expectation of larger steady-state errors.

The robust performance of the closed-loop system is studied also with the aid of the μ-analysis. The closed-loop transfer function matrix has 18 inputs and 17 outputs. The first 15 inputs/outputs correspond to the 15 channels connecting the perturbation matrix Δ, while the last 3 inputs and 2 outputs correspond to the weighted sensitivity of the closed-loop system. Hence, for a μ-analysis of the robust performance the block-structure of the uncertainty should consist of a 15×15 diagonal real uncertainty block and a 3×2 complex uncertainty block.

$$\Delta_P := \left\{ \begin{bmatrix} \Delta & 0 \\ 0 & \Delta_F \end{bmatrix} : \Delta \in \mathcal{C}^{15\times 15},\ \Delta_F \in \mathcal{C}^{3\times 2} \right\}$$

The robust performance (with respect to the uncertainty and performance weighting functions) is achieved if and only if for each frequency $\mu_{\Delta_P}(.)$, computed for the closed-loop frequency response, is less than 1. The robust performance test for all controllers is shown in Figure 10.27. Again, the $\mathcal{H}_\infty$ LSDP controller shows large μ-values over the low-frequency range.

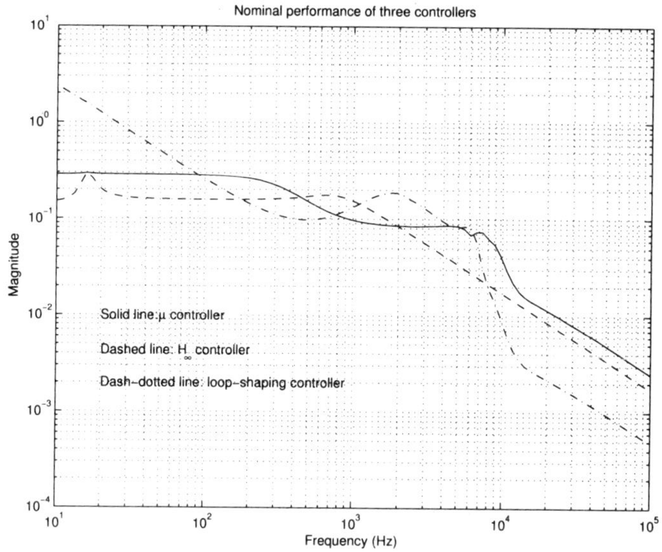

Fig. 10.26. Nominal performance of the closed-loop systems

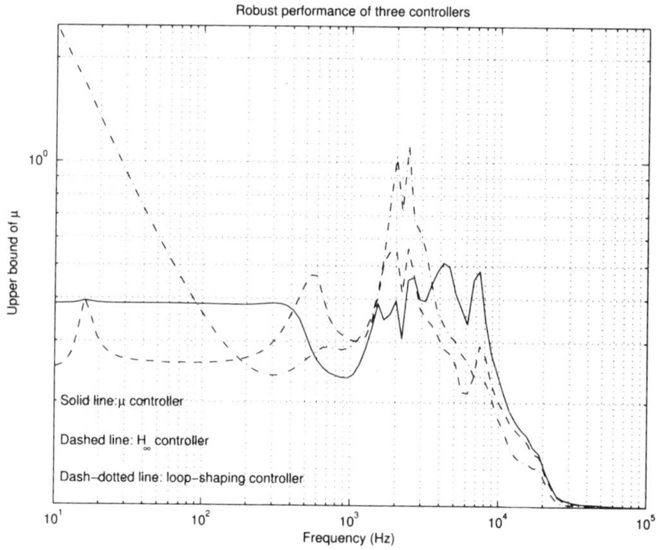

Fig. 10.27. Robust performance of the closed-loop systems

The robust stability and robust performance analysis also shows that the worse results may possibly occur over frequencies around the resonant frequencies.

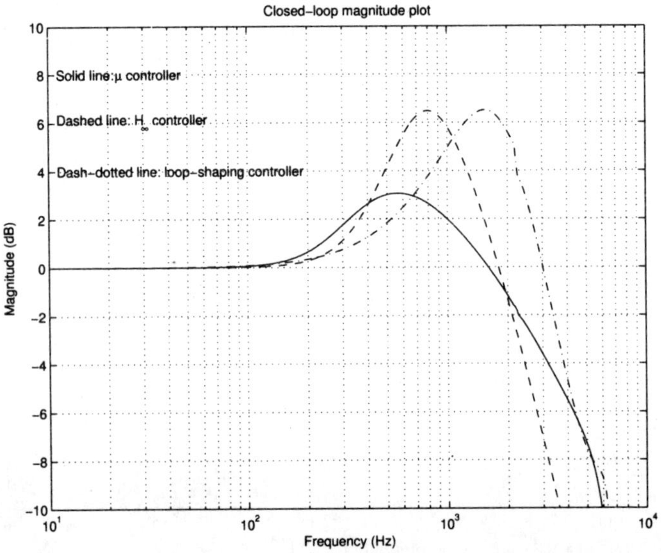

Fig. 10.28. Magnitudes of the closed-loop systems

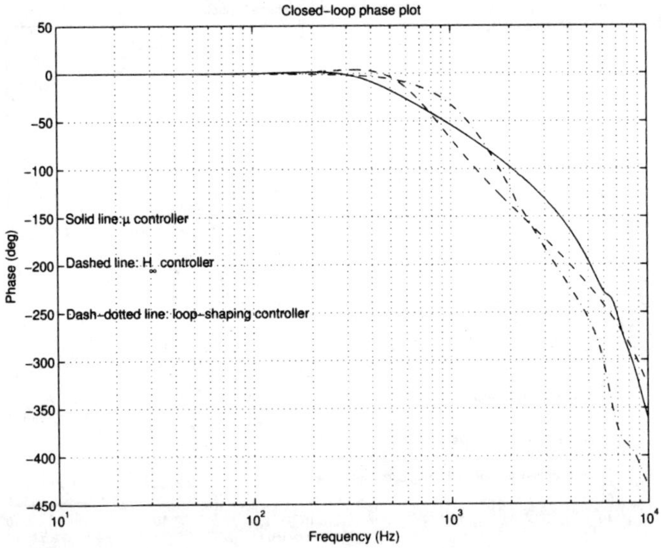

Fig. 10.29. Phase plots of the closed-loop systems

The Bode plots of the closed-loop systems with three controllers are shown in Figures 10.28 and 10.29. It is seen that the system with the $\mathcal{H}_\infty$ LSDP controller has the largest bandwidth that may lead to a fast transient response.

A larger bandwidth, however, may also lead to a larger effect of noises and resonances.

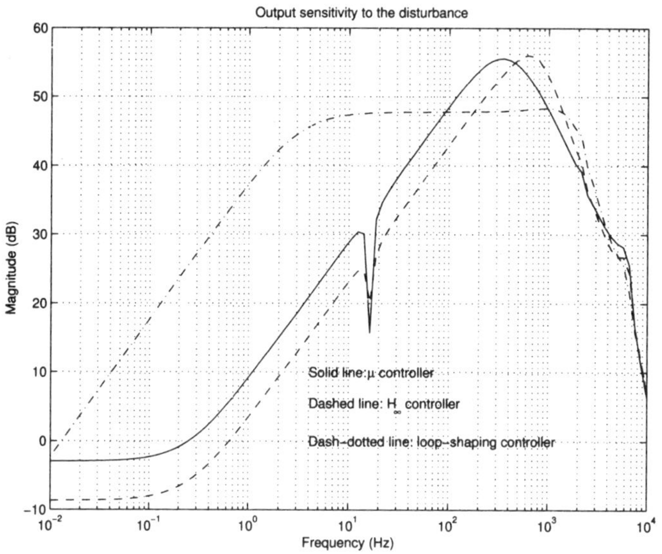

Fig. 10.30. Output sensitivity to disturbance

The plot of the output sensitivity to disturbances (Figure 10.30) shows that in the low-frequency range the influence of the disturbance on the system output in the case of the $\mathcal{H}_\infty$ LSDP controller is the largest. Better disturbance attenuation for this controller may be achieved by choosing higher gain in the prefilter. This will lead, however, to greater overshoot in the transient response. The sensitivity to disturbance in the cases of μ and $\mathcal{H}_\infty$ controllers reaches a maximum value in the frequency range from 300 Hz to 700 Hz, which is inside of the closed-loop bandwidth. This means that the closed-loop system will be susceptible to disturbances over that frequency range. It is interesting to notice that the sensitivity of the $\mathcal{H}_\infty$ LSDP controller over the same range is 8 dB lower.

The output sensitivity to noise is shown in Figure 10.31. (Note that the sensitivity is with respect to the unit-bounded noise, the input of the noise shaping filter.) The lowest sensitivity to noise has been achieved by the system with the μ-controller and the largest sensitivity by the system with the $\mathcal{H}_\infty$ LSDP controller.

The Bode plots of the three controllers are compared in Figures 10.32 and 10.33. It is seen that the $\mathcal{H}_\infty$ LSDP controller has a very low gain in the range from 1 Hz to 500 Hz which is the reason for the weak attenuation of the disturbances over that range.

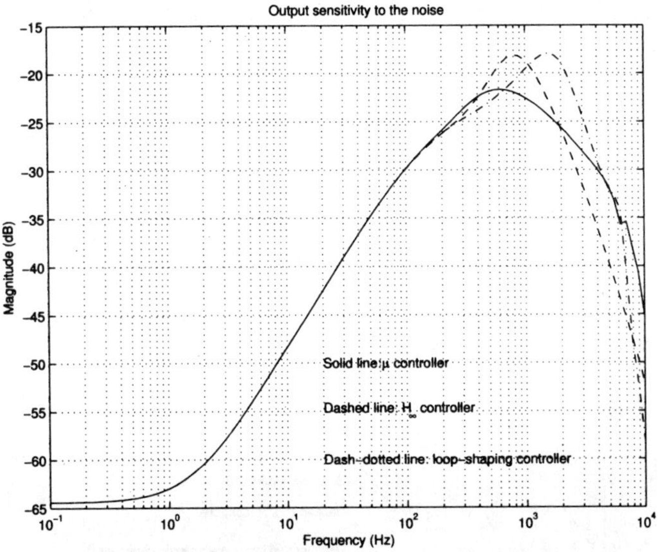

Fig. 10.31. Output sensitivity to noise

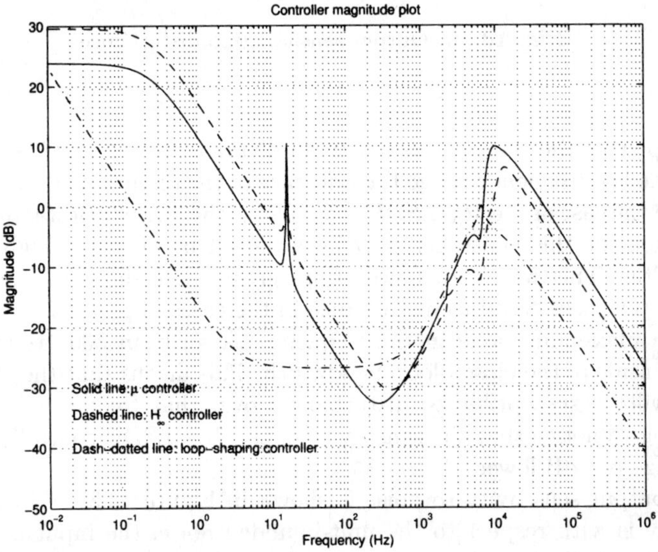

Fig. 10.32. Magnitude plots of three controllers

The transient responses of the closed-loop systems are obtained by using the file `clp_hdd`. In Figure 10.34 we show the transient responses of the closed-loop systems to a reference signal, equivalent to 1 track. While for the $\mathcal{H}_\infty$

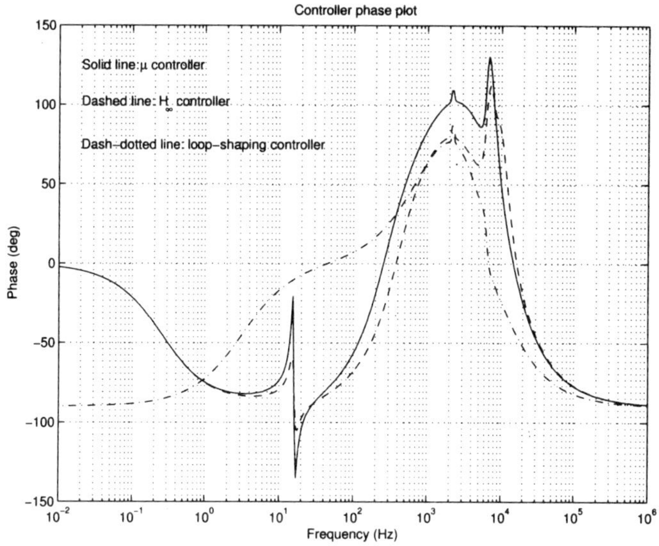

Fig. 10.33. Phase plots of three controllers

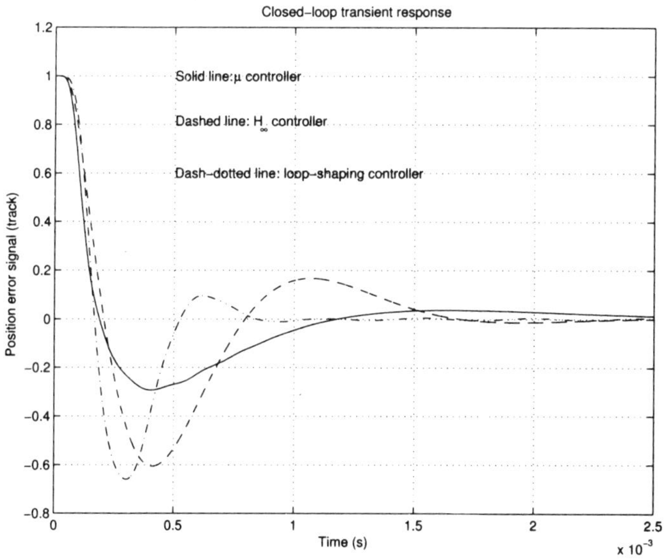

Fig. 10.34. Transient responses with three controllers

and $\mathcal{H}_\infty$ LSDP controllers the undershoot is about 60%, it is only 28% for the μ-controller. The settling time for the $\mathcal{H}_\infty$ LSDP, μ and $\mathcal{H}_\infty$ controllers is 0.8 ms, 1 ms and 1.5 ms, respectively.

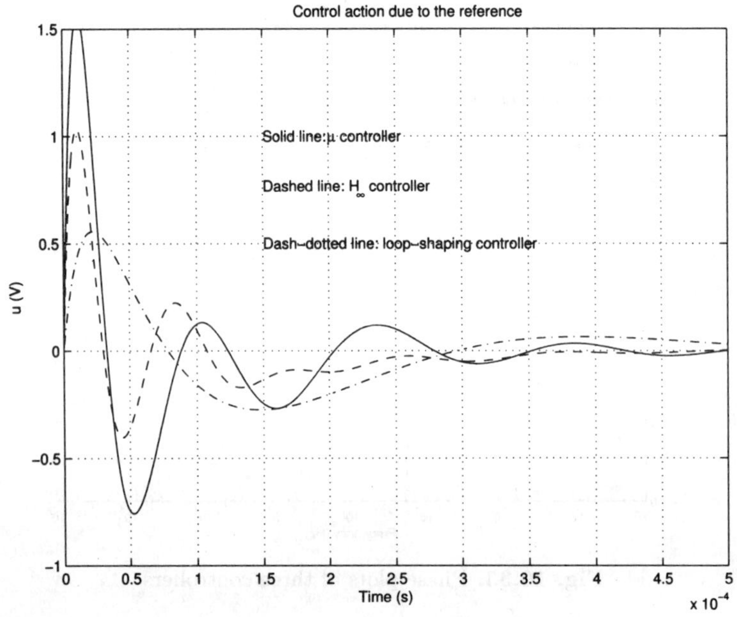

Fig. 10.35. Control actions of three controllers

The control actions of the three controllers are shown in Figure 10.35. For all controllers the control signal amplitude does not exceed 1.2 V, as required. In Figure 10.36 we show the system response to a step torque disturbance $t_d = 0.0005$ N m (equivalent to a force of 9.8×10^{-3} N applied to the disk head sssembly). The transient error for the $\mathcal{H}_\infty$ LSDP controller has the smallest undershoot (11.5%) but has a nonzero steady-state error. This reveals that in the given case the LSDP controller does not have integrating action with respect to the disturbance. The transient error for μ and $\mathcal{H}_\infty$ controllers is less than 17% of the track width and the steady-state error is practically equal to zero.

The output response to the position-sensing noise is simulated for a noise signal with an amplitude that does not exceed 60 mV (5% of the track width). This signal is obtained at the output of the noise shaping filter whose input is chosen as a sequence of uniformly distributed random numbers in the interval $[-1, 1]$. In the case of the μ-controller the output due to noise is less than 1.9% of the track width. The largest output to the noise (2.3% of the track width) is seen in the $\mathcal{H}_\infty$ LSDP controller case due to the largest closed-loop bandwidth of this controller.

The comparison of the robust stability and robust performance for the three controllers, as well as the comparison of the corresponding frequency

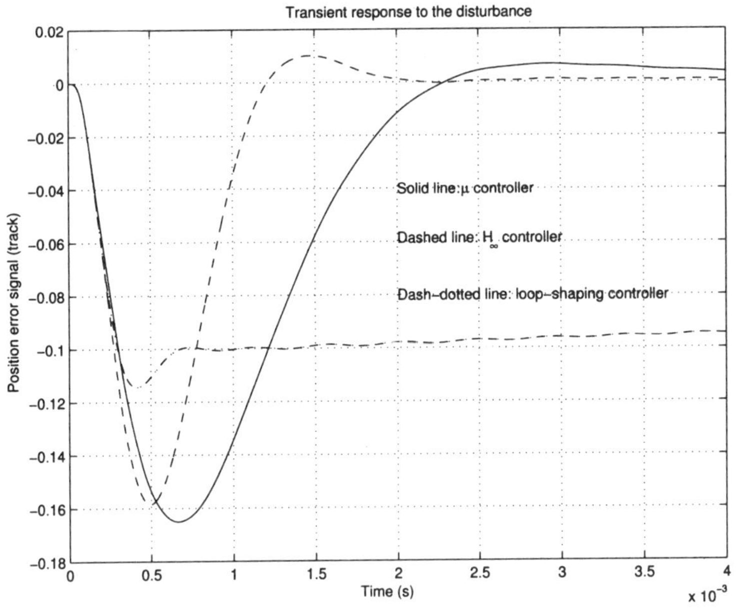

Fig. 10.36. Transient disturbance responses of three controllers

and transient responses shows that it is reasonable to conclude that the μ controller is preferable.

10.7 Controller-order Reduction

The controller obtained by the μ-synthesis is initially of 38th order. It is useful to reduce as much as possible the controller order, which will simplify the implementation and increase the reliability. To do this we use the system balancing followed by optimal Hankel approximation (see Chapter 7). There is a clear gap between the 12th and 13th Hankel singular values. Hence, the order of the μ-controller is reduced to 12. Further reduction of the controller order leads to deterioration of the closed-loop performance.

In Figures 10.37 and 10.38, we compare the Bode plots of the full order and reduced-order controllers. The corresponding plots practically coincide with each other, which implies similar performances in the closed-loop systems. In particular, the transient responses of the closed-loop system with full order and that with the reduced-order controller are practically undistinguishable.

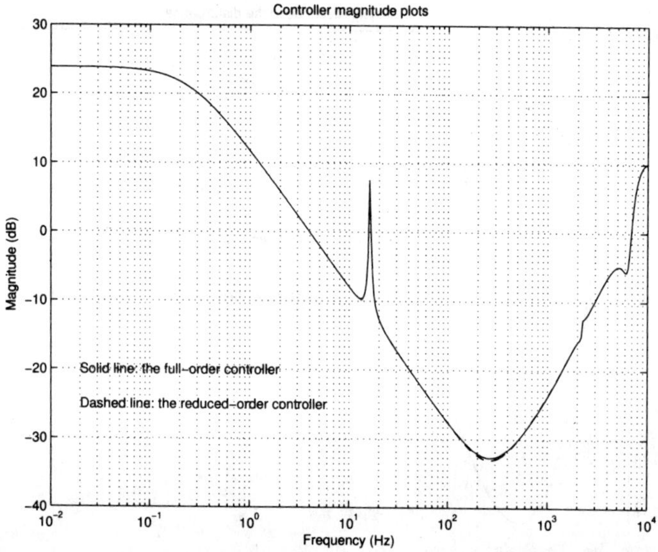

Fig. 10.37. Magnitude plots of full- and reduced-order μ-controllers

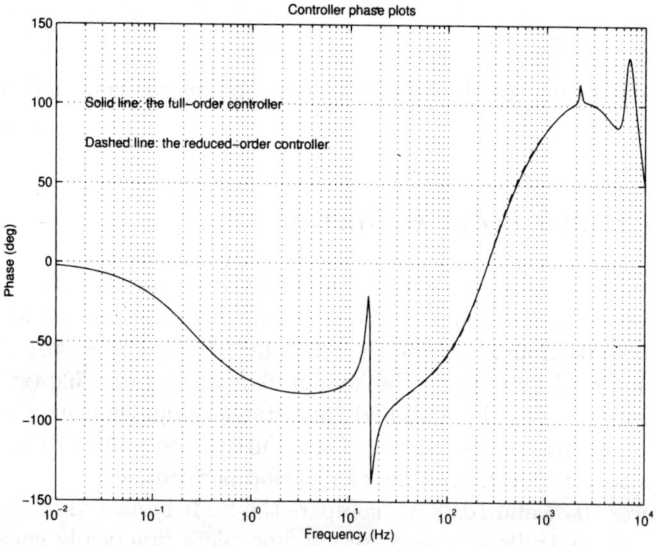

Fig. 10.38. Phase plots of full- and reduced-order μ-controllers

10.8 Design of Discrete-time Controller

In general, there are two approaches to designing a discrete-time servo controller.

The first approach is to sample the already designed continuous-time controller at a given sampling frequency $f_s = 1/T_s$. This may be accomplished by the M-file `dcl_hdd.m` that utilises the Robust Control Toolbox function `samhld`. It is assumed that the control-action calculation requires one sampling period T_s. This introduces a pure delay equal to T_s that is approximated by a rational transfer function using the command `pade`. The resultant sampled-data, closed-loop system is simulated by using the function `sdtrsp`. This approach gives satisfactory results for a sufficiently high sampling frequency (say, 100 kHz in the given case).

The second method is to sample the continuous-time, open-loop system (including the weighting filters) and to design directly a discrete-time controller by using $\mathcal{H}_\infty$ optimisation (implementing the function `dhinfsyn`) or μ-synthesis (implementing the function `dkit`).

The choice of the sampling frequency in the discrete-time case has a strong influence on the closed-loop system performance. A low sampling frequency limits the system bandwidth and would deteriorate the transient performance such as the disturbance rejection. On the other hand, the increase of the sampling frequency would complicate the controller implementation and raise the cost of the HDD.

Later we consider the μ-synthesis of the discrete-time controller for two sampling rates – 24 kHz and 36 kHz. In both cases we use the same performance weighting function

$$W_{p2}(s) = 10^{-4}\frac{s^2 + 4\times 10^5 s + 2.5\times 10^9}{s^2 + 3.9\times 10^5 s + 6.25\times 10^5}$$

utilised already in the continuous-time design. Depending on the sample rate we use two different control weighting functions

$$W_{u1}(s) = 10^{-6}\frac{4s^2 + 2s + 1}{2\times 10^{-3}s^2 + 2\times 10^{-3}s + 1}$$

(for $f_s = 24$ kHz) and

$$W_{u2}(s) = 10^{-6}\frac{1.04s^2 + 2s + 1}{7.5^{-5}s^2 + 2\times 10^{-3}s + 1}.$$

(for $f_s = 36$ kHz). This allows, in both cases, to obtain control signals that do not exceed 1.2 V.

The noise shaping filter is the same as in the continuous-time case.

The sampling of the extended open-loop system for the given sampling rate is conducted by the M-file `dlp_hdd.m`. The discrete-time μ-synthesis is accomplished by the file **dms_hdd.m** that is used in conjunction with the auxiliary file **ddk_hdd.m**. The file **ddk_hdd.m** sets the structure of the uncertainty and the parameter values of the $D-K$ iteration. As in the continuous-time μ-synthesis, only the rigid-body uncertain parameters are taken into account. The frequency is set on the unit circle in the interval $[0, \pi]$. In the discrete-time case it is also necessary to add the operator

```
DISCRETE_DK = 1;
```

in the file **ddk_hdd.m**.

The results from the μ-synthesis for $f_s = 24$ kHz show that for the chosen weighting functions the closed-loop system almost achieves robust performance at the fifth $D-K$ iteration ($\mu_{\max} = 1.06$) but the closed-loop response is relatively slow and the undershoot is large (44%). The maximum control amplitude is 1.19 V. To obtain better results it is necessary to increase the sampling frequency.

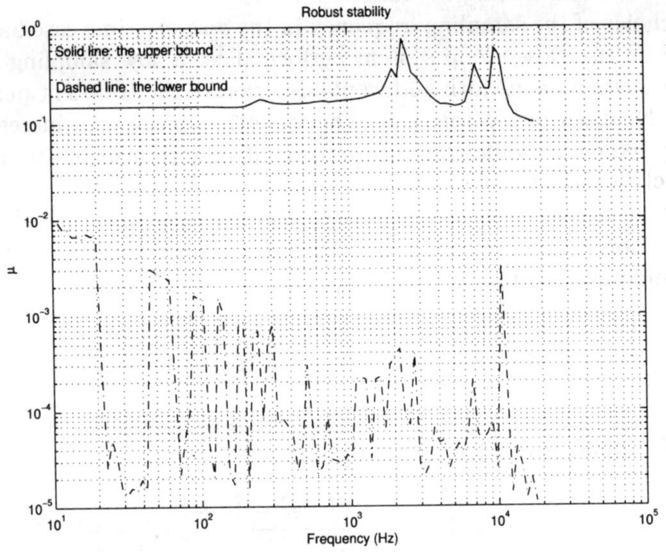

Fig. 10.39. Robust stability of $f_s = 36$ kHz design

We now present in more detail the results from the μ-synthesis at $f_s = 36$ kHz. In this case an appropriate controller is obtained after three $D-K$ iterations and the maximum robust performance achieved is $\mu_{\max} = 1.071$. In Figures 10.39 and 10.40, we show the μ-plots, obtained by the M-file **dmu_hdd.m**, for robust stability analysis and robust performance analysis, respectively. In both plots the worst results are seen around the second resonant frequency of 2.2 kHz.

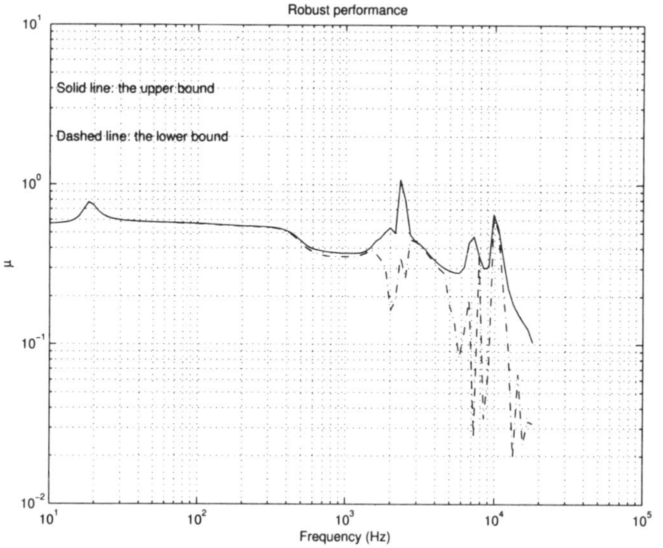

Fig. 10.40. Robust performance of $f_s = 36$ kHz design

The closed-loop transient response is obtained by the file `dsl_hdd.m` that uses the function `sdtrsp`. The function `sdtrsp` also computes the control signal obtained at the output of the digital-to-analogue converter. The closed-loop transient response is shown in Figure 10.41 and the corresponding control action in Figure 10.42. The undershoot of the transient response is less than 36% and the peak amplitude of the control is less than 1.05 V.

The transient response to disturbance is shown in Figure 10.43. Overall, the results obtained are almost as good as the results obtained with the continuous-time, μ-controller.

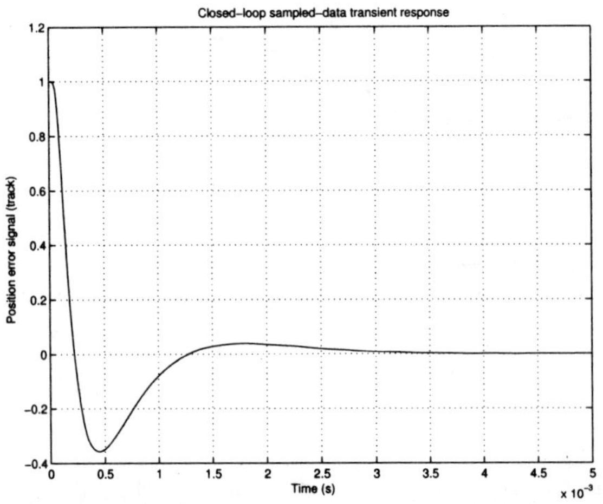

Fig. 10.41. Transient response of $f_s = 36$ kHz design

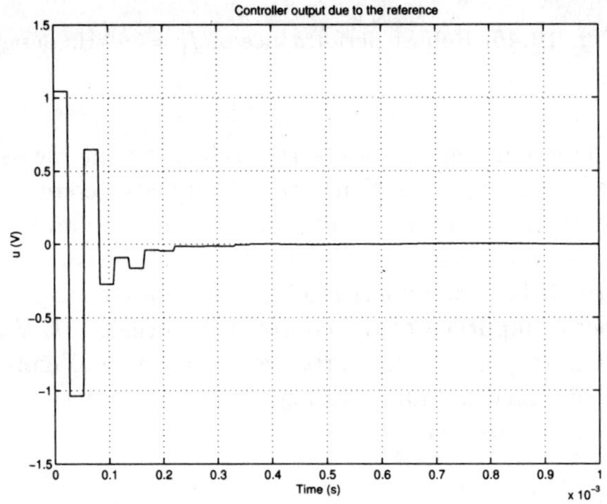

Fig. 10.42. Control action of $f_s = 36$ kHz design

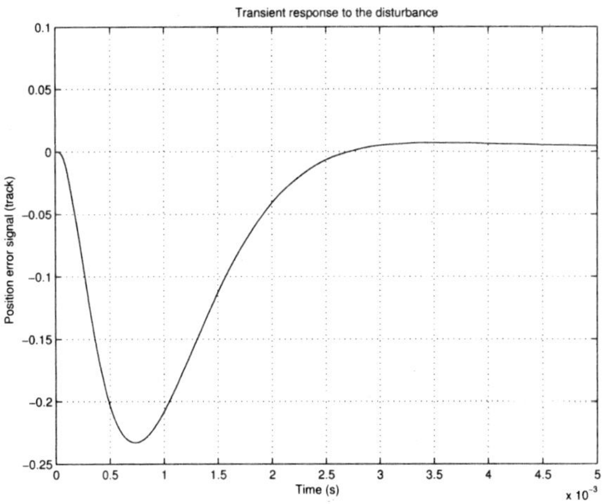

Fig. 10.43. Transient response to disturbance of $f_s = 36$ kHz design

10.9 Nonlinear System Simulation

In order to obtain a realistic idea about the behaviour of the designed system, the nonlinear, closed-loop servo system is simulated by using Simulink®. For this aim, two models are developed, namely `c_hdd.mdl` for the continuous-time system and `d_hdd.mdl` for the sampled-data system. In the simulation we take into account the amplifier saturation, which was neglected in the design so far. Both models allow us to simulate the closed-loop system for different reference, disturbance and noise signals.

Before simulating the continuous-time system it is necessary to assign the model parameters by using the M-file `init_c_hdd.m`.

The sampled-data model involves the discrete-time controller, 16-bit analogue-to-digital converter with maximum input voltage 2.5 V and 16-bit digital-to-analogue converter with maximum output voltage 10 V. It is assumed that the discrete-time controller is implemented on a *digital signal processor(DSP)* with word length of 64 bits. These parameters are set prior to the simulation by using the M-file `init_d_hdd.m`. It is assumed that the control action calculation requires one sampling period T_s.

In Figure 10.44 we show the Simulink® model `d_hdd.mdl` of the nonlinear, sampled-data, closed-loop system.

As in the linear case, the transient responses of the nonlinear closed-loop system are obtained for a simulated runout of 1 track width (1μm) and torque disturbance $t_d = 0.0005$ N m.

In Figure 10.45 and in Figure 10.46 we compare the results from the simulation of the continuous-time and discrete-time nonlinear systems. The continuous-time controller is the reduced-order μ-controller in Section 10.7 and the discrete-time controller is the controller designed at the sampling frequency of 36 kHz.

The transient responses of the nonlinear system are close to the corresponding responses of the linear system due to the small input signals (amplitude less than 1.2 V).

It should be mentioned that the controllers designed are appropriate for small reference signals (equivalent to one track). For larger references the amplifier saturates and it is necessary to implement an appropriate seeking algorithm.

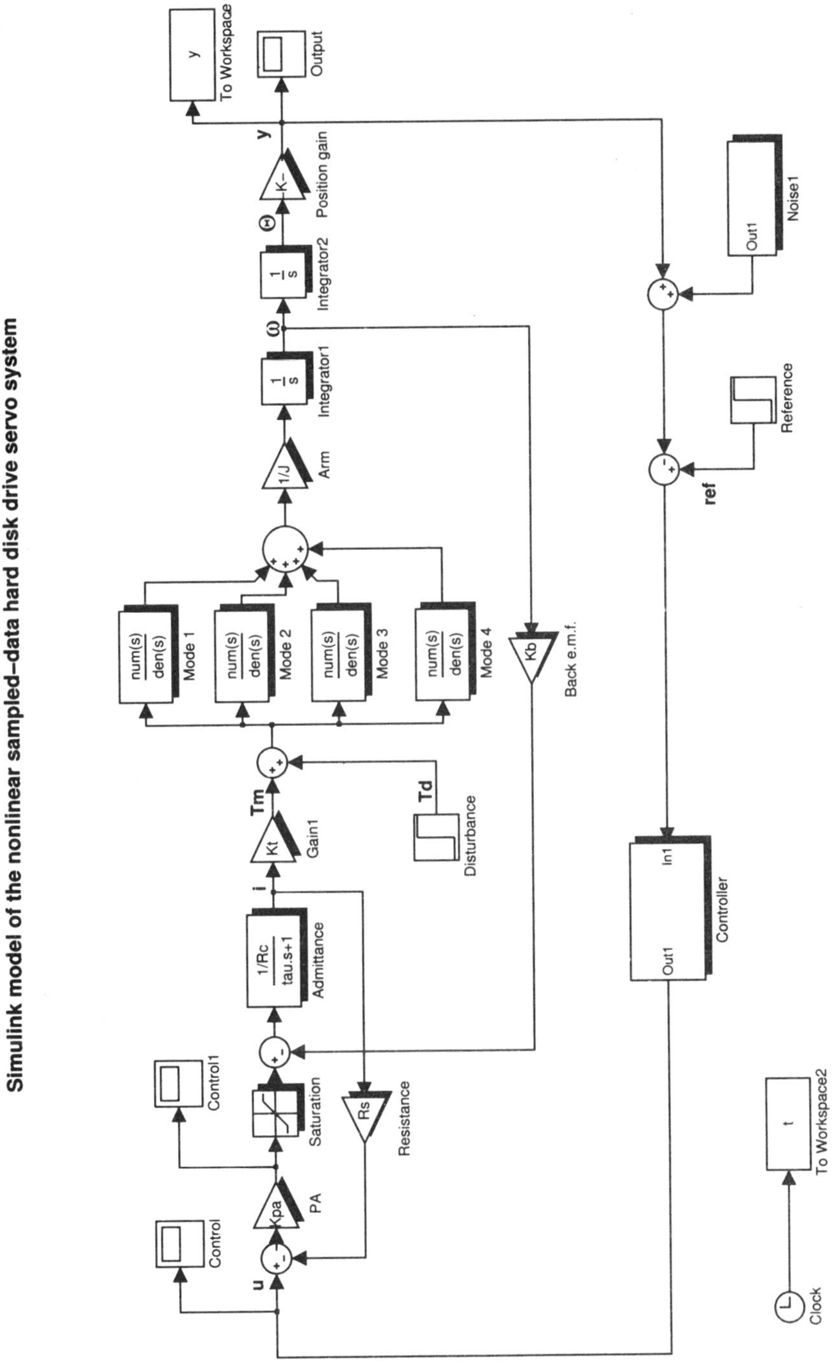

Fig. 10.44. Simulation model of the nonlinear sampled data system

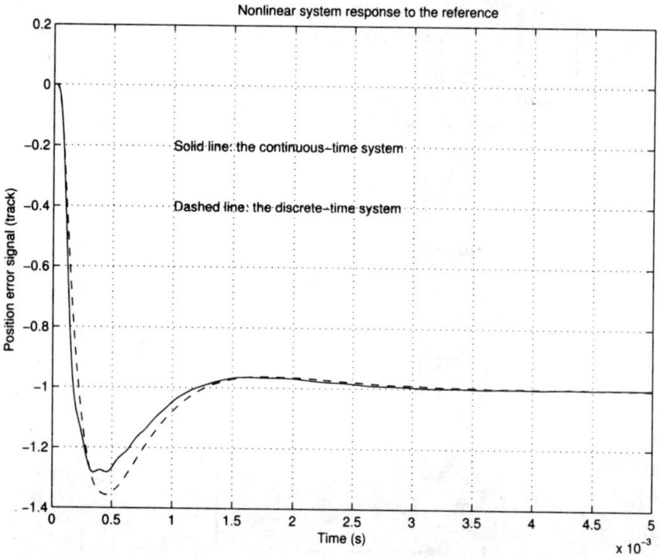

Fig. 10.45. Transient responses of the nonlinear systems

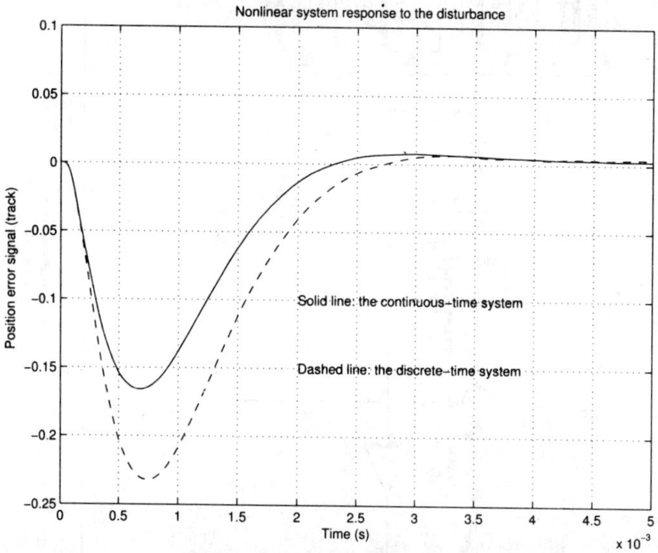

Fig. 10.46. Disturbance responses of the nonlinear systems

10.10 Conclusions

The experience gained in the design of the HDD servo controllers makes it possible to derive the following conclusions:

- The implementation of designed μ, $\mathcal{H}_\infty$ and $\mathcal{H}_\infty$ LSDP controllers in the HDD servo system gives satisfactory results with respect to robustness and performance. All three controllers ensure robust stability of the closed-loop system. The best robust performance is achieved by using the μ and $\mathcal{H}_\infty$ controllers. The implementation of the $\mathcal{H}_\infty$ LSDP controller gives the fastest transient response and the corresponding design is less complicated. This controller, however, leads to the worst performance in the low-frequency range, which results in a large steady-state error. In the given case the best trade-off between the robustness and transient response requirements is achieved by using the μ-controller that is, to some extent, due to the specially chosen weighting functions.
- The number of original uncertain parameters is very large (more than 25 in the given case). This complicates the derivation of the uncertainty model and produces heavy computation demands. This is why it is necessary to investigate the parameter importance with respect to the robustness and performance in order to reduce their number to an acceptable value. However, in the evaluation of the design, it is better to take into account all the possible uncertainties to ensure a satisfactory design in a real case.
- In the μ-synthesis, the order of the resultant controller depends on the order of the plant, of the weighting functions and of the scaling diagonal elements approximations. The designed controllers are usually of high orders, which complicates the implementation of the controller. Hence, an order reduction should usually be considered right after the controller design. Most controllers used in the HDD designs are of order between 8 and 15.
- Good disturbance attenuation requires sufficiently large closed-loop bandwidth. This may, however, lead to difficulties in achieving robust stability and robust performance in the presence of resonant modes. Some resonances whose frequencies are much higher than the closed-loop bandwidth and thus seem innocent may even actually destroy the robust stability of the system. In such cases, it is necessary to increase the damping of these modes by using techniques of passive/active damping.
- The presence of resonant modes may require sufficiently high sampling rates in the case of using a discrete-time controller.
- It is important to stress that better results, with respect to the transient response (overshoot and settling time), are difficult to obtain for the current plant parameters. If higher performance demands are required, it is necessary to change the HDD parameters, for instance to increase the VCM torque constant.

Notes and References

The history of the Hard Disk Drive control is presented in the fascinating papers of Abramovitch and Franklin [1, 2]. An excellent survey on similarities and contrasts in the magnetic and optical disk controls is given in [5]. In [77] one may find an attractive description of the HDD construction and functioning.

The book of Chen *et al.* [15] is the only book up to the moment that is entirely devoted to HDD servo systems and contains rich information related to the design of such systems. A tutorial on HDD control can be found in [106].

A detailed model of the HDD servo system, which has very much influenced the model used in this chapter, is presented in the book of Franklin *et al.* [39, Chapter 14]. Below 100 Hz the rotary actuator dynamics is affected by pivot bearing nonlinearity, which is known under the name "stick-slip" . It has a strong effect particularly in the case of small disk drives with lower actuator inertia. Analysis and simulation of this phenomenon are presented in [4, 156].

An important step in the design of HDD servo systems is the reliable estimation of the various disturbances and noises acting on the system. Methods for such estimations are described in [3, 27, 61].

The harmonic compensation used to reduce RRO is considered in [16, 74].

The track-seeking and track-following modes require different control algorithms. Track-seeking algorithms are described in [39, 123]. The switching of control mode from track-seeking to track-following should be smooth so that the residual vibration of the read/write head suspension is minimal. There are several control algorithms that work for both track-seeking and following, see for instance [70, 63]. Such algorithms utilise two-degree-of-freedom (2DOF) controllers in which case the track seeking is accomplished by using a feedforward controller along with a reference trajectory.

Apart from the track seeking and track following the HDD also contains a spindle velocity control loop. The purpose of this loop is to control the air flow over the disk in order to guarantee the appropriate flying height of the read/write head. This is a low-frequency control loop and its design does not represent a serious difficulty.

Further expansion of the closed-loop bandwith of the HDD control system may be achieved by using the so-called dual-stage servos that consist of a low-bandwidth coarse actuator (the usual VCM) and a high bandwidth fine actuator. The fine actuator has a small stroke and may be implemented as a *piezoelectric transducer(PZT)* [31]. The design of dual-stage servos is considered in [22, 66, 76].

Other important aspects of the analysis and design of HDD servo systems are presented in [50, 69, 83, 99, 164], to name a few.

11

Robust Control of a Distillation Column

In this chapter we present the design of a robust control system for a high-purity distillation column. The original nonlinear model of the column is of high order and it includes parametric gain and time-delay uncertainty. A low-order linearised distillation column model is used in the design of a two-degree-of freedom (2DOF) $\mathcal{H}_\infty$ loop-shaping controller and a μ-controller. Both controllers ensure robust stability of the closed-loop system and fulfillment of a mixture of time-domain and frequency-domain specifications. A reduced order μ-controller is then found that preserves the robust stability and robust performance of the closed-loop system. The simulation of the closed-loop system with the nonlinear distillation column model shows very good performance for different reference and disturbance signals as well as for different values of the uncertain parameters.

11.1 Introduction

Distillation is an important process in the separation and purification of chemicals. The process exploits the difference at boiling points of multicomponent liquids. The control of distillation columns is difficult, because the distillation process is highly nonlinear and the corresponding linearised models are often ill-conditioned around the operating point.

The aim of the design, presented in this chapter, is to find a controller that achieves robust stability and robust performance of the closed-loop control system of a high-purity distillation column. The original nonlinear model of the column is of 82nd order and it includes uncertainties in the form of parametric gains and time delay. The uncertainty model is considered in the form of an input multiplicative complex uncertainty. In our design exercises, we found that it is difficult to achieve the desired performance of the closed-loop system using one-degree-of-freedom controllers. Hence, we turned to the $\mathcal{H}_\infty$ two-degree-of-freedom loop-shaping design procedure and μ-synthesis/analysis method. The designs are based on a 6th-order linearised distillation column model. Both designed controllers ensure robust stability of the closed-loop system and achieve a mixed set of time-domain and frequency-domain specifications. We present several time-domain and frequency-domain characteristics of the corresponding closed-loop systems that makes possible the comparison of controllers efficiency. An 11th-order reduced-order μ-controller is found that preserves the stability and performance of the closed-loop system in the presence of uncertainties. The simulation of the closed-loop system with this μ-controller and with the nonlinear distillation column model is conducted in Simulink® and shows very good performance for different reference and disturbance signals as well as for different values of the uncertain parameters.

11.2 Dynamic Model of the Distillation Column

A typical two-product distillation column is shown in Figure 11.1. The objective of the distillation column is to split the feed F, which is a mixture of a light and a heavy component with composition z_F, into a distillate product D with composition y_D, which contains most of the light component, and a bottom product B with composition z_B, which contains most of the heavy component. For this aim, the column contains a series of trays that are located along its height. The liquid in the columns flows through the trays from top to bottom, while the vapour in the column rises from bottom to top. The constant contact between the vapour and liquid leads to increasing concentration of the more-volatile component in the vapour, while simultaneously increasing concentration of the less volatile component in the liquid. The operation of the column requires that some of the bottom product is reboiled at a rate V to ensure the continuity of the vapor flow and some of the distillate is refluxed to the top tray at a rate L to ensure the continuity of the liquid flow.

The notations used in the derivation of the column model are summarised in Table 11.1 and the column data are given in Table 11.2.

The index i denotes the stages numbered from the bottom ($i = 1$) to the top ($i = N_{\text{tot}}$) of the column. Index B denotes the bottom product and D the distillate product. A particular high-purity distillation column with 40 stages (39 trays and a reboiler) plus a total condensor is considered.

The nonlinear model equations are:

1. Total material balance on stage i

$$\mathrm{d}M_i/\mathrm{d}t = L_{i+1} - L_i + V_{i-1} - V_i$$

2. Material balance for the light component on each stage i

$$\mathrm{d}(M_i x_i)/\mathrm{d}t = L_{i+1}x_{i+1} + V_{i-1}y_{i-1} - L_i x_i - V_i y_i$$

This equation leads to the following expression for the derivative of the liquid mole fraction

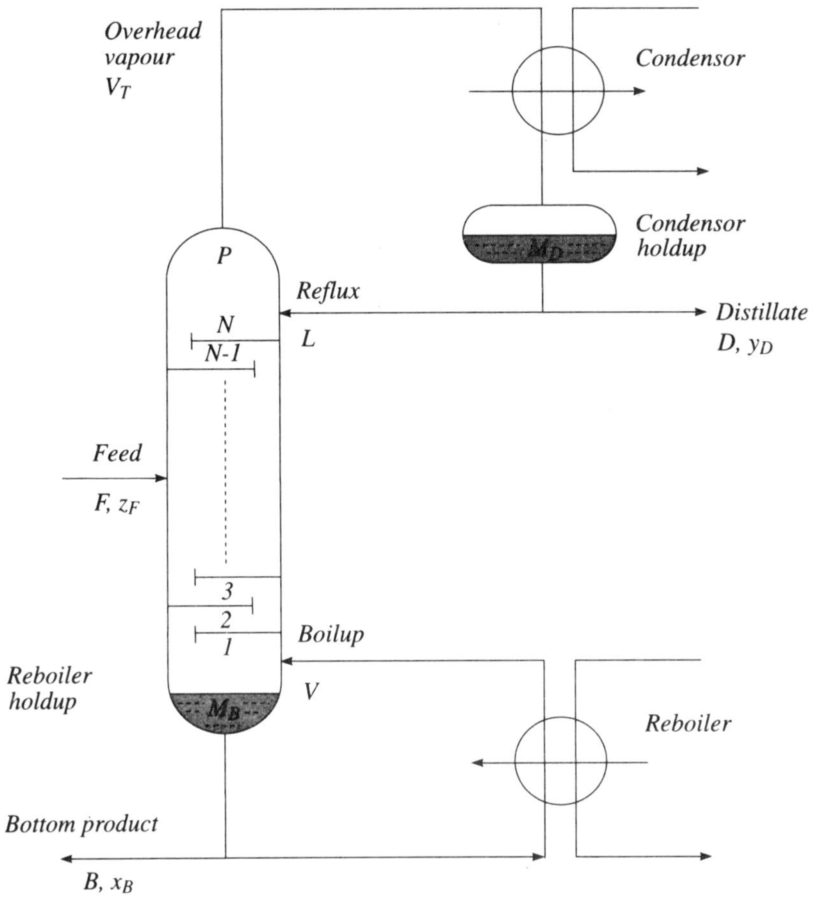

Fig. 11.1. The distillation column system

$$dx_i/dt = (d(M_i x_i)/dt - x_i(dM_i/dt))/M_i$$

3. Algebraic equations
   The vapour composition y_i is related to the liquid composition x_i on the same stage through the algebraic vapour-liquid equilibrium

   $$y_i = \alpha x_i/(1 + (\alpha - 1)x_i)$$

   From the assumption of constant molar flows and no vapour dynamics, one obtains the following expression for the vapour flows

   $$V_i = V_{i-1}$$

   The liquid flows depend on the liquid holdup on the stage above and the vapor flow as follows

Table 11.1. Column nomenclature

Symbol	Description
F	Feed rate [kmol/min]
z_F	feed composition [mole fraction]
q_F	fraction of liquid in feed
D and B	distillate (top) and bottom product flowrate [kmol/min]
y_D and x_B	distillate and bottom product composition (usually of light component) [mole fraction]
L	reflux flow [kmol/min]
V	boilup flow [kmol/min]
N	number of stages (including reboiler)
$N_{\text{tot}} = N + 1$	total number of stages (including condensor)
i	stage number (1 – bottom, N_F – feed stage, N_T – total condensor)
L_i and V_i	liquid and vapour flow from stage i [kmol/min]
x_i and y_i	liquid and vapour composition of light component on stage i
M_i	liquid holdup on stage i [kmol] (M_B – reboiler, M_D – condensor holdup)
α	relative volatility between light and heavy component
τ_L	time constant for liquid flow dynamics on each stage [min]

Table 11.2. Column data

N	N_{tot}	N_F	F	z_F	q_F	D
40	41	21	1	0.5	1	0.5

B	L	V	y_D	x_B	M_i	τ_L
0.5	2.706 29	3.206 29	0.99	0.01	0.5	0.063

$$L_i = L0_i + (M_i - M0_i)/\tau_L + \lambda(V_{i-1} - V0_{i-1})$$

where $L0_i$ [kmol/min] and $M0_i$ [kmol] are the nominal values for the liquid flow and holdup on stage i and $V0_i$ is the nominal boilup flow. If the vapour flow into the stage effects the holdup then the parameter λ is different from zero. For the column under investigation $\lambda = 0$.

The above equations apply at all stages except in the top (condensor), feed stage and bottom (reboiler).

1. For the feed stage, $i = N_F$ (it is assumed that the feed is mixed directly into the liquid at this stage)

$$\mathrm{d}M_i/\mathrm{d}t = L_{i+1} - L_i + V_{i-1} - V_i + F$$

$$\mathrm{d}(M_i x_i)/\mathrm{d}t = L_{i+1}x_{i+1} + V_{i-1}y_{i-1} - L_i x_i - V_i y_i + F z_F$$

2. For the total condensor, $i = N_{\text{tot}}(M_{N_{\text{tot}}} = M_D, L_{N_{\text{tot}}} = L_T)$

$$\mathrm{d}M_i/\mathrm{d}t = V_{i-1} - L_i - D$$

$$\mathrm{d}(M_i x_i)/\mathrm{d}t = V_{i-1} - L_i x_i - D x_i$$

3. For the reboiler, $i = 1(M_i = M_B, V_i = V_B = V)$

$$\mathrm{d}(M_i x_i)/\mathrm{d}t = L_{i+1} x_{i+1} - V_i y_i - B x_i$$

As a result, we obtain a nonlinear model of the distillation column of 82nd order. There are two states per tray, one representing the liquid composition and the other representing the liquid holdup. The model has four manipulated inputs (L_T, V_B, D and B) and three disturbances (F, z_F and q_F).

In order to find a linear model of the distillation column it is necessary to have a steady-state operating point around which the column dynamics is to be linearised. However, the model contains two integrators, because the condensor and reboiler levels are not under control. To stabilise the column, we make use of the so called *LV-configuration of the distillation column* where we use D to control M_D and B to control M_B. This is done by two proportional controllers with both gains equal to 10.

The nonlinear model is linearised at the operating point given in Table 11.2 (the values of F, L, V, D, B, y_D, x_B and z_F). These steady-state values correspond to an initial state where all liquid compositions are equal to 0.5 and the tray holdups are also equal to 0.5 [kmol]. The steady-state vector is obtained for $t = 5000$ min by numerical integration of the nonlinear model equations of the LV-configuration given in the M-file `cola_lv.m`. The linearisation is carried out by implementing the M-file `cola_lin` that makes use of the equations given in the file `cola_lv_lin.m`. The 82nd-order, linear model is stored in the variable G_4u and has four inputs (the latter two are actually disturbances)

$$[L_T \;\; V_B \;\; F \;\; z_F]$$

and two outputs

$$[y_D \;\; x_B]$$

Before reducing the model order, the model G_4u is scaled in order to make all inputs/disturbances and all outputs at about the same magnitude. This is done by dividing each variable by its maximum change, *i.e.*

$$u = U/U_{\max}; \; y = Y/Y_{\max}$$

where U, Y are the input and output of the model G_4u in original units, $U_{\max}$, $Y_{\max}$ are the corresponding maximum values allowed, and u, y are the scaled variables. The scaling is achieved by using the input scaling matrix

$$S_i = \begin{bmatrix} 1 & 0 & 0 & 0 \\ 0 & 1 & 0 & 0 \\ 0 & 0 & 0.2 & 0 \\ 0 & 0 & 0 & 0.1 \end{bmatrix}$$

and output scaling matrix

$$S_o = \begin{bmatrix} 100 & 0 \\ 0 & 100 \end{bmatrix}$$

The scaled model is then found as $G_4 = SoG_4uS_i$.

The final stage in selecting the column model is the order reduction of the scaled model G_4. This is done by using the commands **sysbal** and **hankmr**. As a result, we obtain a 6th-order model saved in the variable G.

All commands for finding the 6th-order linear model of the distillation column are contained in the file `mod_col.m`. The frequency responses of the singular values of G are compared with the singular values of the 82nd order linearised model G_4 in Figure 11.2. It is seen that the behaviour of both models is close until the frequency 2 rad/min.

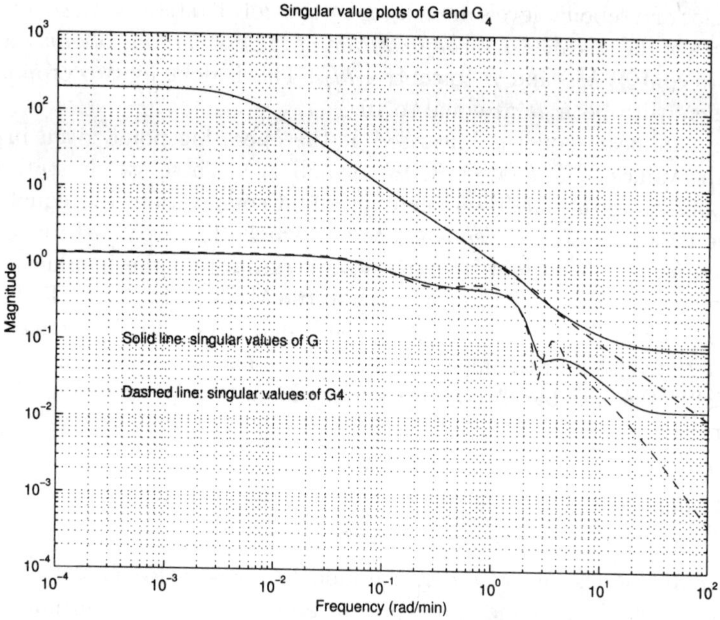

Fig. 11.2. Singular values of G and G_4

11.3 Uncertainty Modelling

The uncertainties considered in the distillation column control systems are a gain uncertainty of $\pm 20\%$ and a time delay of up to 1 min in each input channel. Thus, the uncertainty may be represented by the transfer matrix

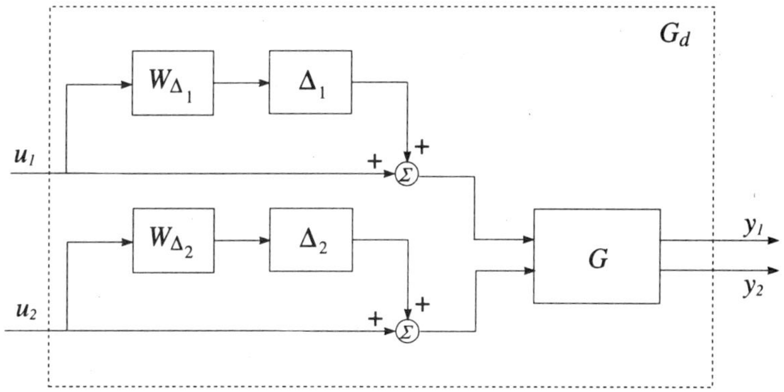

Fig. 11.3. Distillation column with input multiplicative uncertainty

$$W_u = \begin{bmatrix} k_1 e^{-\Theta_1 s} & 0 \\ 0 & k_2 e^{-\Theta_2 s} \end{bmatrix}$$

where $k_i \in [0.8 \;\; 1.2]$; $\Theta_i \in [0.0 \;\; 1.0]$; $i = 1, 2$. It is convenient to represent this uncertainty by an input multiplicative uncertainty, as shown in Figure 11.3, with

$$\Delta = \begin{bmatrix} \Delta_1 & 0 \\ 0 & \Delta_2 \end{bmatrix}$$

where $|\Delta_1| \le 1$, $|\Delta_2| \le 1$. The uncertainty weighting function

$$W_\Delta = \begin{bmatrix} W_{\Delta_1} & 0 \\ 0 & W_{\Delta_2} \end{bmatrix}$$

is determined in the following way.

Denote by $\overline{W}_{u_i} = 1$ the nominal transfer function in the ith channel for $k_i = 1$ and $\Theta_i = 0$; $i = 1, 2$.

According to Figure 11.3 we have that

$$W_{u_i} = (1 + W_{\Delta_i} \Delta_i) \overline{W}_{u_i}, \quad i = 1, 2$$

Taking into account that $|\Delta_i| \le 1$ it follows that the relative uncertainty should satisfy

$$\frac{\left| W_{u_i}(j\omega) - \overline{W}_{u_i}(j\omega) \right|}{\left| \overline{W}_{u_i}(j\omega) \right|} \le \left| W_{\Delta_i}(j\omega) \right|, \; i = 1, 2$$

where $W_{u_i}(j\omega) = k_i e^{j\omega\Theta_i} = k_i(\cos(\omega\Theta_i) + j\sin(\omega\Theta_i))$. In this way, to choose the uncertainty weight W_{Δ_i} is equivalent to determining an upper bound of the frequency response of the relative uncertainty

$$\frac{\left| W_{u_i}(j\omega) - \overline{W}_{u_i}(j\omega) \right|}{\left| \overline{W}_{u_i}(j\omega) \right|} = \sqrt{(k_i \cos(\omega\Theta_i) - 1)^2 + (k_i \sin(\omega\Theta_i))^2}.$$

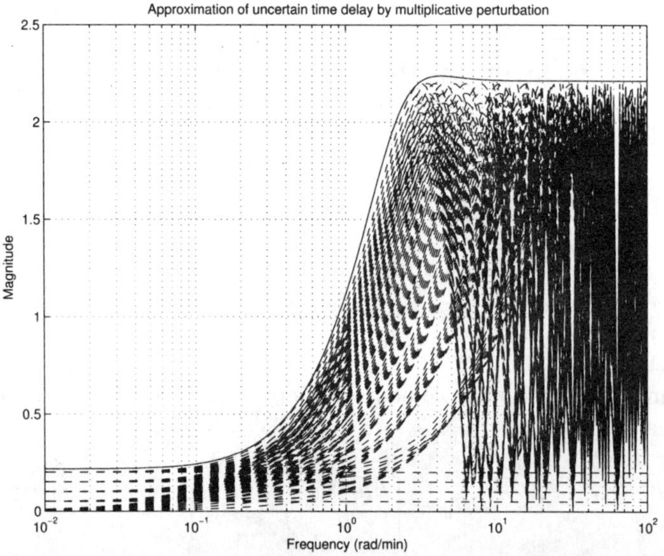

Fig. 11.4. Approximation of the uncertain time delay

The frequency responses of the relative uncertainty

$$\frac{|W_{u_i}(j\omega) - \overline{W}_{u_i}(j\omega)|}{|\overline{W}_{u_i}(j\omega)|}$$

are computed by the file `unc_col.m` and shown in Figure 11.4. These responses are then approximated by 3rd-order transfer functions using the file **`wfit.m`**. As a result, one obtains

$$W_{\Delta_i} = \frac{2.2138s^3 + 15.9537s^2 + 27.6702s + 4.9050}{1.0000s^3 + 8.3412s^2 + 21.2393s + 22.6705}, \ i = 1, 2$$

11.4 Closed-loop System-performance Specifications

The aim of the distillation column control-system design is to determine a controller that meets robust stability and robust performance specifications for the LV configuration. Since these specifications are difficult to satisfy with a one-degree-of-freedom controller, we present the design of two-degree-of-freedom controllers that ensure robust stability and robust performance of the closed-loop system. In the given case, the robust stability means guaranteed closed-loop stability for all $0.8 \leq k_1,\ k_2 \leq 1.2$ and $0 \leq \Theta_1,\ \Theta_2 \leq 1$ min. The time-domain specifications are given in terms of step-response requirements, which must be met for all values of $k_1,\ k_2, \Theta_1$ and Θ_2. Specifically, for a unit step command to the first input channel at $t = 0$, the scaled plant outputs y_1 (tracking) and y_2 (interaction) should satisfy:

- $y_1(t) \geq 0.9$ for all $t \geq 30$ min;
- $y_1(t) \leq 1.1$ for all t;
- $0.99 \leq y_1(\infty) \leq 1.01$;
- $y_2(t) \leq 0.5$ for all t;
- $-0.01 \leq y_2(\infty) \leq 0.01$.

Correspondingly, similar requirements should be met for a unit step command at the second input channel.

In addition, the following frequency-domain specification should be met:

- $\overline{\sigma}(\hat{K}_y\hat{S})(j\omega) < 316$, for each ω, where $\hat{K}_y$ denotes the feedback part of the unscaled controller. (Here and latter, a variable with a hat refers to the case of unscaled plant.) This specification is included mainly to avoid saturation of the plant inputs.
- $\overline{\sigma}(\hat{G}\hat{K}_y)(j\omega) < 1$, for $\omega \geq 150$; or $\overline{\sigma}(\hat{K}_y\hat{S})(j\omega) \leq 1$, for $\omega \geq 150$.

In the above, $\overline{\sigma}$ denotes the largest singular value, and $\hat{S} = (I + \hat{G}\hat{K}_y) < 1$ is the sensitivity function for $\hat{G}$.

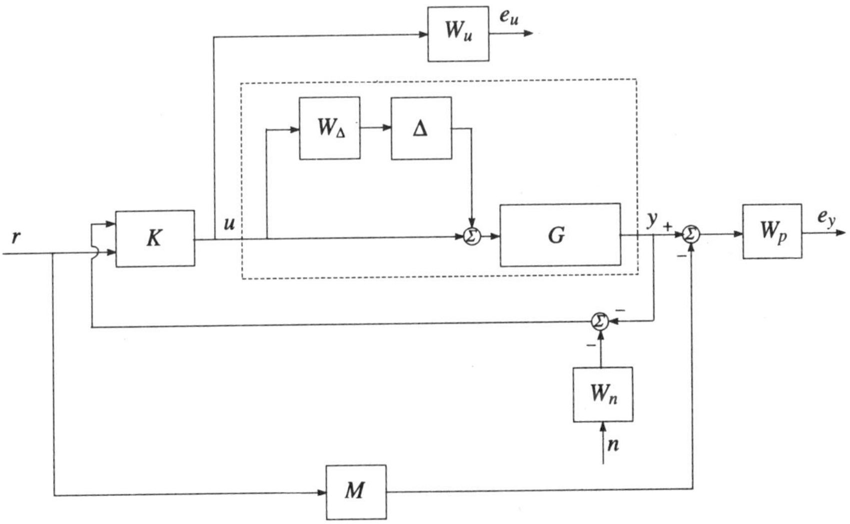

Fig. 11.5. Closed-loop interconnection structure of the distillation column system

The block diagram of the closed-loop system incorporating the design requirements consideration represented by weights is shown in Figure 11.5. The plant enclosed by the dashed rectangle consists of the nominal scaled model G plus the input multiplicative uncertainty. The controller K implements a feedback from outputs y_D and x_B and a feedforward from the reference signal r. The measurement of the distillate and bottom products composition is corrupted by the noise n. The desired dynamics of the closed-loop system is sought by implementation of a suitably chosen model M. The model M represents the desired dynamic behaviour of the closed-loop system from the reference signal to the outputs. The usage of a model of the desired dynamics allows us to take easily into account the design specifications.

The transfer function matrix of the model M is selected as

$$M = \begin{bmatrix} \frac{1}{Ts^2+2\xi Ts+1} & 0 \\ 0 & \frac{1}{Ts^2+2\xi Ts+1} \end{bmatrix}$$

The coefficients of the transfer functions ($T = 6, \ \ \xi = 0.8$) in both channels of the model are chosen such as to ensure an overdamped response with a settling time of about 30 min. The off-diagonal elements of the transfer matrix are set as zeros in order to minimise the interaction between the channels.

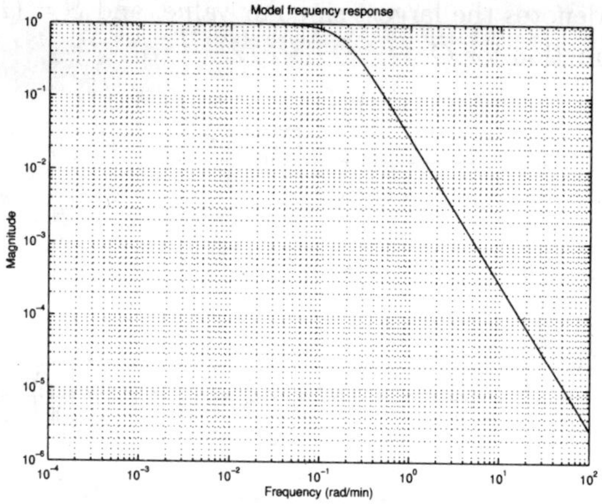

Fig. 11.6. Model frequency response

The frequency response of the model M is shown in Figure 11.6.

Let the scaled, two-degree-of-freedom controller be partitioned as

$$K(s) = [K_y(s) \ K_r(s)]$$

where K_y is the feedback part of the controller and K_r is the prefilter part. It is easy to show that

$$\begin{bmatrix} e_p \\ e_u \end{bmatrix} = \begin{bmatrix} W_p(S\tilde{G}K_r - M) & -W_pTW_n \\ W_u(I + K_y\tilde{G})^{-1}K_r & -W_uK_ySW_n \end{bmatrix} \begin{bmatrix} r \\ n \end{bmatrix}$$

where $S = (I+\tilde{G}K_y)^{-1}$ is the sensitivity function for the scaled plant, $T = (I+\tilde{G}K_y)^{-1}\tilde{G}K_y$ is the complementary sensitivity function and $\tilde{G} = G(I+W_\Delta \Delta)$ is the uncertain, scaled plant model.

The performance objective is to satisfy

$$\left\| \begin{bmatrix} W_p(S\tilde{G}K_r - M) & -W_pTW_n \\ W_u(I+K_y\tilde{G})^{-1}K_r & -W_uK_ySW_n \end{bmatrix} \right\|_\infty < 1 \qquad (11.1)$$

for each uncertain $\tilde{G}$.

The performance and control action weighting functions are chosen as

$$W_p = \begin{bmatrix} 0.55\frac{9.5s+3}{9.5s+10^{-4}} & 0.3 \\ 0.3 & 0.55\frac{9.5s+3}{9.5s+10^{-4}} \end{bmatrix}, \quad W_u = \begin{bmatrix} 0.87\frac{s+1}{0.01s+1} & 0 \\ 0 & 0.87\frac{s+1}{0.01s+1} \end{bmatrix}$$

The implementation of the performance weighting function W_p aims to ensure closeness of the system dynamics to the model over the low-frequency range. Note that this function contains nonzero off-diagonal elements that make it easier to meet the time-domain specifications. A small constant equal to 10^{-4} is added in the denominator in each channel to make the design problem regular.

The usage of the control weighting function W_u allows us to limit the magnitude of control actions over the specified frequency range ($\omega \geq 150$).

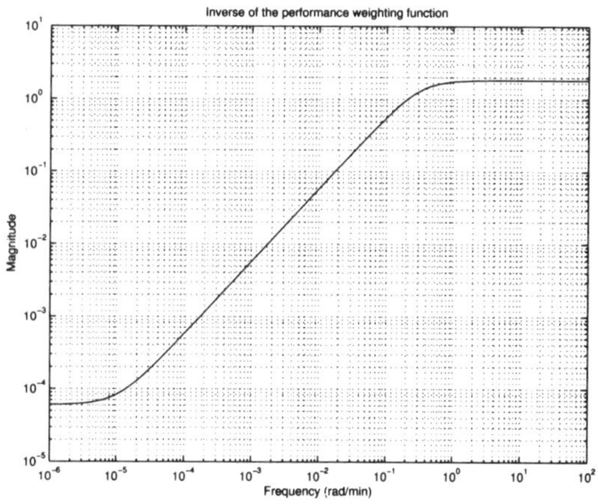

Fig. 11.7. Inverse of performance weighting function

The magnitude plot of the inverse of the performance weighting function W_p is shown in Figure 11.7 and the magnitude plot of the control weighting function is shown in Figure 11.8.

The noise shaping filter

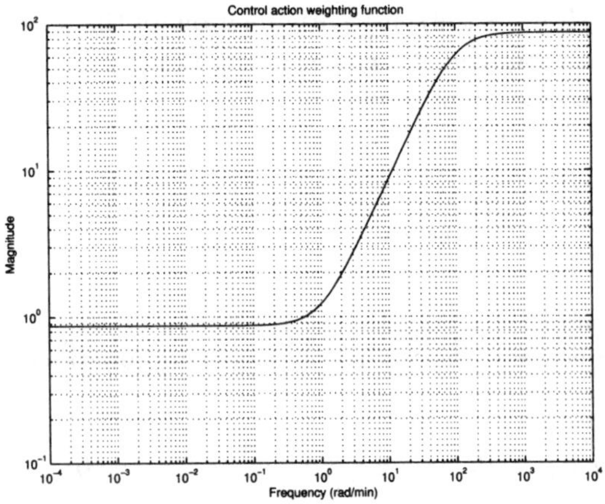

Fig. 11.8. Control-action weighting function

$$W_n = \begin{bmatrix} 10^{-2}\frac{s}{s+1} & 0 \\ 0 & 10^{-2}\frac{s}{s+1} \end{bmatrix}$$

is determined according to the spectral contents of the sensor noises accompanying the measurement of the distillate and bottom product composition.

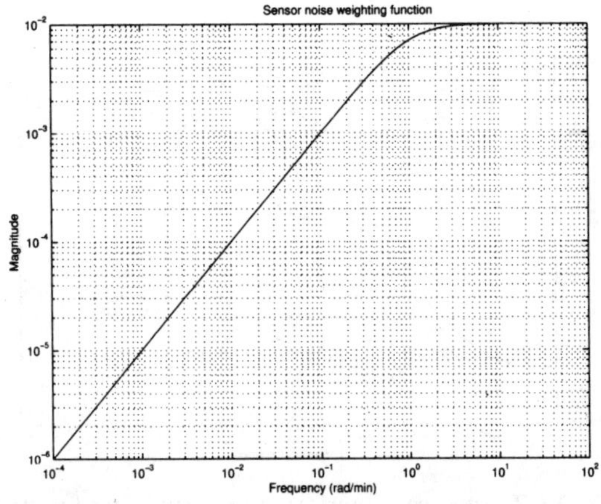

Fig. 11.9. Noise weighting function

The magnitude plot of the noise shaping filter is shown in Figure 11.9.

The model transfer function, the performance and control weighting functions as well as the noise shaping filter are all set in the file `wts_col.m`.

11.5 Open-loop and Closed-loop System Interconnections

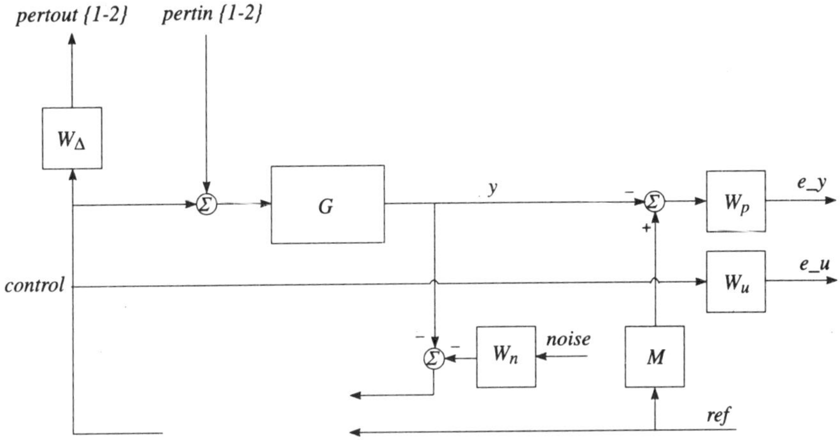

Fig. 11.10. Open-loop interconnection structure of the distillation column system

The open-loop system interconnection is obtained by the M-file `olp_col`. The internal structure of the eight-input, ten-output open-loop system, which is saved as the variable `sys_ic`, is shown in Figure 11.10. The inputs and outputs of the uncertainties are saved as the variables `pertin` and `pertout`, the references and the noises – as the variables `ref` and `noise`, respectively, and the controls – as the variable `control`.

All variables have two elements (*i.e.* 2-dimensional vectors).

The schematic diagram showing the specific input/output ordering for the variable `sys_ic` is given in Figure 11.11.

The block-diagram used in the simulation of the closed-loop system is shown in Figure 11.12. The corresponding closed-loop system interconnection, which is saved as the variable `sim_ic`, is obtained by the M-file `sim_col.m`.

The schematic diagram showing the specific input/output ordering for the variable `sim_ic` is shown in Figure 11.13.

11.6 Controller Design

Successful design of the distillation column control system may be obtained by using the $\mathcal{H}_\infty$ loop-shaping design procedure (LSDP) and the μ-synthesis. Note that in the case of LSDP we do not use the performance specifications implemented in the case of μ-synthesis. Instead of these specifications we use a prefilter W_1 and a postfilter W_2 in order to shape appropriately the open-loop transfer function $W_1 G W_2$.

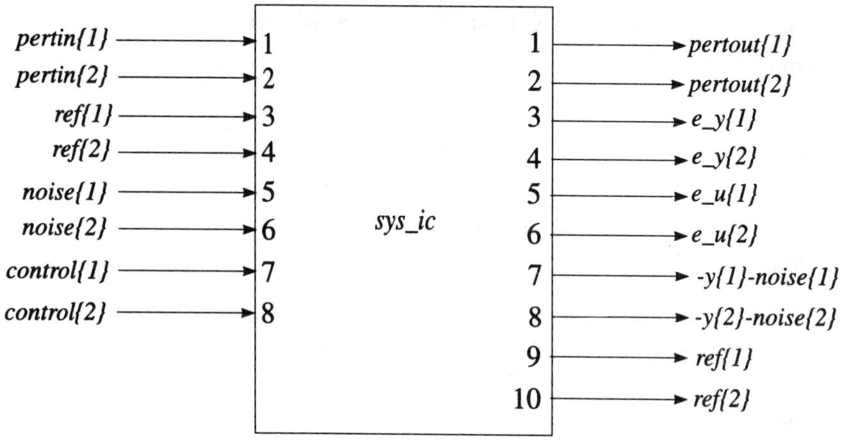

Fig. 11.11. Schematic diagram of the open-loop interconnection

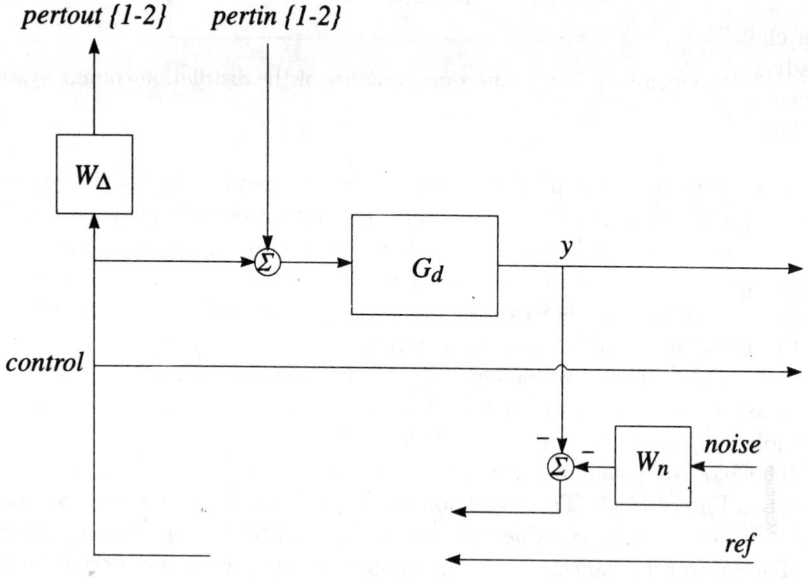

Fig. 11.12. Closed-loop interconnection structure of the distillation column system

11.6.1 Loop-shaping Design

In the present case, we choose a prefilter with transfer function

$$W_1 = \begin{bmatrix} 1.7\frac{1.1s+1}{10s} & 0 \\ 0 & 1.7\frac{1.1s+1}{10s} \end{bmatrix}$$

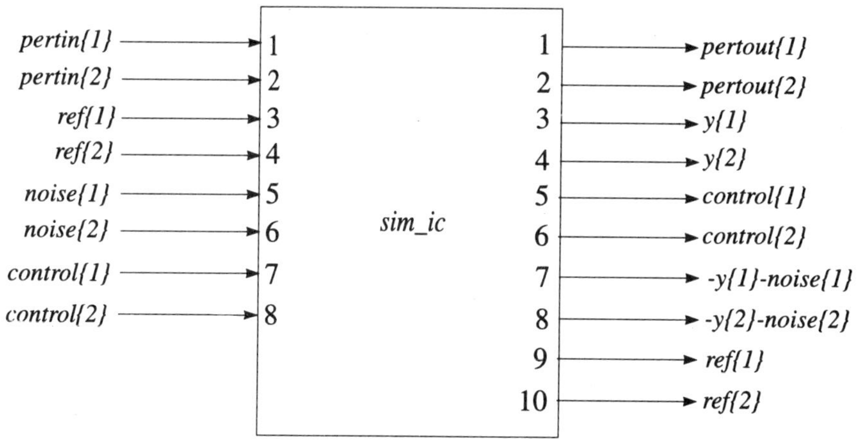

Fig. 11.13. Schematic diagram of the closed-loop interconnection

The choice of the gain equal to 1.7 is done to ensure a sufficiently small steady-state error. Larger gain leads to smaller steady-state errors but worse transient response.

The postfilter is taken simply as $W_2 = I_2$.

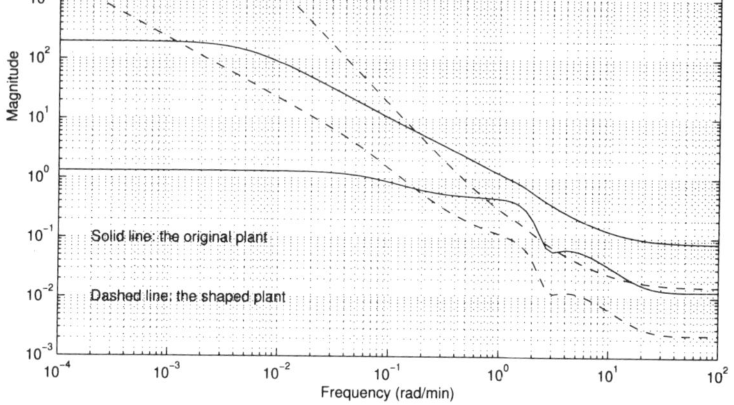

Fig. 11.14. Singular values of the original system and shaped system

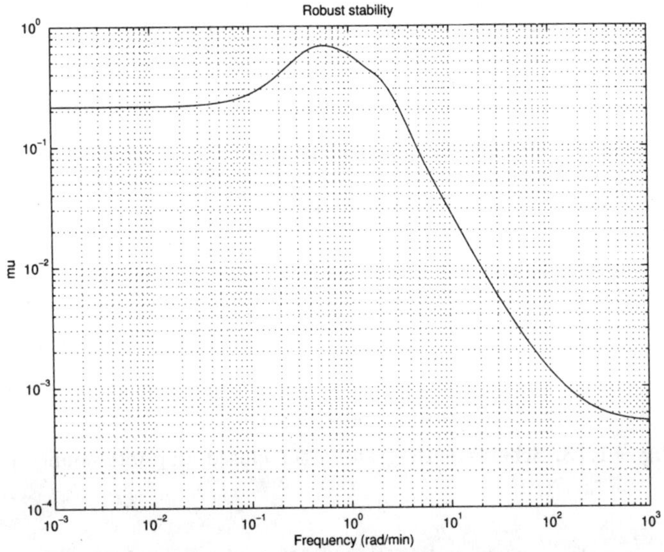

Fig. 11.15. Robust stability for loop-shaping controller

The singular value plots of the original and shaped systems are shown in Figure 11.14. The design of the two-degree-of-freedom LSDP controller is done by using the M-file `lsh_col.m` that implements the function `ncfsyn`. The controller obtained is of order 10.

The robust stability analysis of the closed-loop system is done by the file `mu_col` the frequency response plot of the structured value μ shown in Figure 11.15. According to this plot the closed-loop system preserves stability for all perturbations with norm less than 1/0.6814. As usual, the requirements for nominal performance and robust performance are not fulfilled with this controller.

The closed-loop frequency responses are obtained by using the file `frs_col.m`.

The singular value plot of the unscaled closed-loop system transfer function is shown in Figure 11.16. Both low-frequency gains are equal to 1 that ensures zero steady-state errors in both channels.

The singular value plots of the transfer function matrix with respect to the noises (Figure 11.17) show that the noises are attenuated by at least a factor of 10^4 times at the system output.

The singular-value plots of the transfer function matrices $\hat{G}\hat{K}_y$ and $\hat{K}_y\hat{S}$ are shown in Figures 11.18 and 11.19, respectively. The maximum of the largest singular value of $\hat{G}\hat{K}_y$ is less than 1 for $\omega \geq 150$ and the maximum of the largest singular value of $\hat{K}_y\hat{S}$ is less than 200 so that the corresponding frequency-domain specification is met.

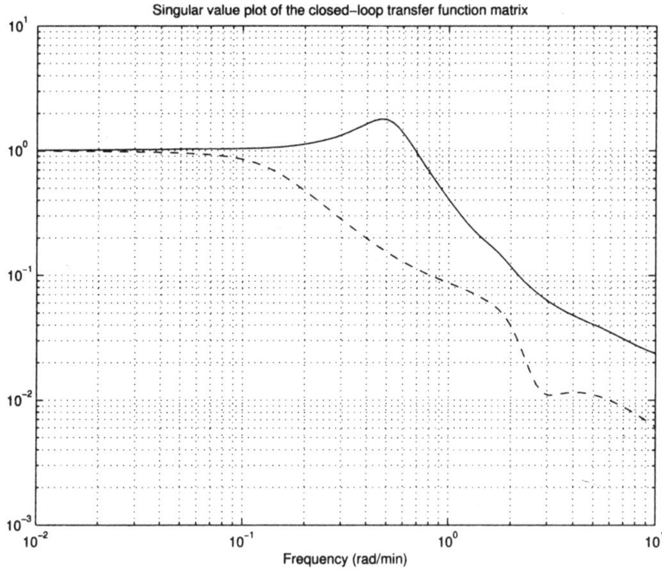

Fig. 11.16. Frequency response of the closed-loop system with loop-shaping controller

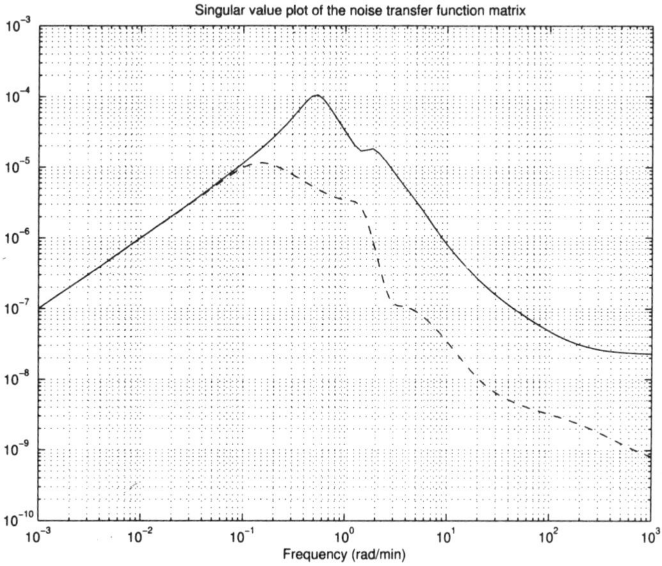

Fig. 11.17. Frequency response to the noises

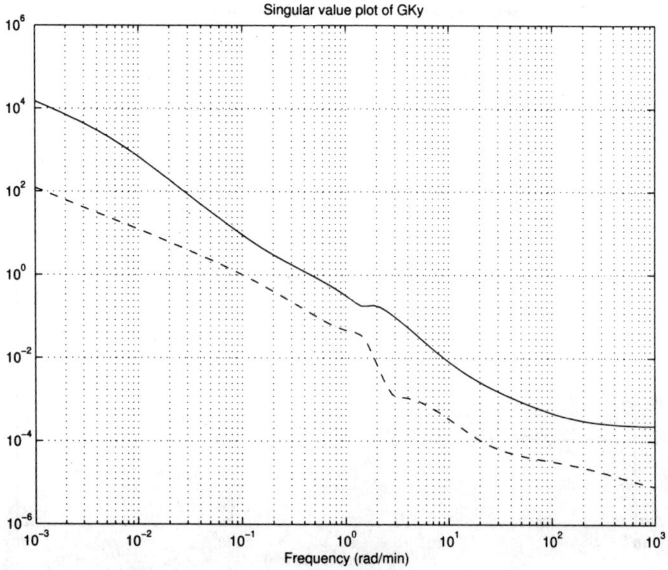

Fig. 11.18. Singular-value plot of $\hat{G}\hat{K}_y$

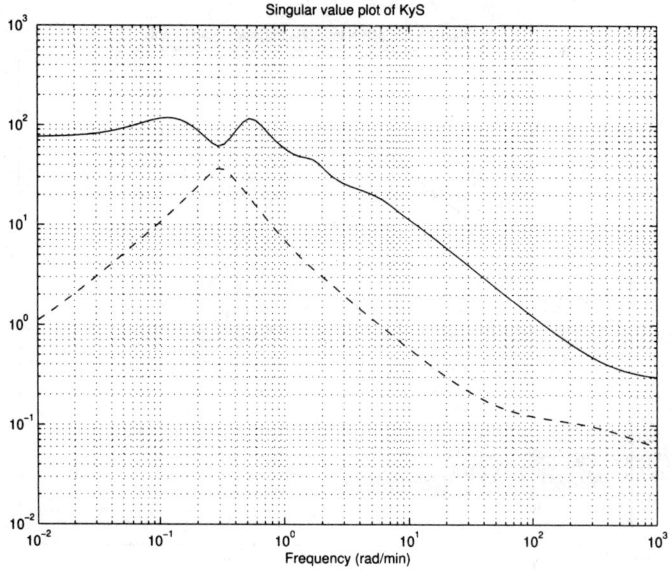

Fig. 11.19. Singular-value plot of $\hat{K}_y\hat{S}$

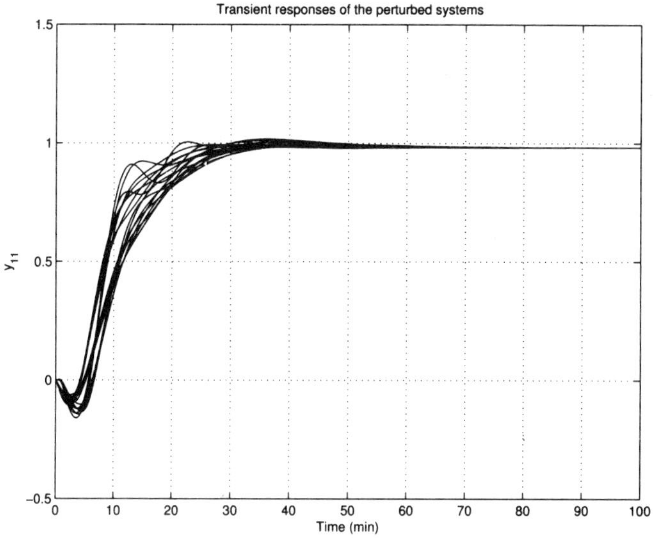

Fig. 11.20. Perturbed transient response y_{11} for loop-shaping controller

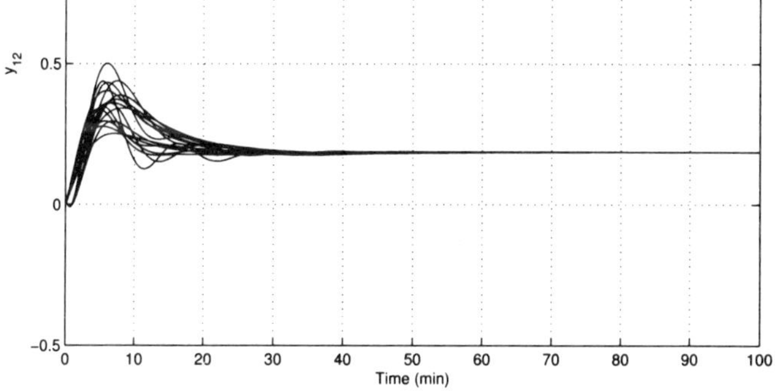

Fig. 11.21. Perturbed transient response y_{12} for loop-shaping controller

In Figures 11.20 – 11.23 we show the transient responses of the scaled closed-loop system obtained by the file `prtcol.m` for different values of the

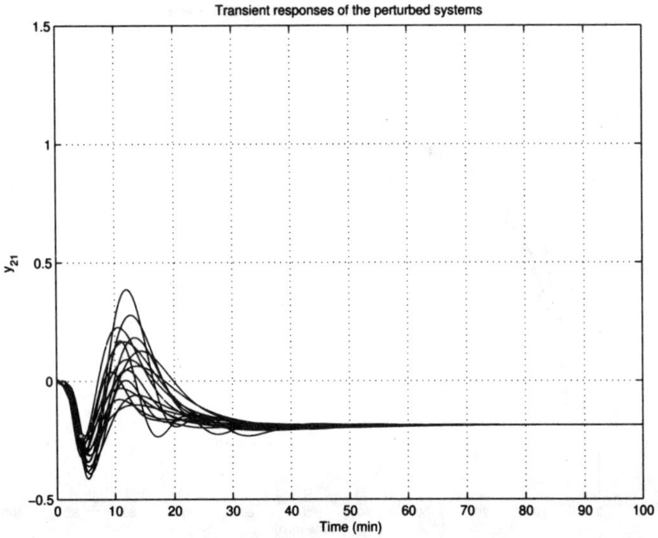

Fig. 11.22. Perturbed transient response y_{21} for loop-shaping controller

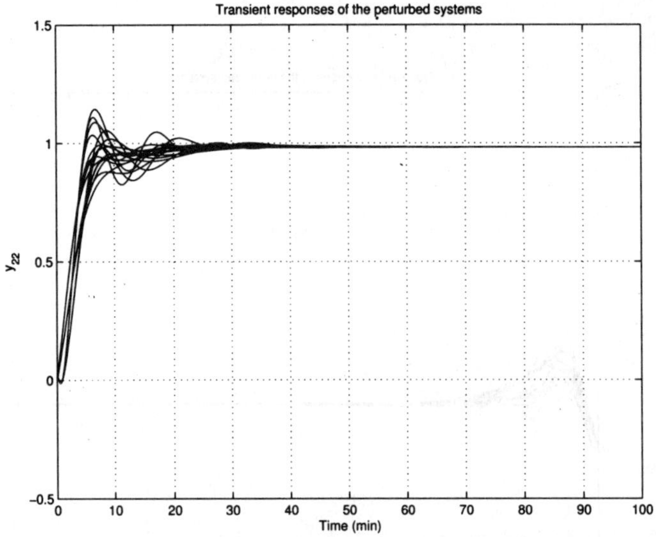

Fig. 11.23. Perturbed transient response y_{22} for loop-shaping controller

uncertain gain and time delay. The time-domain specification is met and the closed-loop system transient response has a small settling time.

The control action in the closed-loop system for the same variations of the uncertain parameters is shown in Figures 11.24 – 11.27.

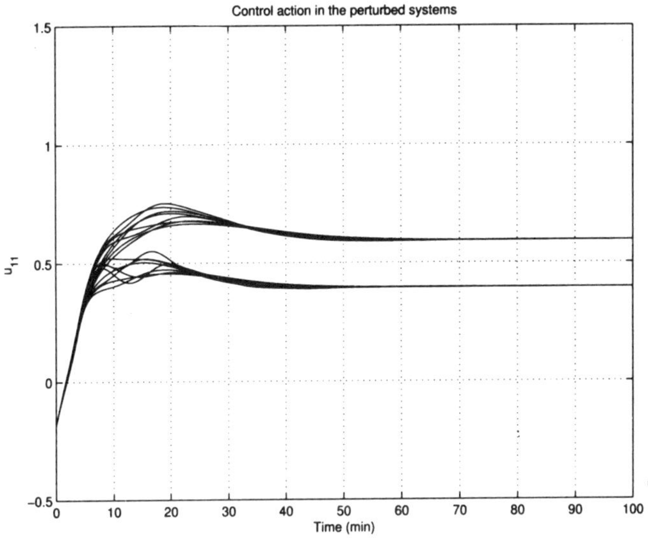

Fig. 11.24. Perturbed control action u_{11} for loop-shaping controller

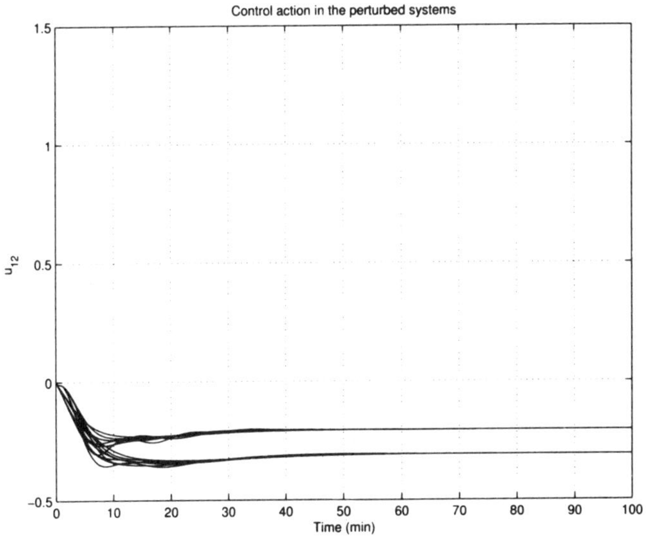

Fig. 11.25. Perturbed control action u_{12} for loop-shaping controller

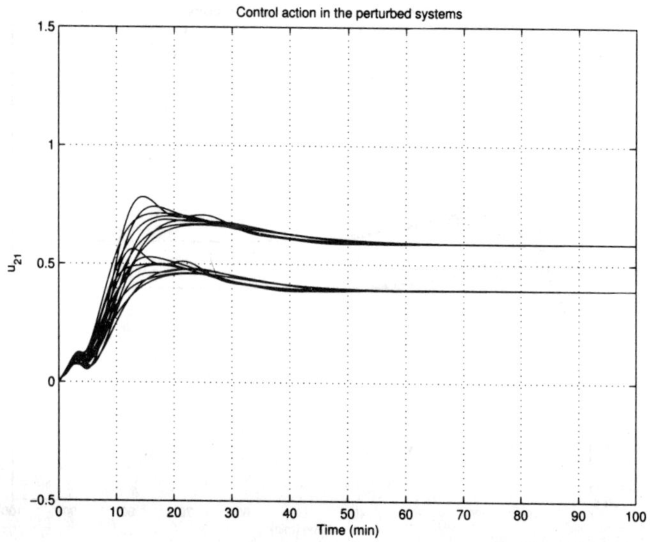

Fig. 11.26. Perturbed control action u_{21} for loop-shaping controller

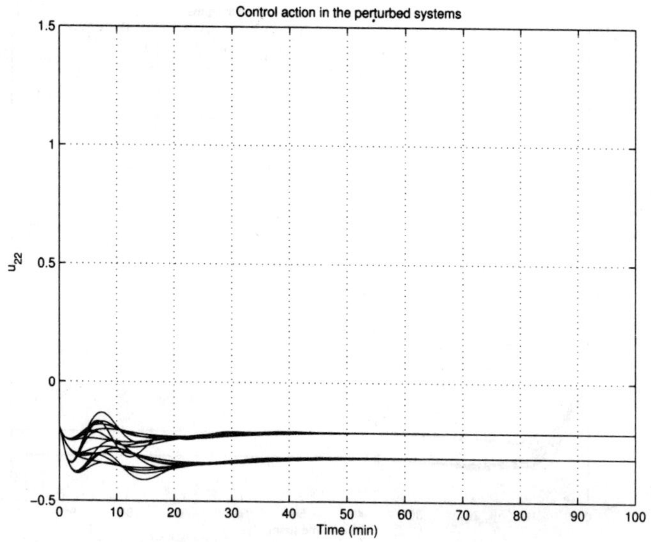

Fig. 11.27. Perturbed control action u_{22} for loop-shaping controller

11.6.2 μ-Synthesis

Let us denote by $P(s)$ the transfer function matrix of the eight-input, ten-output open-loop system consisting of the distillation column model plus the weighting functions and let the block structure Δ_P is defined as

$$\Delta_P := \left\{ \begin{bmatrix} \Delta & 0 \\ 0 & \Delta_F \end{bmatrix} : \Delta \in \mathcal{C}^{2\times 2},\ \Delta_F \in \mathcal{C}^{4\times 4} \right\}$$

The first block of this matrix corresponds to the uncertainty block Δ, used in modelling the uncertainty of the distillation column. The second block Δ_F is a fictitious uncertainty 4×4 block, introduced to include the performance objectives in the framework of the μ-approach. The inputs to this block are the weighted error signals e_p and e_u the outputs being the exogenous inputs r and n.

To meet the design objectives a stabilising controller K is to be found such that, at each frequency $\omega \in [0, \infty]$, the structured singular value satisfies the condition

$$\mu_{\Delta_P}[F_L(P, K)(j\omega)] < 1$$

The fulfillment of this condition guarantees robust performance of the closed-loop system, *i.e.*,

$$\left\| \begin{bmatrix} W_p(S\tilde{G}K_r - M) & -W_pTW_n \\ W_u(I + K_y\tilde{G})^{-1}K_r & -W_uK_ySW_n \end{bmatrix} \right\|_\infty < 1 \qquad (11.2)$$

The μ-synthesis is done by using the M-file `ms_col.m`. The uncertainty structure and other parameters used in the D-K iteration are set in the auxiliary file `dk_col.m`.

Table 11.3. Results of the μ-synthesis

Iteration	Controller order	Maximum value of μ
1	22	1.072
2	28	0.980
3	30	0.984
4	28	0.975

The progress of the D-K iteration is shown in Table 11.3.

In the given case an appropriate controller is obtained after the fourth D-K iteration. The controller is stable and its order is equal to 28.

It can be seen from Table 11.3 that after the fourth iteration the maximum value of μ is equal to 0.975.

The μ-analysis of the closed-loop system is done by the file `mu_col`.

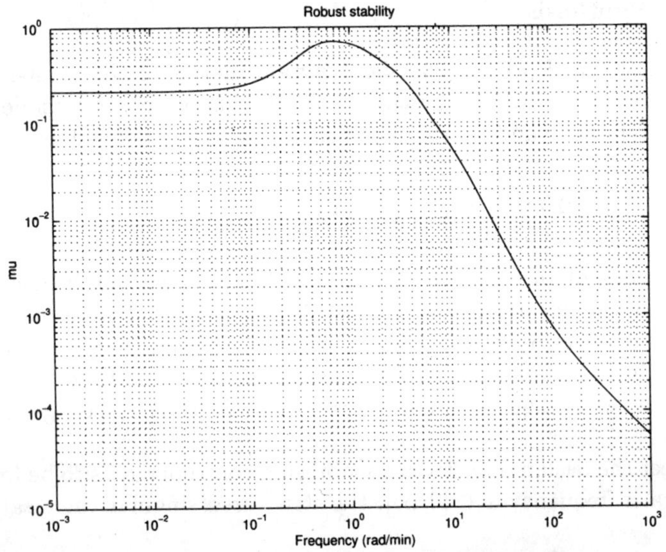

Fig. 11.28. Robust stability for μ-controller

The frequency-response plot of the structured singular value for the case of robust stability is shown in Figure 11.28. The maximum value of μ is 0.709,

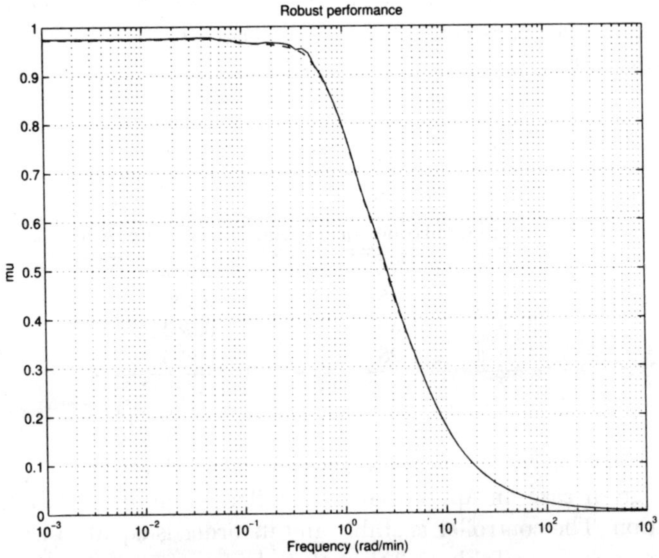

Fig. 11.29. Robust performance for μ-controller

which means that the stability of the system is preserved under perturbations that satisfy $\|\Delta\|_\infty < \frac{1}{0.709}$.

The frequency response of μ for the case of robust performance, analysis is shown in Figure 11.29. The closed-loop system achieves robust performance, the maximum value of μ being equal to 0.977.

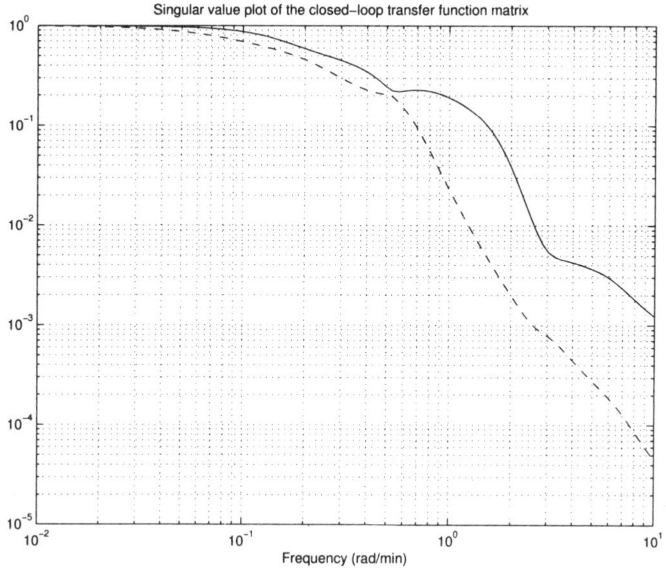

Fig. 11.30. Closed-loop singular-value plots

The unscaled closed-loop system singular-value plot is shown in Figure 11.30. The closed-loop bandwidth is about 0.1 rad/min.

The frequency responses with respect to the noise are shown in Figure 11.31. It is seen from the figure that the noises in measuring the distillate and bottom-product composition have a relatively small effect on the system output.

In Figure 11.32 we show the singular-value plot of the unscaled sensitivity function $\hat{S}$. The singular-value plots of the unscaled μ-controller are shown in Figure 11.33.

The singular-value plots of $\hat{G}\hat{K}_y$ and $\hat{K}_y\hat{S}$ are shown in Figures 11.34 and 11.35, respectively. The maximum of the largest singular value of $\hat{G}\hat{K}_y$ is less than 1 for $\omega \geq 150$ and the maximum of the largest singular value of $\hat{K}_y\hat{S}$ is less than 300, thus the frequency-domain specification is met.

Consider now the effect of variations of uncertain parameters on the system dynamics.

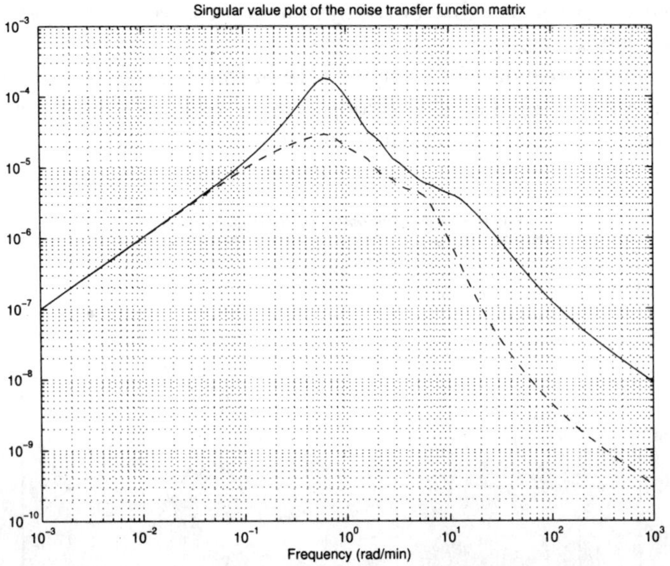

Fig. 11.31. Frequency responses with respect to noises

The frequency responses of the perturbed sensitivity function $\hat{S}$ obtained by the file `pfr_col.m` are shown in Figure 11.36.

The frequency responses of the perturbed transfer function matrix $\hat{K}_y\hat{S}$ are shown in Figure 11.36. The maximum of the largest singular value of this matrix does not exceed 300 for all values of the uncertain parameters.

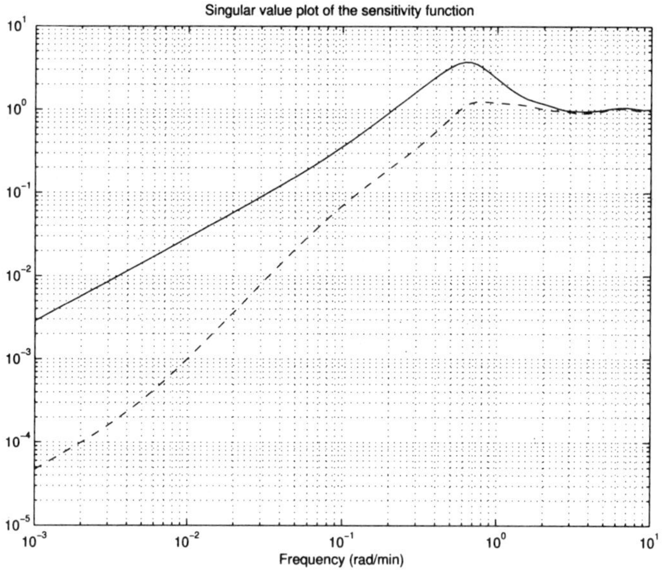

Fig. 11.32. Frequency responses of the sensitivity function

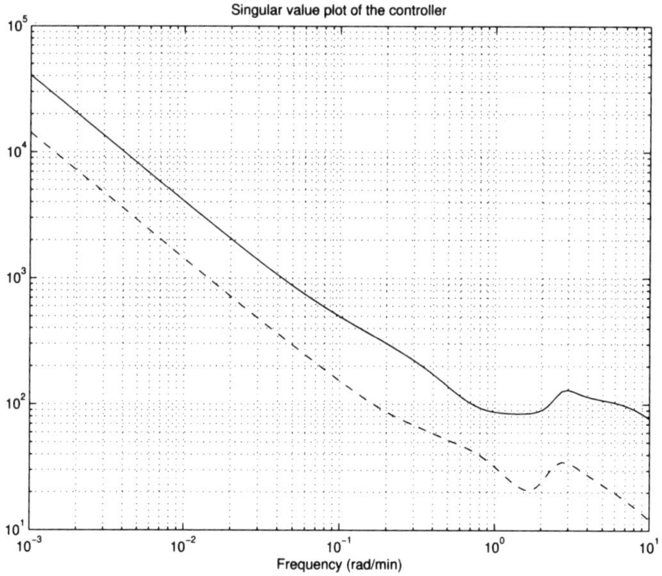

Fig. 11.33. Singular values of the controller

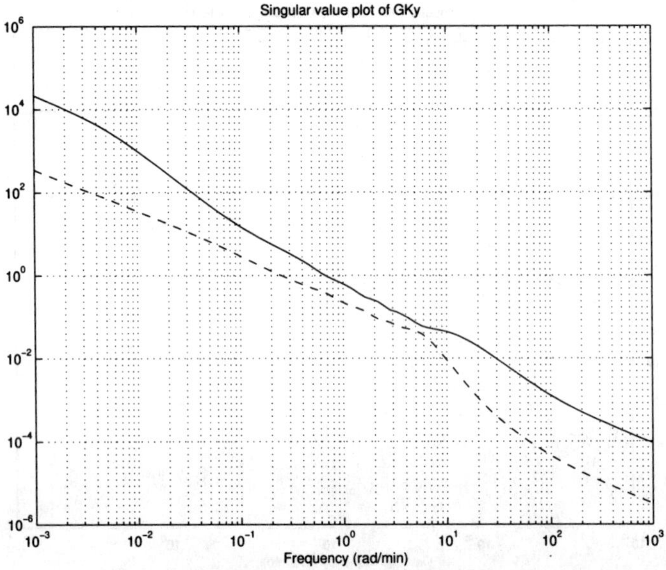

Fig. 11.34. Frequency responses of $\hat{G}\hat{K}_y$

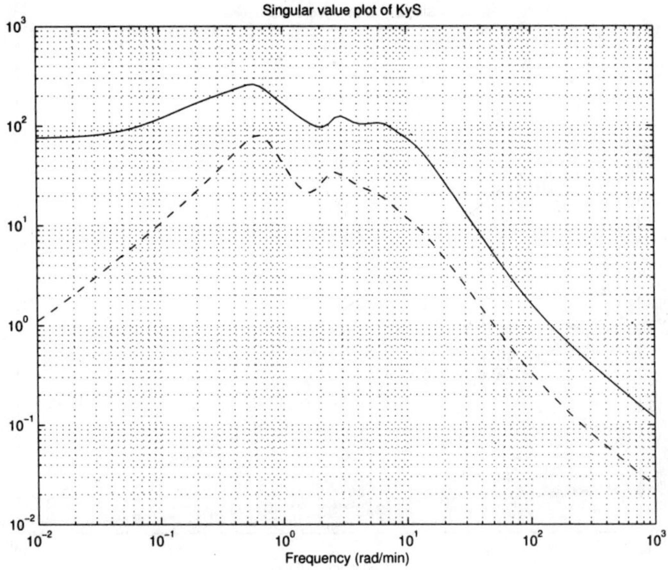

Fig. 11.35. Frequency responses of of $\hat{K}_y\hat{S}$

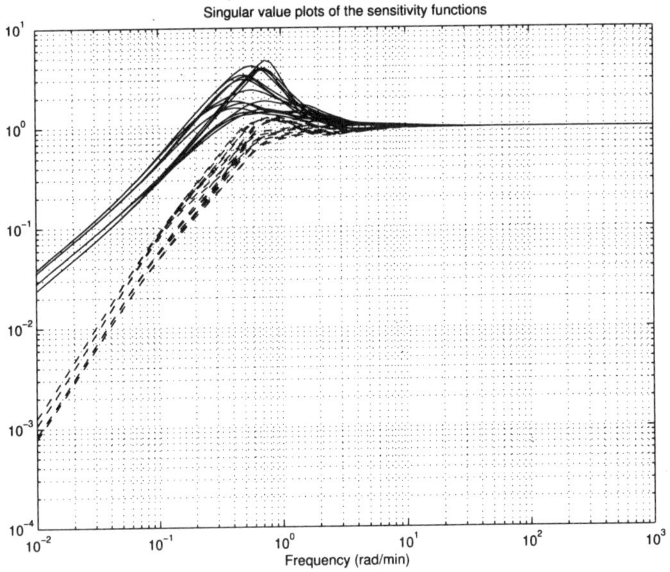

Fig. 11.36. Frequency responses of the perturbed sensitivity function

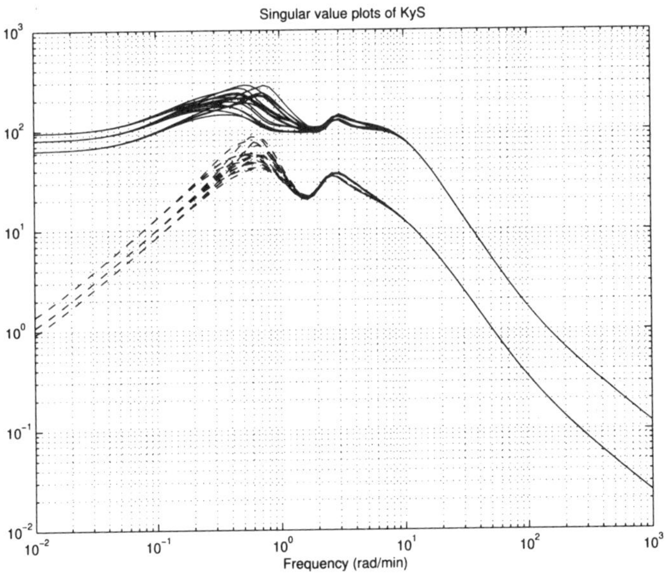

Fig. 11.37. Perturbed frequency responses of $\hat{K}_y\hat{S}$

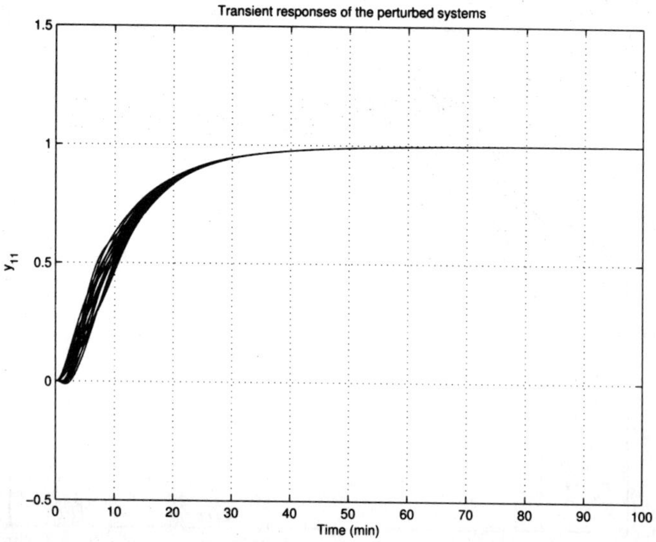

Fig. 11.38. Perturbed transient response y_{11} for μ-controller

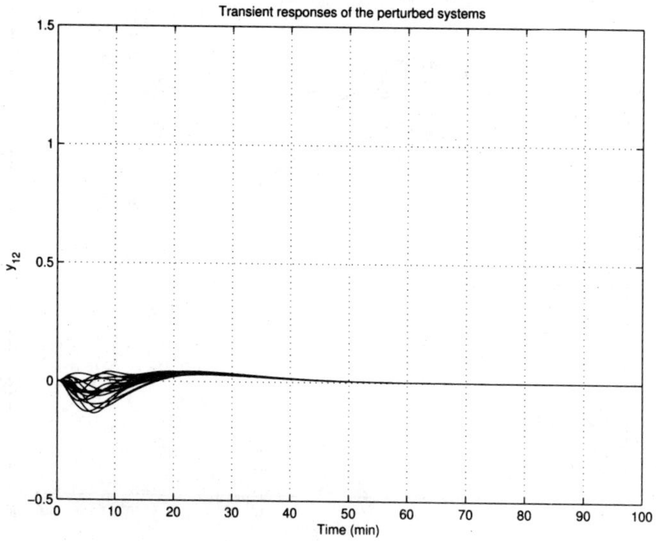

Fig. 11.39. Perturbed transient response y_{12} for μ-controller

The perturbed transient responses of the scaled closed-loop system with a μ-controller are shown in Figures 11.38 – 11.41. The responses to the cor-

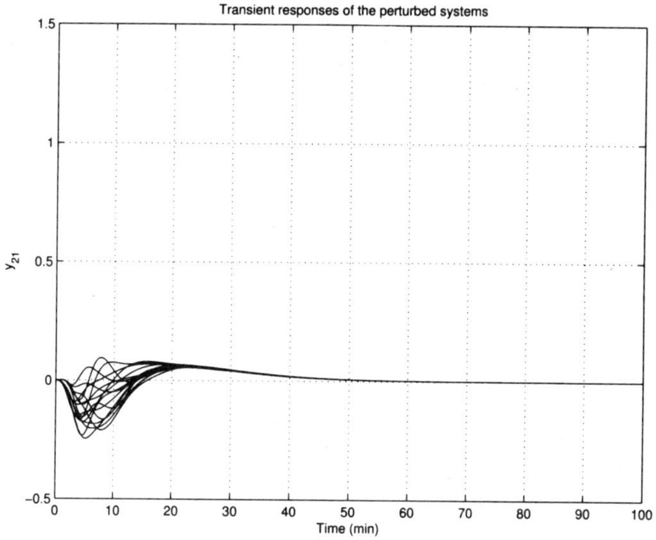

Fig. 11.40. Perturbed transient response y_{21} for μ-controller

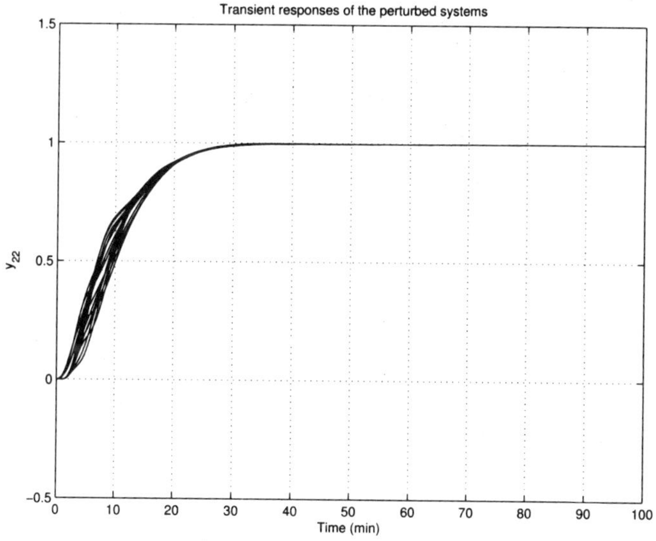

Fig. 11.41. Perturbed transient response y_{22} for μ-controller

responding references have no overshoots and the interaction of channels is weaker than in the case of using loop-shaping controller.

The control actions in the case of perturbed system with the μ-controller is shown in Figures 11.42 – 11.45.

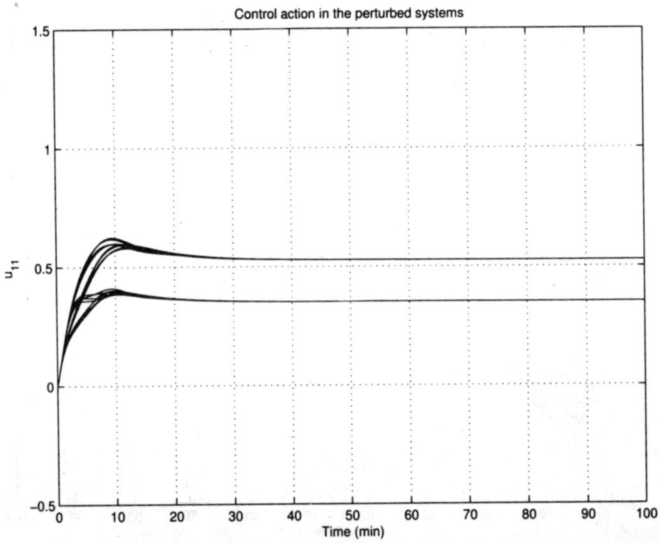

Fig. 11.42. Perturbed control action u_{11} for μ-controller

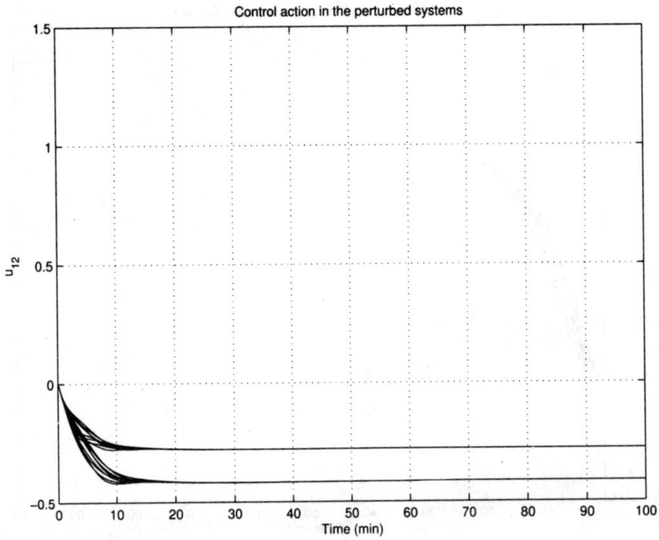

Fig. 11.43. Perturbed control action u_{12} for μ-controller

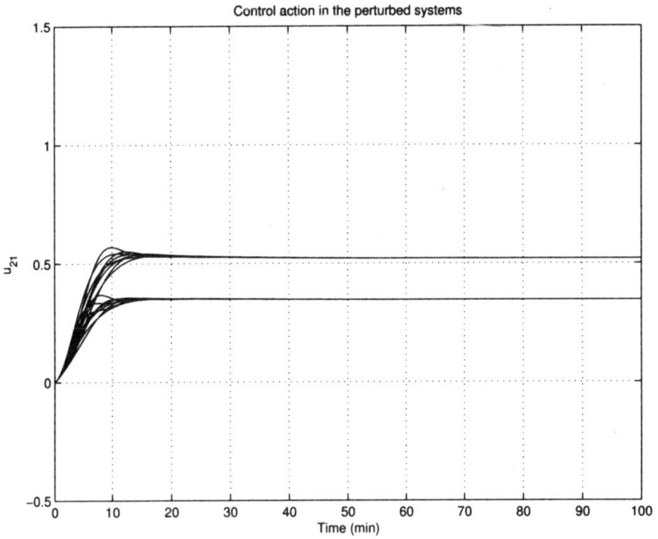

Fig. 11.44. Perturbed control action u_{21} for μ-controller

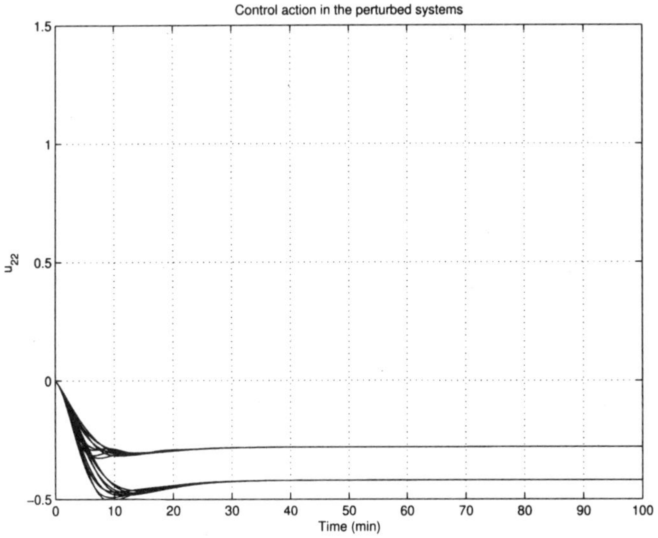

Fig. 11.45. Perturbed control action for μ-controller

Consider now the reduction of controller order. For this aim we implement the M-file `red_col.m`. After balancing of the controller and neglecting the small Hankel singular values its order is reduced to 11.

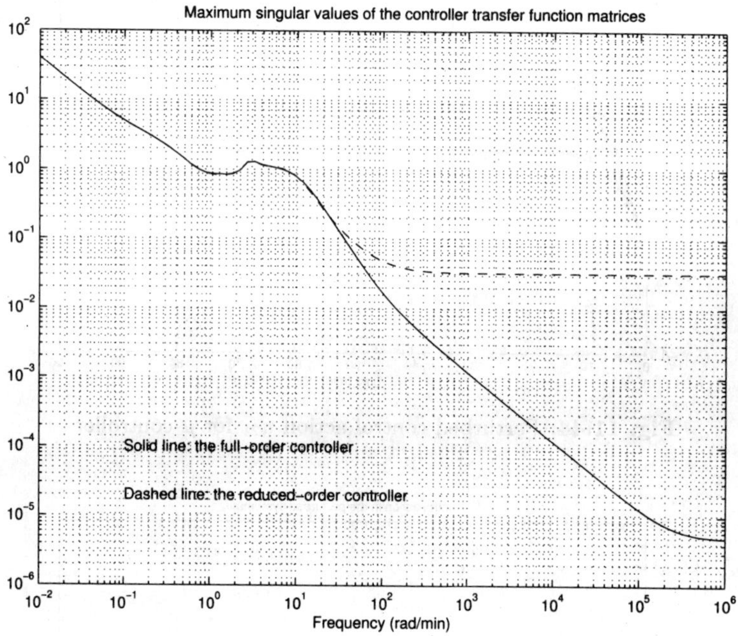

Fig. 11.46. Frequency responses of the full-order and reduced-order controllers

In Figure 11.46 we compare the frequency responses of the maximum singular values of the scaled full-order and reduced-order controllers. The frequency responses of both full-order and reduced-order controllers coincide up to 23 rad/min that is much more than the closed-loop bandwidth of the system. This is why the transient responses of the closed-loop system with full-order and with reduced-order controllers are practically undistinguishable.

11.7 Nonlinear System Simulation

The LSDP controller and μ-controller designed are investigated by simulation of the corresponding nonlinear closed-loop system. The simulation is carried out by the Simulink® model `nls_col.mdl` that implements the nonlinear plant model given in Section 11.2. To simulate the nonlinear plant we use the M-files `colamod` and `colas` by kind permission of the author, Sigurd Skogestad.

The Simulink® model of the distillation column control system shown in Figure 11.47 allows us to carry out a number of simulations for different set points and disturbances. Note that the inputs to the controller are formed as differences between the values of the corresponding variables and their nominal (steady-state) values used in the linearisation. In contrast, the controller outputs are added to the corresponding nominal inputs in order to obtain the full inputs to the nonlinear model of the column.

Before simulation of the system it is necessary to set the model parameters by using the M-file `init_col.m`. Also, the controller is rescaled so as to implement the unscaled input/output variables.

The nonlinear system simulation is done for the following reference and disturbance signals. At $t = 10$ min the feed rate F increases from 1 to 1.2, at $t = 100$ min the feed composition z_F increases from 0.5 to 0.6 and at $t = 200$ min the set point in y_D increases from 0.99 to 0.995.

The time response of the distillate y_D for the case of the reduced-order μ-controller is given in Figure 11.48. It is seen from the figure that the disturbances are attenuated well and the desired set point is achieved exactly.

The time response of the bottom-product composition x_B for the same controller is given in Figure 11.49.

The simulation results show that the robust design method is appropriately chosen and confirm the validity of the uncertain model used.

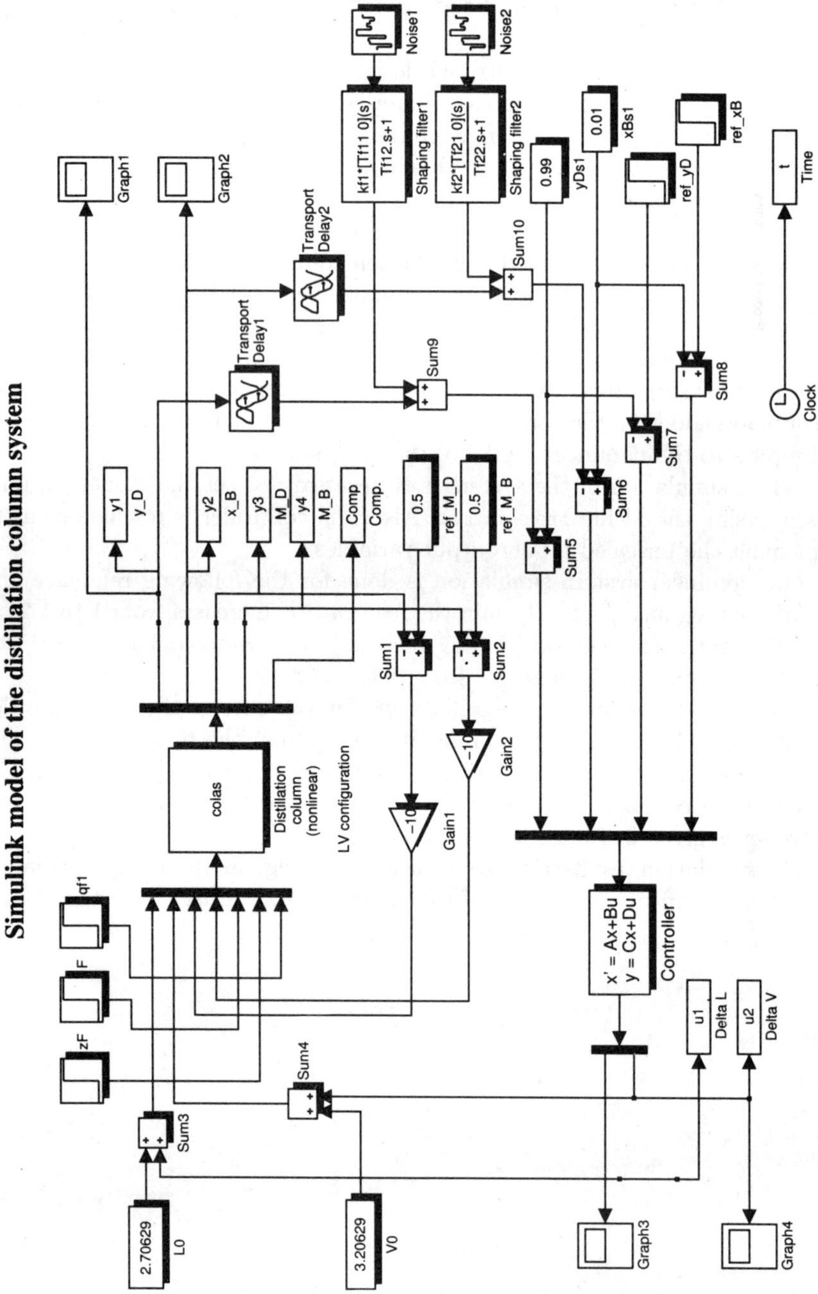

Fig. 11.47. Simulation model of the nonlinear system

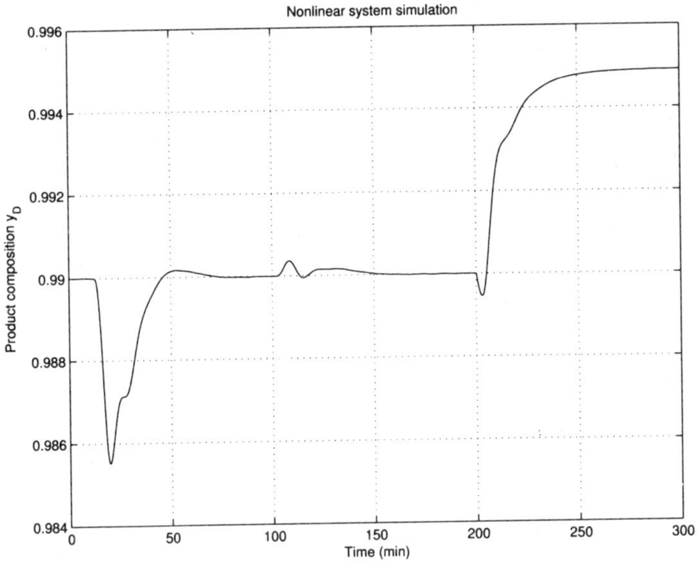

Fig. 11.48. Transient response of the nonlinear system - y_D

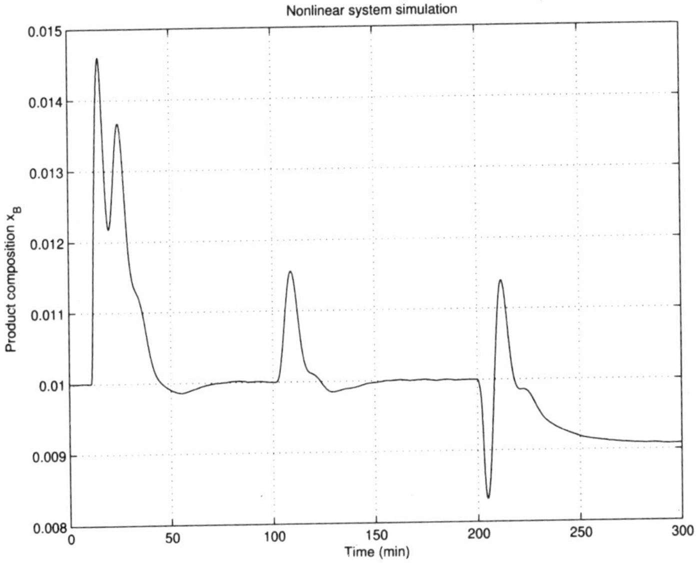

Fig. 11.49. Transient response of the nonlinear system - x_B

11.8 Conclusions

The results from the analysis and design of a distillation column control system may be summarised as follows.

- It is possible to use a sufficiently low-order linearised model of the given nonlinear plant, so that the designed linear controllers allow to be achieved satisfactory dynamics of the nonlinear closed-loop system. The linearised model is scaled in order to avoid very small or very large signals.
- The one-degree-of-freedom controller does not allow us to meet the time-domain and frequency-domain specifications, which makes it necessary to use two-degree-of-freedom controllers. Two controllers are designed – one by using the $\mathcal{H}_\infty$ loop-shaping design method and the other by using the μ-synthesis method. Both controllers satisfy the time-domain and frequency-domain specifications and ensure robust stability of the corresponding closed-loop systems. It is impressive how the low-order, easily designed loop-shaping controller allows us to obtain practically the same characteristics of the closed-loop systems as the μ-controller, while the latter requires much more experiments for tuning the weighting functions.
- The nonlinear system simulation results confirm the ability of the loop-shaping controller and the reduced-order μ-controller to achieve disturbance attenuation and good responses to reference signals. The simulation confirms the validity of the uncertain model used.

Notes and References

The distillation column control problem presented in this chapter was introduced by Limebeer [86] as a benchmark problem at the 1991 *Conference on Decision and Control.* In [86] the uncertainty is defined in terms of parametric gain and delay uncertainty and the control objectives are a mixture of time-domain and frequency-domain specifications. The problem originates from Skogestad *et al.* [141] where a simple model of a high-purity distillation column was used and uncertainty and performance specifications were given as frequency-dependent weighting functions. A tutorial introduction to the dynamics of the distillation column is presented in [140].

A design of a two-degree-of-freedom loop-shaping controller for the distillation column is presented in [53] where an 8th-order model of the column is used. A two-degree-of-freedom controller for the distillation column system is proposed in [95] with a reference model and using μ-synthesis. In that paper, one may find a selection procedure for the weighting functions described in details. Our design differs from the design in [95] in several respects. First, instead of a 2nd-order model with time delay we use a 6th-order model that is justified by the results from nonlinear system simulation. Second, we use modified weighting functions in order to obtain better results. In particular, we use a performance weighting transfer function matrix with nonzero off-diagonal elements that meets the time-domain specifications much better. Also, the control weighting functions are taken as first-order, low-pass filters.

Various design methods have been reported, in addition to the above, to tackle this distillation column problem ([127, 161, 147, 113, 142]). In [161], the design problem is formulated as a mixed optimisation problem. It is well known that control-system design problems can be formulated as constrained optimisation problems. Design specifications in both the time and frequency domains as well as stability can be naturally formulated as constraints. Numerical optimisation approaches can be used directly and a solution obtained, if there is one, will characterise an acceptable design. However, the optimisation problems so derived are usually very complicated with many unknowns, many nonlinearities, many constraints, and in most cases, they are multiobjective with several conflicting design aims that need to be simultaneously achieved. Furthermore, a direct parameterization of the controller will increase the complexity of the optimisation problem. In [161], the $\mathcal{H}_\infty$ loop-shaping design procedure is followed. Instead of direct parameterization of controllers, the pre- and postweighting functions used to shape the open-loop, augmented system are chosen as design (optimisation) parameters. The low order and simple structure of such weighting functions make the numerical optimisation much more efficient. The $\mathcal{H}_\infty$ norm requirement is also included in the cost/constraint set. The stability of the closed-loop system is naturally met by such designed controllers. Satisfactory designs are reported in that paper. Reference [147] further extends the optimisation approach in [161] by using a Genetic Algorithm to choose the weighting function parameters.

12

Robust Control of a Rocket

In this chapter we consider the design of a robust system for attitude stabilisation of a winged supersonic rocket flying at altitudes between 1000 m and 10000 m. The plant to be controlled is timevariant, which makes the controller design difficult. To simplify the design, we first derive linearised equations of the longitudinal motion of the rocket. Variations of the aerodynamic coefficients of this motion are considered as parametric uncertainties in the design. In this study, both continuous-time and discrete-time μ-controllers are designed that are implemented for pitch and yaw control and ensure the desired closed-loop dynamics in the presence of uncertainty, disturbances and noises. Robust stability and robust performance of the closed-loop systems with the implementation of each controller are investigated, respectively, and the nonlinear closed-loop, sampled-data system simulation results are given.

12.1 Rocket Dynamics

We consider a winged rocket that has a canard aerodynamic configuration and is equipped with a solid propellant engine. The actuators of the attitude-stabilization system in the longitudinal and lateral motions are four control surfaces (fins) that may rotate in pairs about their axes. The roll angle stabilisation is realised by auxiliary surfaces (ailerons). The control aim is to ensure accurate tracking of required acceleration maneuvers in the presence of uncertainties in the aerodynamic characteristics, disturbances (wind gusts) and sensor noises. The controller produces the inputs to two servo-actuators that rotate the fins. The normal acceleration in each plane and the pitch (yaw) rate are measured, respectively, by an accelerometer and a rate gyro, and are feedback signals to the controller.

To describe the rocket motion in space we shall need three orthogonal reference frames: the vehicle-carried vertical reference frame, the body-fixed reference frame and the flight-path reference frame. These three reference frames all have their origin in the rocket's mass centre.

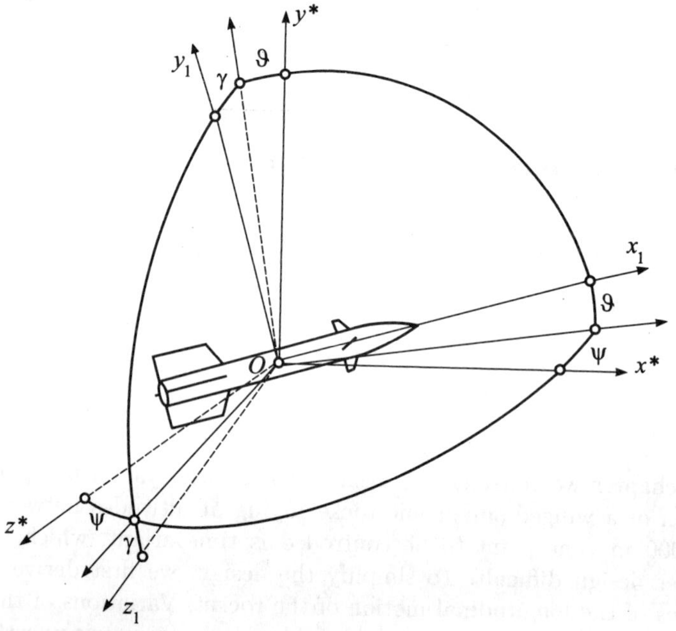

Fig. 12.1. Relationship between the vehicle-carried vertical reference frame and the body-fixed reference frame

The x^*-axis of the vehicle-carried vertical reference frame is directed to the North, the y^*-axis to the East, and the z^*-axis points downwards along the local direction of the gravity. The x_1-axis of the body-fixed reference frame is directed towards to the nose of the rocket, the y_1-axis points to the top wing, and the z_1-axis points to the right wing. In Figure 12.1 we show the relationship between the vehicle-carried vertical reference frame $O^*x^*y^*z^*$ and the body-fixed reference frame $Ox_1y_1z_1$. The rocket attitude (*i.e.* the position of the frame $Ox_1y_1z_1$ with respect to $O^*x^*y^*z^*$) is characterised by the angles ϑ, ψ, γ that are called pitch angle, yaw angle, and roll angle, respectively. The value of the roll angle in the nominal motion is usually very small.

The x-axis of the flight-path reference frame is aligned with the velocity vector V of the rocket and the y-axis lies in the plane Ox_1y_1. The relationship between the body-fixed reference frame and the flight-path reference frame is shown in Figure 12.2. The orientation of the body-fixed reference frame with respect to the flight-path reference frame is determined by the angle of attack α and the sideslip angle β.

Finally, the relationship between the vehicle-carried vertical reference frame $O^*x^*y^*z^*$ and the flight-path reference frame $Oxyz$ is shown in Figure 12.3. The position of the frame $Oxyz$ with respect to $O^*x^*y^*z^*$ is determined

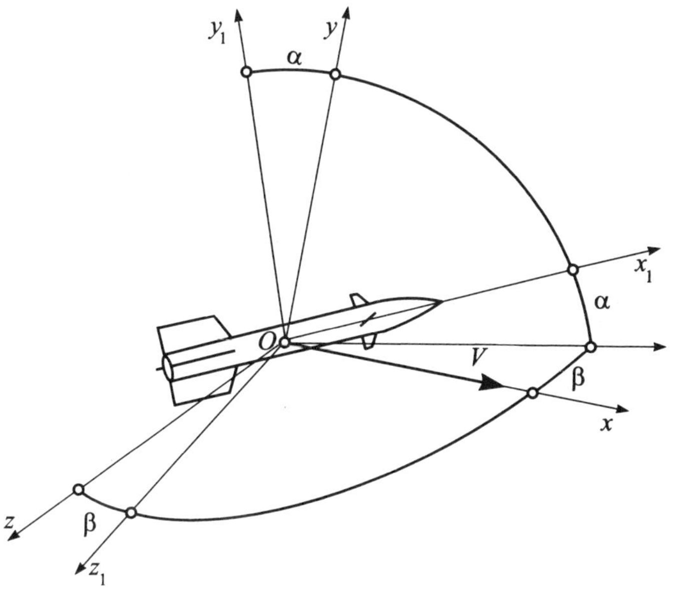

Fig. 12.2. Relationship between the body-fixed reference frame and the flight-path reference frame

by the flight-path angle Θ, the bank angle Ψ and the aerodynamic angle of roll γ_c.

The characteristic points along the longitudinal body axis of the rocket are shown in Figure 12.4, where

x_G is the coordinate of the mass centre of the rocket;

x_C is the coordinate of the aerodynamic centre of pressure (the point where the aerodynamic forces are applied upon);

x_R is the coordinate of the fins rotation axis.

The rocket will be statically stable or unstable depending on the location of the centre of pressure relative to the centre of mass. If $x_G < x_C$ then the rocket is statically stable.

The control of the lateral acceleration of the rocket is carried out in the following way. The moments about the mass centre, due to the fins deflections, create the corresponding angle of attack and sideslip angle. These angles, in turn, lead to the lifting forces and accelerations in the corresponding planes. The control problem consists of generation of the fins deflections by the autopilot that produce angle of attack and sideslip angle, corresponding to a maneuver called for by the guidance law, while stabilising the rocket rotational motion.

The nonlinear differential and algebraic equations describing the six degree-of-freedom motion of the rocket are as follows.

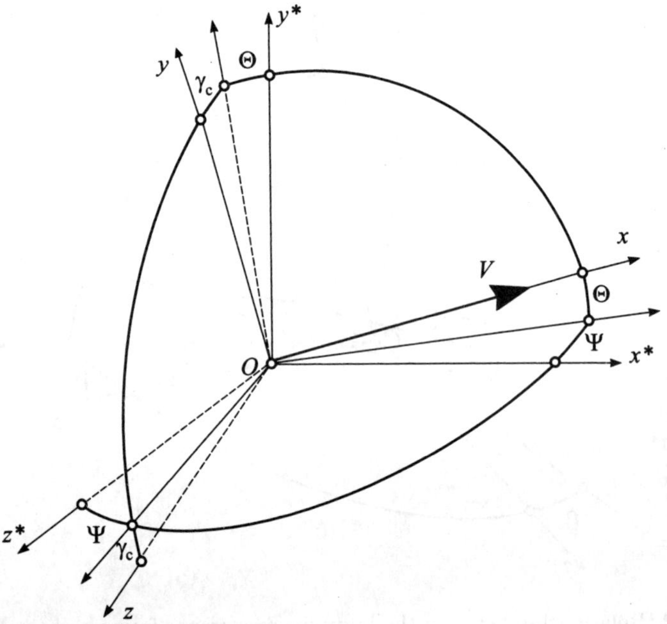

Fig. 12.3. Relationship between the vehicle-carried vertical reference frame and the flight-path reference frame

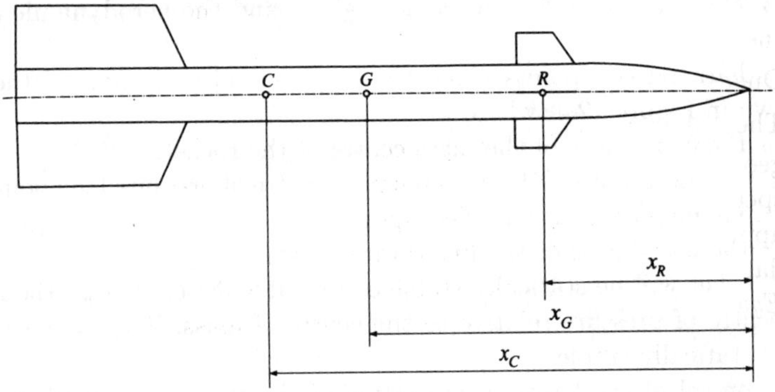

Fig. 12.4. Coordinates of the characteristic points of the body

1. *Equations describing the motion of the mass centre*

$$
\begin{aligned}
m\dot{V} &= P\cos\alpha\cos\beta - Q - G\sin\Theta + F_x(t) \\
mV\dot{\Theta} &= P(\sin\alpha\cos\gamma_c + \cos\alpha\sin\beta\sin\gamma_c) \\
&\quad + Y\cos\gamma_c - Z\sin\gamma_c - G\cos\Theta + F_y(t) \qquad (12.1) \\
-mV\cos\Theta\dot{\Psi} &= P(\sin\alpha\sin\gamma_c - \cos\alpha\sin\beta\cos\gamma_c)
\end{aligned}
$$

$$+Y \sin \gamma_c + Z \cos \gamma_c + F_z(t)$$

In these equations, P is the engine thrust, Q is the drag force, Y and Z are the lift forces in directions y and z, respectively, $G = mg$ is the rocket weight ($g = 9.80665$ m/s^2 is the acceleration of gravity at sea level), $F_x(t)$, $F_y(t)$, $F_z(t)$ are generalised disturbance forces in directions x, y, z, respectively. Also, m is the mass of the rocket in the current moment of the time.

The engine thrust is given by $P = P_0 + (p_0 - p)S_a$, where P_0 is the engine thrust at the sea level, $p_0 = 101325$ N/m^2 is the atmospheric pressure at the sea level, p is the pressure at the flight altitude, and S_a is the area of the engine nozzle output section. It is assumed that the mass consumption rate μ of the propellant remains constant, *i.e.* $P_0 = const.$

Further, let S and L be the reference area and the length of the rocket body, respectively and let $q = \frac{\rho V^2}{2}$ be the dynamic pressure, where ρ is the air density at the corresponding altitude. Then,

$Q = Q_a + Q_c$, where $Q_a = c_x q S$ is the drag force of the body and wings, $Q_c = (c_x^{\delta_y}|\delta_y| + c_x^{\delta_z}|\delta_z|)qS$ is the drag force due to the fins deflections, δ_y, δ_z are the angles of the fins deflection in the longitudinal and lateral motion, respectively, and c_x, $c_x^{\delta_y}$, $c_x^{\delta_z}$ are dimensionless coefficients;

$Y = Y_a + Y_c$, where $Y_a = c_y q S$ is the lift force of the body and wings, $Y_c = c_y^{\delta_z}\delta_z q S$ is the lift force due to the deflection of the horizontal fins, $c_y = c_y^{\alpha}\alpha$ and c_y^{α} is a dimensionless coefficient. Similar expressions hold for the lift force Z: $Z = Z_a + Z_c$, $Z_a = c_z q S$, $Z_c = c_z^{\delta_y}\delta_y q S$, where $c_z = c_z^{\beta}\beta$. Due to the rocket symmetry $c_z^{\beta} = -c_y^{\alpha}$, $c_z^{\delta_y} = -c_y^{\delta_z}$.

The aerodynamic coefficients c_x, $c_x^{\delta_y}$, $c_x^{\delta_z}$, c_y^{α}, c_z^{β} depend on the rocket geometry as well as on the Mach number $M = \frac{V}{a}$, where a is the sound speed at the corresponding altitude. These coefficients are determined by approximate formulae and are defined more precisely using experimental data.

2. *Equations, describing the rotational motion about the mass centre*

$$\begin{aligned}
I_x\dot{\omega}_x + (I_z - I_y)\omega_y\omega_z &= M_x^f + M_x^d + M_x^c + M_x(t) \\
I_y\dot{\omega}_y + (I_x - I_z)\omega_z\omega_x &= M_y^f + M_y^d + M_y^c + M_y(t) \\
I_z\dot{\omega}_z + (I_y - I_x)\omega_x\omega_y &= M_z^f + M_z^d + M_z^c + M_z(t)
\end{aligned} \quad (12.2)$$

$$\begin{aligned}
\dot{\psi} &= (\omega_y \cos\gamma - \omega_z \sin\gamma)/\cos\theta \\
\dot{\theta} &= \omega_y \sin\gamma + \omega_z \cos\gamma \\
\dot{\gamma} &= \omega_x - \tan\theta(\omega_y \cos\gamma - \omega_z \sin\gamma)
\end{aligned} \quad (12.3)$$

In these equations, I_x, I_y, I_z are the rocket moments of inertia; M_x^f, M_y^f, M_z^f are aerodynamic moments due to the angle of attack α and the sideslip angle β; M_x^d, M_y^d, M_z^d are aerodynamic damping moments due to the roll, pitch and yaw rates ω_x, ω_z, ω_y, respectively; M_x^c, M_y^c, M_z^c are control moments due to the fins deflections δ_x, δ_y, δ_z, respectively, and $M_x(t)$, $M_y(t)$, $M_z(t)$ are generalised disturbance moments about corresponding axes. We have that

- $M_x^f = (m_x^\alpha \alpha + m_x^\beta \beta)qSL$; $M_y^f = m_y^\beta \beta qSL$; $M_z^f = m_z^\alpha \alpha qSL$, where m_x^α, m_x^β, m_y^β, m_z^α are dimensionless coefficients;
- $M_x^d = m_x^{\omega_x}\omega_x qSL^2/V$, $M_y^d = m_y^{\omega_y}\omega_y qSL^2/V$, $M_z^d = m_z^{\omega_z}\omega_z qSL^2/V$, where $m_x^{\omega_x}$, $m_y^{\omega_y}$, $m_z^{\omega_z}$ are dimensionless coefficients;
- $M_x^c = m_x^{\delta_x}\delta_x qSL$, $M_y^c = m_y^{\delta_y}\delta_y qSL$, $M_z^c = m_z^{\delta_z}\delta_z qSL$, where $m_x^{\delta_x}$, $m_y^{\delta_y}$, $m_z^{\delta_z}$ are dimensionless coefficients. For the given rocket configuration it is fulfilled that $M_y^c = Z_c(x_G - x_R)\cos\beta$, $M_z^c = Y_c(x_G - x_R)\cos\alpha$ so that $m_y^{\delta_y} = -c_z^{\delta_y}(x_G - x_R)/L\cos\beta$, $m_z^{\delta_z} = c_y^{\delta_z}(x_G - x_R)/L\cos\alpha$.

3. *Equations providing the relationships between the angles ψ, θ, γ, α, β, Ψ, Θ, γ_c*

$$
\begin{aligned}
\sin\beta &= (\sin\theta\sin\gamma\cos(\Psi - \psi) - \cos\gamma\sin(\Psi - \psi))\cos\Theta \\
&\quad - \cos\theta\sin\gamma\sin\Theta \\
\sin\alpha &= ((\sin\theta\cos\gamma\cos(\Psi - \psi) + \sin\gamma\sin(\Psi - \psi))\cos\Theta \\
&\quad - \cos\theta\cos\gamma\sin\Theta)/\cos\beta \\
\sin\gamma_c &= (\cos\alpha\sin\beta\sin\theta - (\sin\alpha\sin\beta\cos\gamma - \cos\beta\sin\gamma)\cos\theta)/\cos\Theta
\end{aligned}
\tag{12.4}
$$

4. *Equations providing the accelerations n_{x1}, n_{y1}, n_{z1} in the body-fixed reference frame*

$$
\begin{aligned}
n_{x1} &= n_x\cos\alpha\cos\beta + n_y\sin\alpha - n_z\cos\alpha\sin\beta \\
n_{y1} &= -n_x\sin\alpha\cos\beta + n_y\cos\alpha + n_z\sin\alpha\sin\beta \\
n_{z1} &= n_x\sin\beta + n_z\cos\beta
\end{aligned}
\tag{12.5}
$$

where the accelerations n_x, n_y, n_z in the flight-path reference frame are given by

$$
\begin{aligned}
n_x &= (P\cos\alpha\cos\beta - Q)/G \\
n_y &= (P\sin\alpha + Y)/G \\
n_z &= (-P\cos\alpha\sin\beta + Z)/G
\end{aligned}
$$

Note that the accelerations n_{y1}, n_{z1} are measured by accelerometers fixed with the axes y_1, z_1, respectively. For brevity, later instead of n_{y1}, n_{z1} we shall write n_y, n_z.

Equations (12.1) – (12.5) are used later in the simulation of the nonlinear rocket stabilization system.

For small deviations from the nominal(unperturbed) motion, the three-dimensional motion of the rocket can be decomposed with sufficient accuracy in three independent motions – pitch, yaw and roll. In the controller design, we consider only the pitch perturbation motion. In this case the rocket attitude is characterised by the angles Θ and θ (or, equivalently, α and θ). Due to the rocket symmetry the yaw stabilisation system is analogous to the pitch stabilisation system.

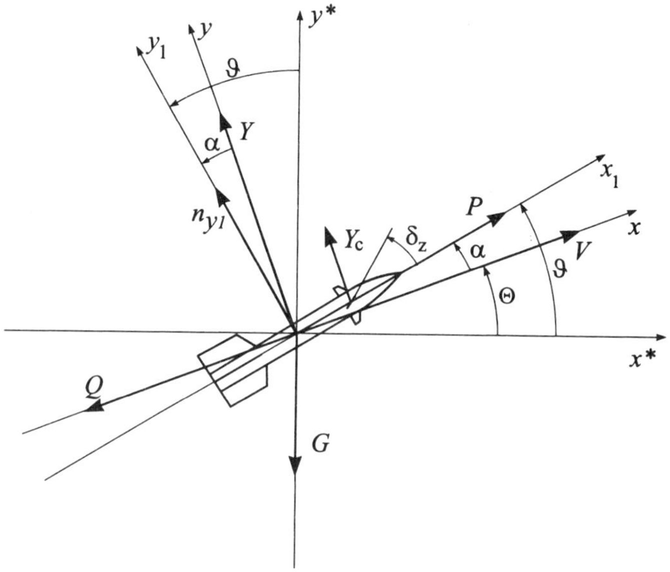

Fig. 12.5. Force diagram in the vertical plane

The equations describing the rocket's longitudinal motion have the form (Figure 12.5).

1. *Equations describing the motion of the mass centre*
   These equations are obtained from (12.1) by substituting $\beta = 0,\ \gamma_c = 0$.

$$m\dot{V} = P\cos\alpha - Q - G\sin\Theta + F_x(t) \tag{12.6}$$

$$mV\dot{\Theta} = P\sin\alpha + Y - G\cos\Theta + F_y(t) \tag{12.7}$$

2. *Equations, describing the rotational motion about the mass centre*
   These equations are obtained from (12.2) and (12.3) by substituting $\omega_x = 0,\ \omega_y = 0, \gamma = 0$.

$$I_z\dot{\omega}_z = M_z^f + M_z^d + M_z^c + M_z(t) \tag{12.8}$$
$$\dot{\theta} = \omega_z$$

3. *Equation giving the relationships between the angles α, θ, Θ*
   This equation is obtained from (12.4) for $\Psi = 0,\ \psi = 0$.

$$\alpha = \theta - \Theta \tag{12.9}$$

4. *Equation giving the normal acceleration n_y*
   This equation is obtained from (12.5) for $\beta = 0$.

$$n_y = -\frac{P\cos\alpha - Q}{G}\sin\alpha + \frac{P\sin\alpha + Y}{G}\cos\alpha \tag{12.10}$$

Later, it is assumed that the rocket velocity V (or Mach number) is constant and the nonlinear differential equation (12.6) associated with V is dropped from the design model. This assumption is justified by the results from the nonlinear system simulation presented later. Also, as is often customary, we shall neglect the effect of the gravitational force in (12.7).

In order to obtain a linear controller (12.7) – (12.10) are linearised about trim operating points ($\Theta = \Theta^o$, $\theta = \theta^o$, $\alpha = \alpha^o$, $\omega_z = \omega_z^o$, $\delta_z = \delta_z^o$) under the assumptions that the variations $\Delta\Theta = \Theta - \Theta^o$, $\Delta\theta = \theta - \theta^o, \Delta\alpha = \alpha - \alpha^o$, $\Delta\omega_z = \omega_z - \omega_z^o$, $\Delta\delta_z = \delta_z - \delta_z^o$ are sufficiently small. In such a case it is fulfilled that $\sin\Delta\alpha \approx \Delta\alpha$, $\cos\alpha \approx 1$.

As a result, the linearised equations of the perturbed motion of the rocket take the form (for convenience the symbol Δ is omitted)

$$\begin{aligned}
\dot{\Theta} &= \frac{P + Y^\alpha}{mV}\alpha + \frac{Y^{\delta_z}}{mV}\delta_z + \frac{F_y(t)}{mV}\\
\dot{\omega}_z &= \frac{M_z^\alpha}{J_z}\alpha + \frac{M_z^{\omega_z}}{J_z}\omega_z + \frac{M_z^{\delta_z}}{J_z}\delta_z + \frac{M_z(t)}{J_z}\\
\dot{\theta} &= \omega_z\\
\alpha &= \theta - \Theta\\
n_y &= \frac{Q + Y^\alpha}{G}\alpha + \frac{Y^{\delta_z}}{G}\delta_z
\end{aligned} \tag{12.11}$$

where $Y^\alpha = c_y^\alpha qS$, $Y^{\delta_z} = c_y^{\delta_z} qS$, $M_z^\alpha = m_z^\alpha qSL$, $M_z^{\omega_z} = m_z^{\omega_z} qSL^2/V$, $M_z^{\delta_z} = m_z^{\delta_z}\delta_z qSL$.

Equations (12.11) are represented as

$$\begin{aligned}
\dot{\Theta} &= a_{\Theta\Theta}\alpha + a_{\Theta\delta_z} + \overline{F}_y(t)\\
\ddot{\theta} &= a_{\theta\dot{\theta}}\dot{\theta} + a_{\theta\theta}\alpha + a_{\theta\delta_z}\delta_z + \overline{M}_z(t)\\
\alpha &= \theta - \Theta\\
n_y &= a_{n_y\alpha}\alpha + a_{n_y\delta_z}\delta_z
\end{aligned} \tag{12.12}$$

where

$$\begin{aligned}
a_{\Theta\Theta} &= \tfrac{P+Y^\alpha}{mV}, \quad a_{\Theta\delta_z} = \tfrac{Y^{\delta_z}}{mV}\\
a_{\theta\dot{\theta}} &= \tfrac{M_z^{\omega_z}}{J_z}, \quad a_{\theta\theta} = \tfrac{M_z^\alpha}{J_z}\\
a_{\theta\delta_z} &= \tfrac{M_z^{\delta_z}}{J_z}, \quad a_{n_y\alpha} = \tfrac{Q+Y^\alpha}{G}\\
a_{n_y\delta_z} &= \tfrac{Y^{\delta_z}}{G}
\end{aligned}$$

In (12.12) we used the notation

$$\overline{F_y(t)} = \frac{F_y(t)}{mV}$$

and

$$\overline{M}_z(t) = \frac{M_z(t)}{J}$$

Equations (12.12) are extended by the equation describing the rotation of the fins

$$\ddot{\delta}_z + 2\xi_{\delta_z}\omega_{\delta_z}\dot{\delta}_z + \omega_{\delta_z}^2\delta_z = \omega_{\delta_z}^2\delta_z^o \tag{12.13}$$

where δ_z^o is the desired angle of fin's deflection (the servo-actuator reference), ω_{δ_z} is the natural frequency and ξ_{δ_z} the damping coefficient of the servo-actuator.

The set of equations (12.12) and (12.13) describes the perturbed rocket longitudinal motion.

The block diagram of the stabilisation system, based on (12.12) and (12.13), is shown in Figure 12.6.

In the following we consider the attitude stabilisation of a hypothetical rocket whose parameters are given in Table 12.1.

Table 12.1. Rocket parameters

Parameter	Description	Value	Dimension
L	rocket length	1.6	m
d	rocket diameter	0.12	m
S	reference area	0.081	m^2
S_a	engine nozzle output section area	0.011	m^2
m_o	initial rocket mass	45	kg
μ	propellant consumption per second	0.3	kg/s
J_{x0}	initial rocket moment of inertia about x-axis	8.10×10^{-2}	kg m^2
J_{y0}	initial rocket moment of inertia about y-axis	9.64	kg m^2
J_{z0}	initial rocket moment of inertia about z-axis	9.64	kg m^2
P_0	engine thrust at the sea level	$740g$	N
x_{G0}	initial rocket mass centre coordinate	0.8	m
x_{C0}	initial rocket pressure centre coordinate	1.0	m
x_R	fins rotation axis coordinate	0.3	m
ω_{δ_z}	natural frequency of the servo-actuator	150	rad/s
ξ_{δ_z}	servo-actuator damping	0.707	
t_f	duration of the active stage of the flight	40	s

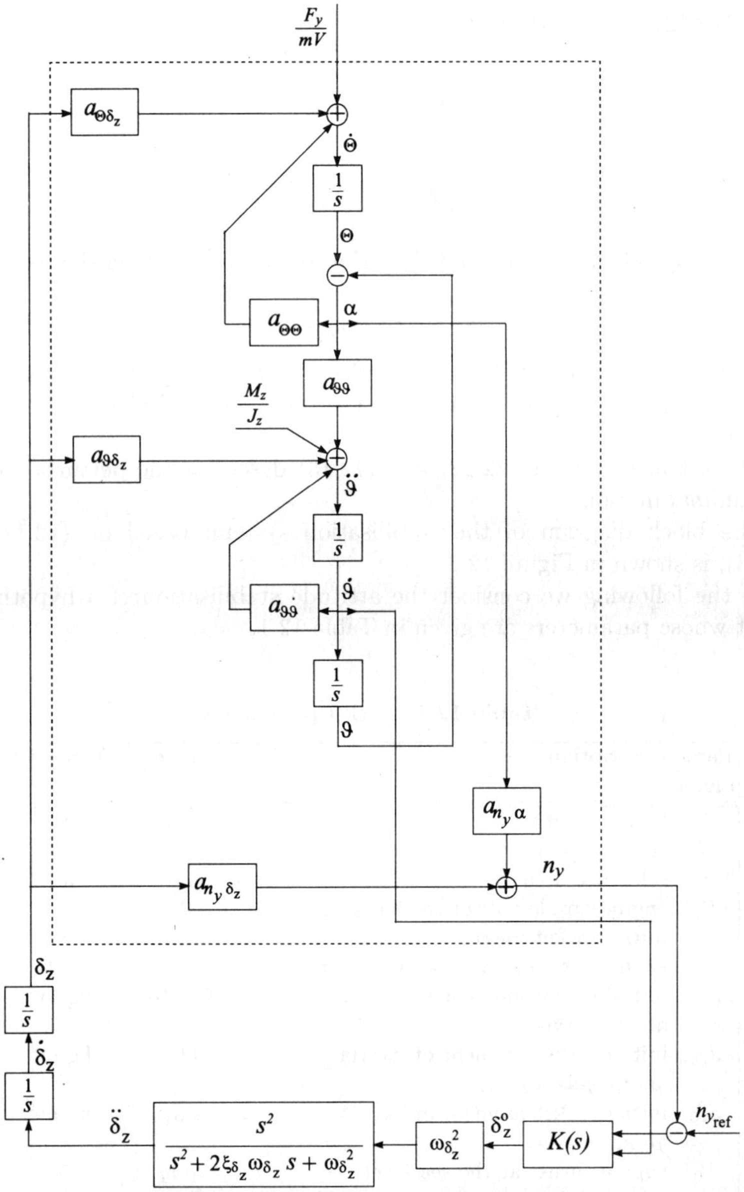

Fig. 12.6. Block diagram of the stabilisation system

During the flight the coefficients of the rocket motion equations vary due to the decreasing of the vehicle mass and also due to variation of the aerodynamic coefficients as a result of the velocity and altitude change.

The coefficients in the motion equations are to be determined for the nominal (unperturbed) rocket motion. Nominal values of the parameters are assumed in the absence of disturbance forces and moments.

The unperturbed longitudinal motion is described by the equations

$$m\dot{V}^* = P\cos\alpha^* - Q^* - mg_0\sin\Theta^*$$

$$mV^*\dot{\Theta} = P\sin\alpha^* + Y^* - mg_0\cos\Theta^*$$

$$\dot{H} = V^*\sin\Theta^*$$

$$\dot{m} = -\mu^*$$

where

$$\alpha^* = \theta^* - \Theta^*$$

$$\delta_z^* = -(m_z^\alpha/m_z^{\delta_z})\alpha^*$$

$$Q^* = c_x qS + c_x^{\delta_z}|\delta_z^*|qS$$

$$Y^* = c_y^\alpha\alpha^* qS + c_y^{\delta_z}\delta_z^* qS$$

$$q = \frac{\rho V^{*2}}{2},\ \rho = \rho(H)$$

$$c_x = c_x(M),\ c_y^\alpha = c_y^\alpha(M),\ c_y^{\delta_z} = c_y^{\delta_z}(M)$$

$$M = V^*/a^*,\ a^* = a^*(H)$$

$$m_z^\alpha = (c_x + c_y^\alpha)(x_G - x_C)/L,\ m_z^{\delta_z} = c_y^{\delta_z}(x_G - x_R)/L$$

$$\theta^* = \theta^*(t)$$

In these equations, $\theta^*(t)$ is the desired time program for changing the pitch angle of the vehicle.

We now consider the dynamics of the stabilisation system with coefficients calculated for the unperturbed motion in which the rocket performs a horizontal flight in the atmosphere, *i.e.* $\vartheta^* = 0^\circ$ with initial velocity $V^* = 300$ m/s and different altitude H^*. The parameters of the unperturbed motion are computed by numerical integration of the equations given above, taking into account the change of the vehicle mass during the flight, the dependence of the air density on the flight altitude and the dependence of the aerodynamic coefficients on the Mach number. The parameters of the unperturbed motion are obtained by the function `sol_rock` that invokes the function `dif_rock`, setting the differential equations, and the MATLAB® function `ode23tb` intended to solve stiff differential equations. The dependence of the aerodynamic coefficients on the Mach number for the specific vehicle is described in the files `cx_fct.m` and `cy_fct.m`. The flight parameters p, ρ and M are computed as functions of the altitude by the command `air_data` that implements an approximate model of the *international standard atmosphere(ISA)*. The time program of the pitch angle is set by the function `theta_rock`.

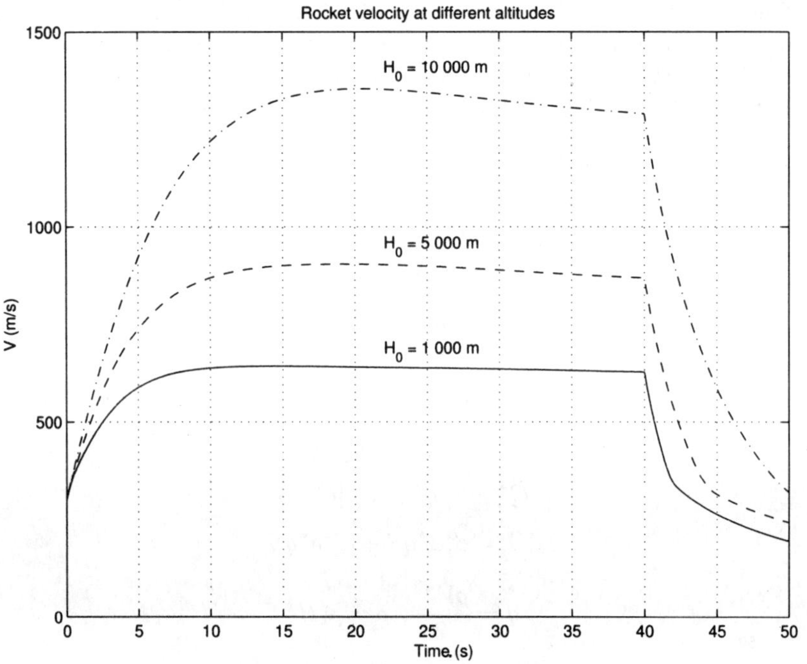

Fig. 12.7. Rocket velocity at different altitudes

The rocket velocity as a function of the time for initial velocity $V = 300$ m/s and flights at altitudes 1000, 5000 and 10000 m is shown in Figure 12.7. It is seen that between the 15th second and 40th second of the flight the velocity may be assumed constant.

The flight parameters corresponding to the unperturbed motion with initial altitude $H^* = 5000$ m are given in Table 12.2 for different time periods. The coefficients in the differential equations of the perturbed motion are computed on the basis of the parameters of unperturbed motion by using the function `cfn_rock`.

We consider the following as a rocket transfer function

$$G_{n_y\delta_z^o}(s) = \frac{n_y(s)}{\delta_z^o(s)}$$

In Figure 12.8 we show a family of magnitude frequency responses (logarithmic) of the vehicle for initial altitude $H_0 = 5000$ m and various time instants. The magnitude response of the open-loop system has a large peak and varies with time.

In the following sections, we will consider the controller design of the stabilisation system for the rocket dynamics at the 15th second of flight at an initial altitude $H^* = 5000$ m.

Table 12.2. Flight parameters for the unperturbed motion

t (s)	10	15	20	25	30	35	40
V (m/s)	869.8	902.4	907.2	905.0	901.5	898.0	894.7
M	2.710	2.809	2.822	2.814	2.801	2.789	2.778
H (m)	4873	4815	4763	4716	4674	4637	4604
x_G (m)	0.75	0.725	0.70	0.675	0.65	0.625	0.60

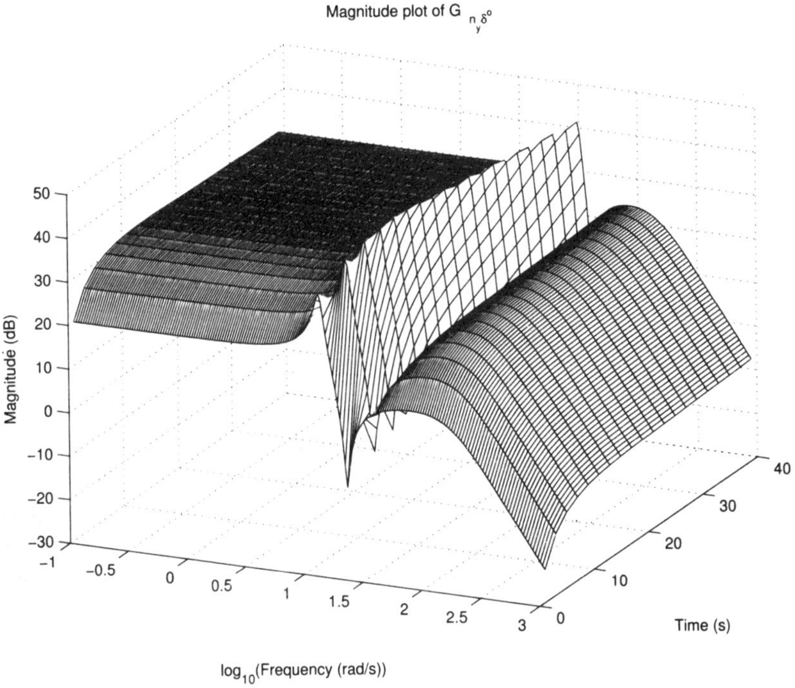

Fig. 12.8. Family of logarithmic magnitude frequency responses of the rocket

12.2 Uncertainty Modelling

The main variation of coefficients of perturbed motion happens in the aerodynamics coefficients c_x, c_y^α, m_z^α, $m_z^{\omega_z}$. As noted above, these coefficients are usually determined experimentally as functions of the Mach number and may vary in sufficiently wide intervals in practice. This is why it is supposed that the construction parameters of the vehicle are known exactly and the aerodynamic coefficients lead to 30% variations in the perturbed motion case. This uncertainty is actually much larger than the real one but it makes it possible for the controller designed to work satisfactorily under the changes of the rocket mass and velocity. In this way, the increased robustness of the closed-loop system may help overcome (to some extent) the timevariance of the plant parameters.

In deriving the uncertain model of the rocket dynamics we shall eliminate the angle Θ in the system (12.12) by using the relationship $\Theta = \theta - \alpha$. This allows us to avoid the use of the angle θ itself working only with its derivatives $\ddot{\theta}$ and $\dot{\theta}$. This, in turn, makes it possible to avoid the usage of an additional integrator in the plant dynamics that leads to violation of the conditions for controller existence in the $\mathcal{H}_\infty$ design. As a result, we obtain the equations

$$\begin{aligned} \dot{\alpha} &= -a_{\Theta\Theta}\alpha + \dot{\theta} - a_{\Theta\delta_z}\delta_z \\ \ddot{\theta} &= a_{\theta\dot{\theta}}\dot{\theta} + a_{\theta\theta}\alpha + a_{\theta\delta_z}\delta_z \\ n_y &= a_{n_y\alpha}\alpha + a_{n_y\delta_z}\delta_z \end{aligned} \tag{12.14}$$

The seven uncertain coefficients of the perturbed motion equations are given in Table 12.3.

Table 12.3. Uncertainty in rocket coefficients

Coefficients	Uncertainty range (%)
$a_{\Theta\Theta}$	30
$a_{\Theta\delta_z}$	30
$a_{\vartheta\dot{\vartheta}}$	30
$a_{\vartheta\vartheta}$	30
$a_{\vartheta\delta_z}$	30
$a_{n_y\alpha}$	30
$a_{n_y\delta_z}$	30

Each uncertain coefficient (c) may be represented in the form

$$c = \bar{c}(1 + p_c\delta_c)$$

where $\bar{c}$ is the nominal value of the coefficient c (at a given time instant), p_c is the relative uncertainty ($p_c = 0.3$ for uncertainty 30%) and $-1 \leq \delta_c \leq 1$. The uncertain coefficient $c = \bar{c}(1 + p_c\delta_c)$ may be further represented as an upper linear fractional transformation (LFT) in δ_c,

$$c = F_U(M_c, \delta_c)$$

where

$$M_c = \begin{bmatrix} 0 & \bar{c} \\ p_c & \bar{c} \end{bmatrix}$$

as is shown in Figure 12.9.

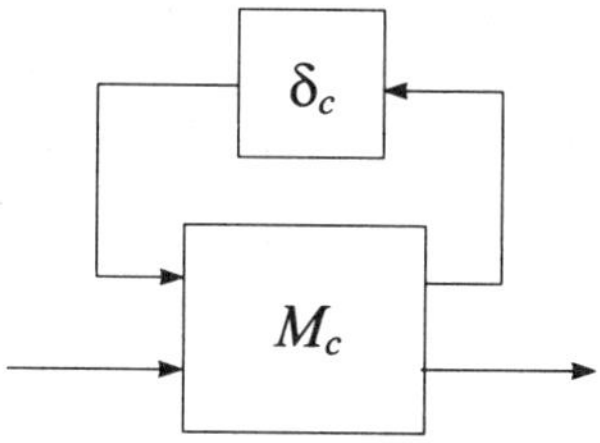

Fig. 12.9. Uncertain parameter as LFT

The uncertainty model corresponding to the system of (12.14) is difficult to obtain directly. This is why we shall derive the uncertain models corresponding to the individual equations in the system (12.14) and then we shall combine them in a common model.

Consider first the equation

$$\dot{\alpha} = -a_{\Theta\Theta}\alpha + \dot{\theta} - a_{\Theta\delta_z}\delta_z$$

The uncertain model corresponding to this equation is shown in Figure 12.10

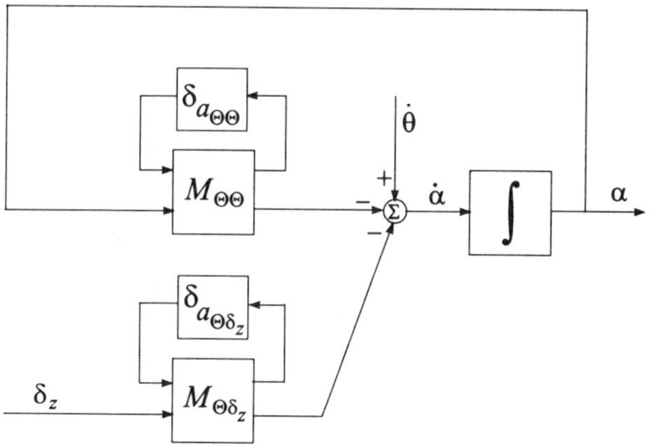

Fig. 12.10. Block diagram of the uncertain model for α

where

$$M_{\Theta\Theta} = \begin{bmatrix} 0 & \bar{a}_{\Theta\Theta} \\ p_{\Theta\Theta} & \bar{a}_{\Theta\Theta} \end{bmatrix}$$

$$M_{\Theta\delta_z} = \begin{bmatrix} 0 & \bar{a}_{\Theta\delta_z} \\ p_{\Theta\delta_z} & \bar{a}_{\Theta\delta_z} \end{bmatrix}$$

$p_{\Theta\Theta} = 0.3$, $p_{\Theta\delta_z} = 0.3$, $|\delta_{a_{\Theta\Theta}}| \leq 1$, $|\delta_{a_{\Theta\delta_z}}| \leq 1$ and the nominal values of the coefficients are denoted by a bar.

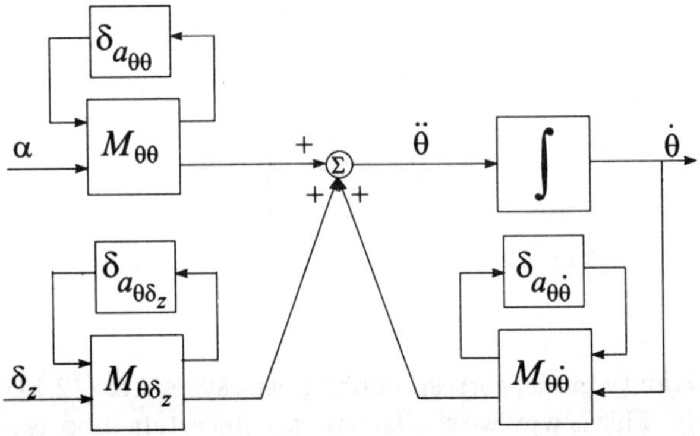

Fig. 12.11. Block diagram of the uncertain model for θ

The uncertain model corresponding to the equation

$$\ddot{\theta} = a_{\theta\dot{\theta}}\dot{\theta} + a_{\theta\theta}\alpha + a_{\theta\delta_z}\delta_z$$

is shown in Figure 12.11, where

$$M_{\theta\dot{\theta}} = \begin{bmatrix} 0 & \bar{a}_{\theta\dot{\theta}} \\ p_{\theta\dot{\theta}} & \bar{a}_{\theta\dot{\theta}} \end{bmatrix}$$

$$M_{\theta\theta} = \begin{bmatrix} 0 & \bar{a}_{\theta\theta} \\ p_{\theta\theta} & \bar{a}_{\theta\theta} \end{bmatrix}$$

$$M_{\theta\delta_z} = \begin{bmatrix} 0 & \bar{a}_{\theta\delta_z} \\ p_{\theta\delta_z} & \bar{a}_{\theta\delta_z} \end{bmatrix}$$

and $p_{\theta\dot{\theta}} = 0.3$, $p_{\theta\theta} = 0.3$, $p_{\theta\delta_z} = 0.3$, $|\delta_{a_{\theta\dot{\theta}}}| \leq 1$, $|\delta_{a_{\theta\theta}}| \leq 1$, $|\delta_{a_{\theta\delta_z}}| \leq 1$.

Finally, the uncertain model corresponding to the equation

$$n_y = a_{n_y\alpha}\alpha + a_{n_y\delta_z}\delta_z$$

is shown in Figure 12.12, where

$$M_{n_y\alpha} = \begin{bmatrix} 0 & \bar{a}_{n_y\alpha} \\ p_{n_y\alpha} & \bar{a}_{n_y\alpha} \end{bmatrix}$$

$$M_{n_y\delta_z} = \begin{bmatrix} 0 & \bar{a}_{n_y\delta_z} \\ p_{n_y\delta_z} & \bar{a}_{n_y\delta_z} \end{bmatrix}$$

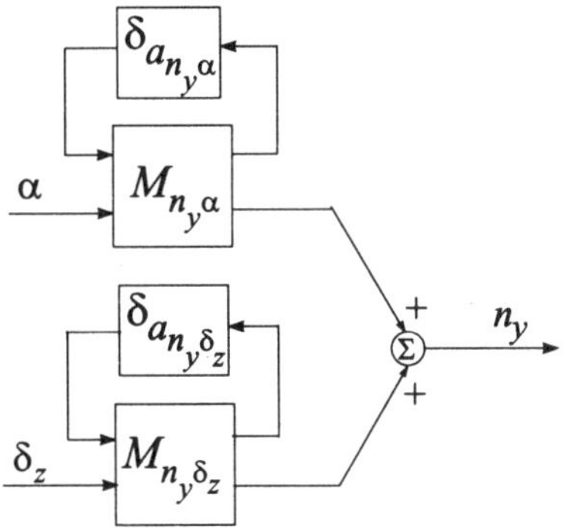

Fig. 12.12. Block diagram of the uncertain model for n_y

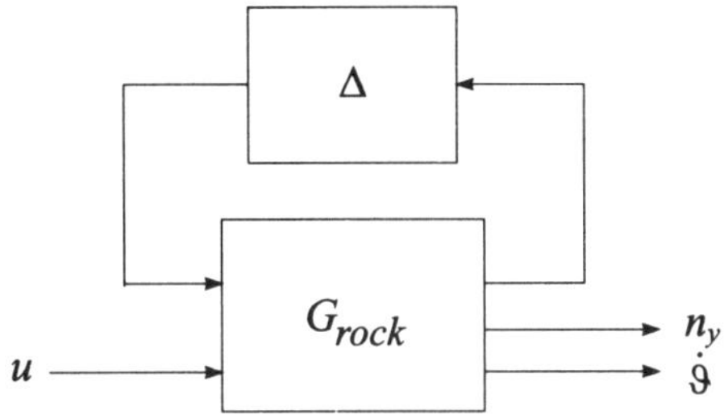

Fig. 12.13. Plant model in the form of upper LFT

and $p_{n_y\alpha} = 0.3$, $p_{n_y\delta_z} = 0.3$, $|\delta_{a_{n_y\alpha}}| \leq 1$, $|\delta_{a_{n_y\delta_z}}| \leq 1$.

By "pulling out"the uncertain parameters from the known part of the model one obtains an uncertain model in the form of upper LFT, shown in Figure 12.13 with a 7×7 matrix Δ of uncertain parameters,

$$\Delta = \mathrm{diag}(\delta_{a_{\Theta\Theta}}, \delta_{a_{\Theta\delta_z}}, \delta_{a_{\theta\dot{\theta}}}, \delta_{a_{\theta\theta}}, \delta_{a_{\theta\delta_z}}, \delta_{a_{n_y\alpha}}, \delta_{a_{n_y\delta_z}})$$

Due to the complexity of the plant, the easiest way in simulation and design to define the uncertainty model is to implement the command `sysic`. In this case, the plant input is considered as the reference signal $u(t) = \delta_\vartheta^o$ to the fin's servo-actuator, and the plant outputs are the normal acceleration n_y and the derivative $\dot{\theta}$ of the pitch angle. A model thus obtained is of 4th order.

The rocket uncertainty model is implemented by the M-file `mod_rock.m`. For a given moment of the time, the nominal model coefficients are computed by using the file `cfn_rock.m`. The values of the rocket parameters, as given in Table 12.1, are set in the file `prm_rock.m`.

12.3 Performance Requirements

The aim of the rocket stabilisation system is to achieve and maintain the desired normal acceleration in the presence of disturbances and of sensor noises.

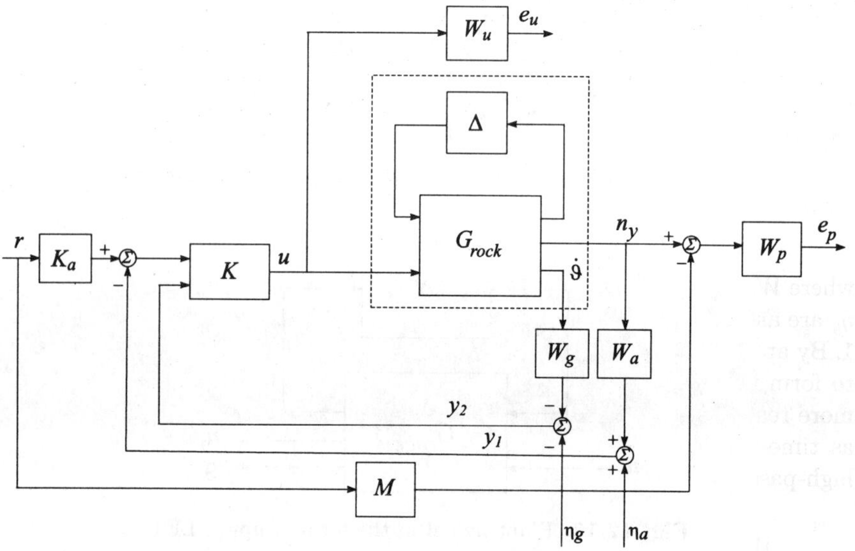

Fig. 12.14. Block diagram of the closed-loop system with performance requirements

A block-diagram of the closed-loop system including the feedback and controller as well as the elements reflecting the model uncertainty and weighting functions related to performance requirements is shown in Figure 12.14.

This system has a reference signal r, noises η_a and η_g on measurement of n_y and $\dot{\theta}$, respectively, and two weighted outputs e_p and e_u that characterise performance requirements. The transfer functions W_a and W_g represent the dynamics of the accelerometer measuring n_y, and the dynamics of the rate gyro measuring $\dot{\theta}$, respectively. The coefficient K_a is the accelerometer gain. The system M is the ideal model to be matched by the designed closed-loop system. The rectangular box, shown with dashed lines, represents the perturbed plant model $G = F_U(G_{\text{rock}}, \Delta)$. Inside the rectangular box is the nominal model G_{rock} of the rocket and the block Δ parametrising the model uncertainty. The matrix Δ is unknown but has a diagonal structure and is norm bounded, $\|\Delta\|_\infty < 1$. It is required for performance that the transfer function matrix from r, η_a and η_g to e_p and e_u should be small in the sense of $\|.\|_\infty$, for all possible uncertain matrices Δ. The transfer function matrices W_p and W_u are employed to represent the relative significance of performance requirements over different frequency ranges. The measured, output feedback signals are

$$y_1 = W_a n_y + \eta_a, \quad y_2 = W_g \dot{\theta} + \eta_g$$

The transfer functions W_a and W_g in this design are chosen as

$$W_a = \frac{0.4}{1.0 \times 10^{-6}s^2 + 2.0 \times 10^{-3}s + 1}$$

$$W_g = \frac{5.5}{2.56 \times 10^{-6}s^2 + 2.3 \times 10^{-3}s + 1}$$

The high-frequency noises η_a and η_g, which occurred in measuring the normal acceleration and the derivative of the pitch angle, may be represented in the form

$$\eta_a = W_{an}\tilde{\eta}_a \quad \eta_g = W_{gn}\tilde{\eta}_g$$

where W_{an} and W_{gn} are weighting transfer functions (shaping filters) and $\tilde{\eta}_a$, $\tilde{\eta}_g$ are arbitrary (random) signals satisfying the condition $\|\tilde{\eta}_a\|_2 \leq 1, \ \|\tilde{\eta}_g\|_2 \leq 1$. By appropriate choice of the weighting functions W_{an} and W_{gn} it is possible to form the desired spectral contents of the signals η_a and η_g. (Note that a more realistic approach to represent the noises η_a and η_g is to describe them as time-varying stochastic processes.) In the given case W_{an} and W_{gn} are high-pass filters whose transfer functions are

$$W_{an}(s) = 2 \times 10^{-4}\frac{0.12s + 1}{0.001s + 1}, \quad W_{gn}(s) = 6 \times 10^{-5}\frac{0.18s + 1}{0.002s + 1}$$

The synthesis problem of the rocket-attitude stabilisation system considered here is to find a linear, stabilising controller $K(s)$, with feedback signals y_1 and y_2, to ensure the following properties of the closed-loop system.

Nominal performance:

The closed-loop system achieves nominal performance. That is, the performance criterion is satisfied for the nominal plant model. In this case, the perturbation matrix Δ is zero.

Denote by $\Phi = \Phi(G, K)$ the transfer function of the closed-loop system from r, $\tilde{\eta}_a$ and $\tilde{\eta}_g$ to e_p and e_u,

$$\begin{bmatrix} e_p(s) \\ e_u(s) \end{bmatrix} = \Phi(s) \begin{bmatrix} r(s) \\ \tilde{\eta}_a \\ \tilde{\eta}_g \end{bmatrix}$$

The criterion for nominal performance is to satisfy the inequality

$$\|\Phi_{rock}(s)\|_\infty < 1 \tag{12.15}$$

where $\Phi_{\text{rock}}(s)$ is the transfer function matrix of the closed-loop system for the case $\Delta = 0$.

This criterion is a generalisation of the mixed sensitivity optimisation problem and includes performance requirements by matching an "ideal system" M.

Robust stability:

The closed-loop system achieves robust stability if the closed-loop system is internally stable for all possible, perturbed plant dynamics $G = F_U(G_{\text{rock}}, \Delta)$.

Robust performance:

The closed-loop system must remain internally stable for each $G = F_U(G_{\text{rock}}, \Delta)$ and, in addition, the performance criterion

$$\|\Phi(s)\|_\infty < 1 \tag{12.16}$$

must be satisfied for each $G = F_U(G_{\text{rock}}, \Delta)$.

For the problem under consideration here, it is desired to design a controller to track commanded acceleration maneuvers up to $15g$ with a time constant no greater than 1 s and a command following accuracy no worse than 5%. The controller designed should also generate control signals that do not violate the constraints $|\delta_z| < 30\,\text{deg}$ (≈ 0.52 rad), $|\alpha| < 20\,\text{deg}$ (≈ 0.35 rad) where δ_z, α are the angle of the control fin's deflection and angle of attack, respectively. In addition to these requirements it is desirable that the controller should have acceptable complexity, *i.e.* it is of sufficiently low order.

The ideal system model to be matched with, which satisfies the requirements to the closed-loop dynamics, is chosen as

$$M = \frac{1}{0.0625s^2 + 0.20s + 1}$$

and the performance weighting functions are

$$W_p(s) = \frac{1.0s^2 + 3.0s + 5.0}{1.0s^2 + 2.9s + 0.005}, \quad W_u(s) = 10^{-5}$$

The magnitude frequency response of the model is shown in Figure 12.15.

The performance weighting functions are chosen so as to ensure an acceptable trade-off between the nominal performance and robust performance of the closed-loop system with control action that satisfies the constraint imposed. These weighting functions are obtained iteratively by trial and error.

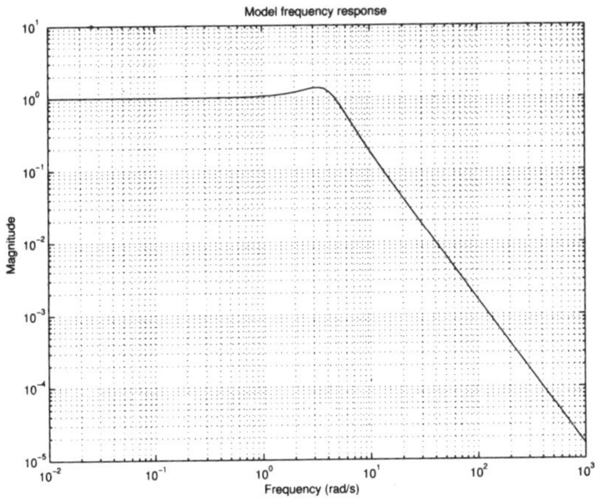

Fig. 12.15. Model frequency response

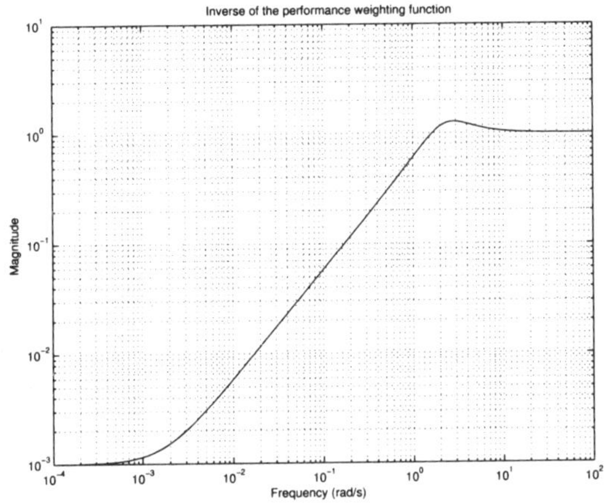

Fig. 12.16. Frequency response of the inverse weighting function

In Figure 12.16 we show the frequency response of the inverse of the performance weighting function W_p^{-1}. As can be seen from the figure, over the low-frequency range we require a small difference between the system and model and a small effect on the system output due to the disturbances. This ensures good reference tracking and a small error in the case of low-frequency disturbances.

The open-loop interconnection for the rocket stabilisation system is generated by the file `olp_rock.m`. The internal structure of the system with 11 inputs and 11 outputs is shown in Figure 12.17 and is computed by using the file `sys_rock.m` that saves the open-loop system as the variable `sys_ic`. The sensors transfer functions and weighting performance functions are defined in

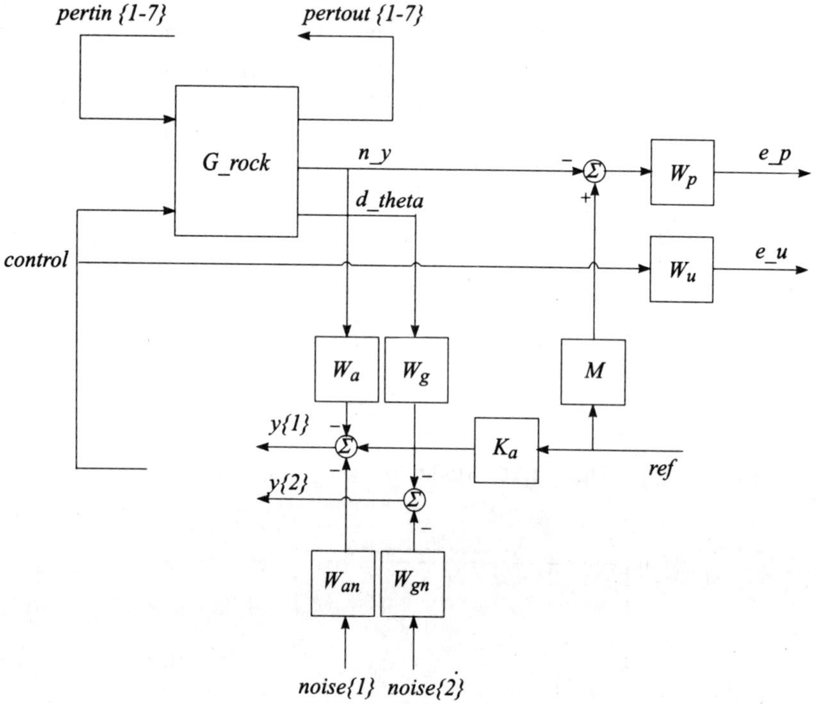

Fig. 12.17. Block diagram of the open-loop system with performance requirements

the files `wsa_rock.m` and `wts_rock.m`, respectively. The open-loop system is of 14th order. The inputs and outputs of the uncertainty are stored as the variables `pertin` and `pertout`. The reference signal is the variable `ref` and the noises at the inputs of the shaping filters are `noise{1}` and `noise{2}`. The control action is the variable `control` and the measured outputs are the variables `y{1}` and `y{2}`.

Both variables `pertin` and `pertout` are 7-dimensional. Variables `noise` and `y` are 2-dimensional. `ref, n_y, d_theta, e_p` and `e_u` are all scalar variables.

A schematic diagram showing the specific input/output ordering for the system variable `sys_ic` is shown in Figure 12.18.

12.4 $\mathcal{H}_\infty$ Design

In this section, the $\mathcal{H}_\infty$ (sub)optimal design method is employed to find an output feedback controller K for the connection shown in Figure 12.19. It is noted that in this setup the uncertainty inputs and outputs (*i.e.* signals

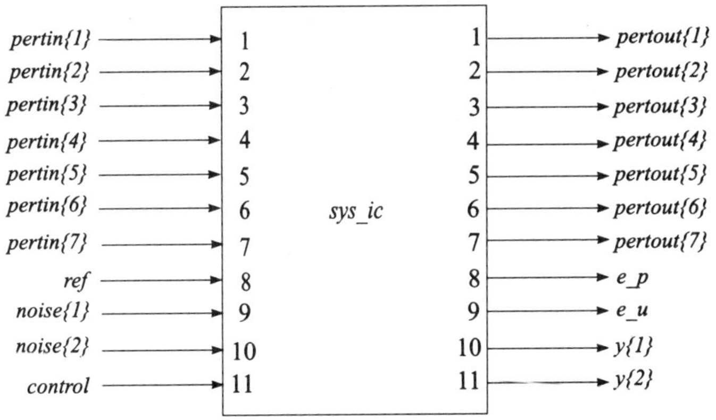

Fig. 12.18. Schematic diagram of the open-loop system structure

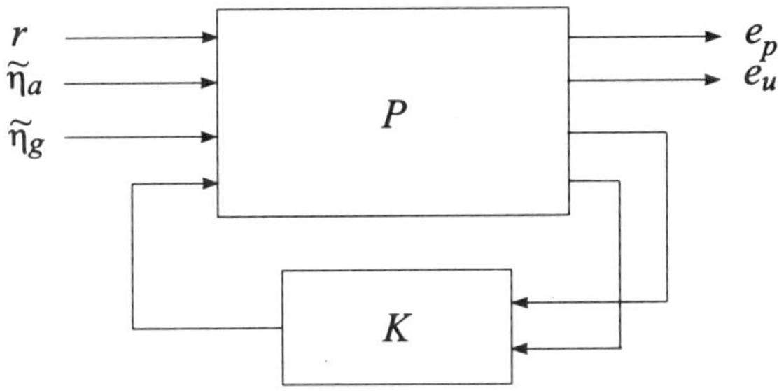

Fig. 12.19. Closed-loop system with $\mathcal{H}_\infty$ controller

around Δ) are excluded. The variable `hin_ic`, corresponding to the open-loop transfer function P in Figure 12.19, is obtained by the command line

```
hin_ic = sel(sys_ic,[8:11],[8:11])
```

The $\mathcal{H}_\infty$ optimal control synthesis minimises the $\|.\|_\infty$ norm of $\Phi_{\text{rock}} = F_L(P, K)$ in terms of all stabilising controllers (K). According to the definition given above, Φ_{rock} is the nominal transfer function matrix of the closed-loop system from the reference and noises (the variables `ref` and `noise`) to the weighted outputs `e_p` and `e_u`. The $\mathcal{H}_\infty$ controller is computed by the M-file `hin_rock.m`. It utilises the function `hinfsyn`, which for a given open-loop system calculates an $\mathcal{H}_\infty$ (sub)optimal control law. In the given case the final value of $\|\Phi_{\text{rock}}\|_\infty$ (γ) achieved is 4.53×10^5 that shows a poor $\mathcal{H}_\infty$ performance. The controller obtained is of 14th order.

Robust stability and robust performance analysis of the closed-loop system are carried out by the file `mu_rock.m`. The robust stability test is related to the upper-left 7×7 sub-block of the closed-loop system transfer matrix formed by `sys_ic` and the designed controller K. To achieve robust stability it is necessary for each frequency considered an upper bound of the structured singular value μ of that sub-block transfer function matrix must be less than 1. In the computation of μ the structured uncertainty is defined as

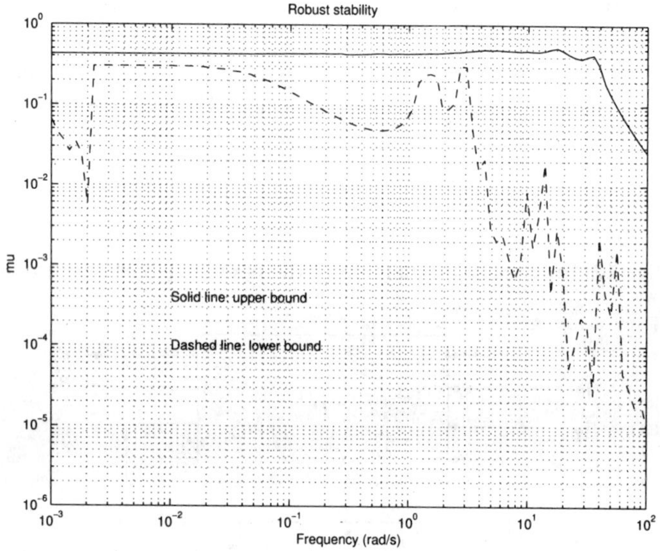

Fig. 12.20. Robust stability of the closed-loop system with K_{hin}

```
blkrsR = [-1 1;-1 1;-1 1;-1 1;-1 1;-1 1;-1 1]
```

The frequency response of the structured singular value for the case of robust stability analysis is shown in Figure 12.20. The maximum value of μ is 0.52, which means that the stability of the closed-loop system is preserved under all perturbations that satisfy $\|\Delta\|_\infty < \frac{1}{0.52}$.

The nominal performance is tested on the lower 2×3 block of the closed-loop system transfer matrix. From the frequency response of the nominal performance shown in Figure 12.21 it is seen that the nominal performance is achieved within a large margin. The obtained peak value of γ is 0.43, which is much less than 1.

The robust performance of the closed-loop system with the $\mathcal{H}_\infty$ controller is also investigated by means of the μ-analysis. The closed-loop transfer function matrix has 10 inputs and 9 outputs. The first 7 inputs and outputs correspond to the 7 channels that connect to the uncertainty matrix Δ, while the last 3 inputs and 2 outputs correspond to the weighted sensitivity of the closed-loop system and connect to a fictitious uncertainty matrix. Hence, for μ-analysis of the robust performance the block-structure of the uncertainty must comprise a 7×7 diagonal, real uncertainty block Δ and a 2×3 complex

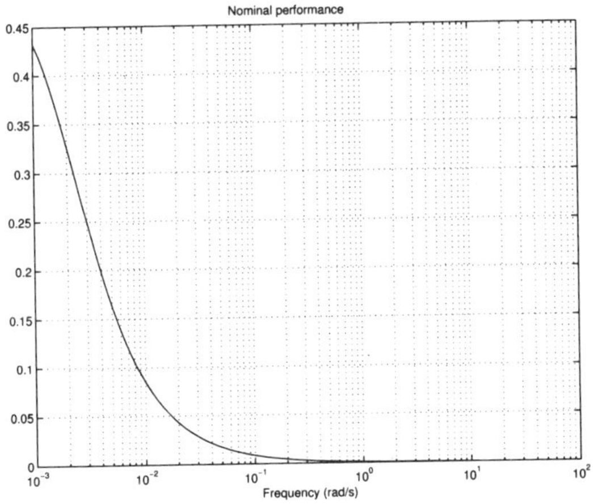

Fig. 12.21. Nominal performance of the closed-loop system for K_{hin}

uncertainty block Δ_F

$$\Delta_P := \left\{ \begin{bmatrix} \Delta & 0 \\ 0 & \Delta_F \end{bmatrix} : \Delta \in \mathcal{R}^{7\times 7},\ \Delta_F \in \mathcal{C}^{2\times 3} \right\}$$

The robust performance (with respect to the uncertainty and performance weighting functions) is achieved if and only if, over a range of frequency under consideration, the structured singular value $\mu_{\Delta_P}(j\omega)$ at each ω is less than 1.

The frequency response of μ for the case of robust performance analysis is given in Figure 12.22. The peak value of μ is 1.153, which shows that the robust performance has not been achieved. Or in other words, the system does not preserve performance under all relative parameter changes shown in Table 12.3.

The results obtained are valid for $t = 15$ s. To check if the controller designed achieves robust stability and robust performance of the closed-loop system at other time instants of flight, further analysis should be conducted with corresponding dynamics.

The simulation of the closed-loop system is implemented by using the file `clp_rock.m` that corresponds to the structure shown in Figure 12.23. In this structure the performance weighting functions W_p and W_u, used in the system design and performance analysis, are absent. The simulation shows the transient responses with respect to the reference signal for $t = 15$ s. The transient responses are computed by using the function `trsp` under the assumption of "frozen"model parameters. (Note that all transient responses are shown with a time offset of 15 s.)

In Figure 12.24 we show the transient responses of the closed-loop system with the designed $\mathcal{H}_\infty$ controller for a step signal with magnitude $r = \pm 15g$,

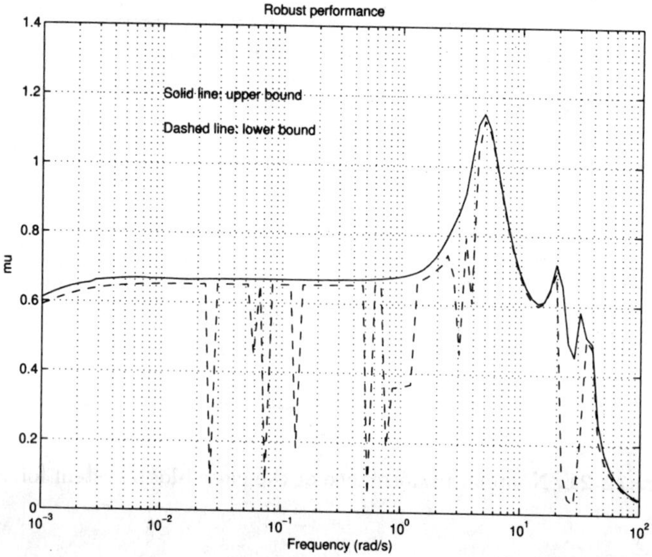

Fig. 12.22. Robust performance of the closed-loop system for K_{hin}

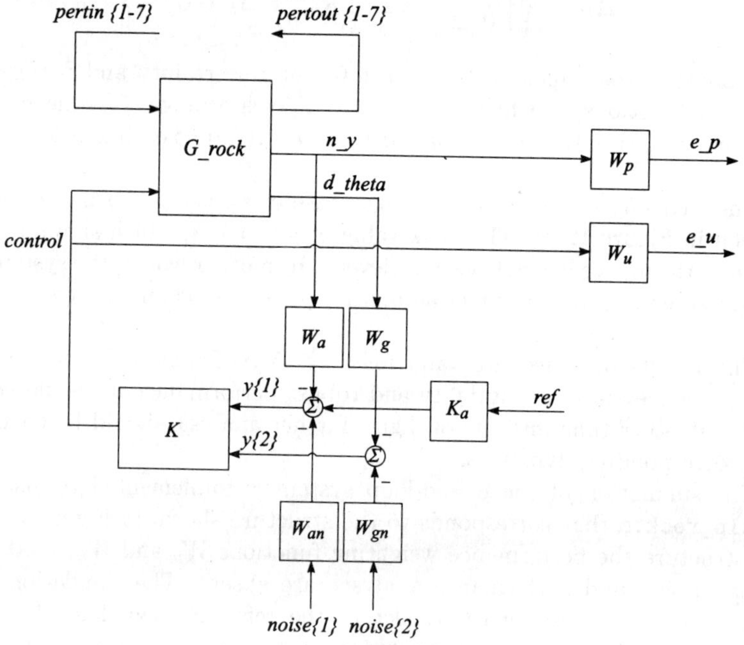

Fig. 12.23. Structure of the closed-loop system

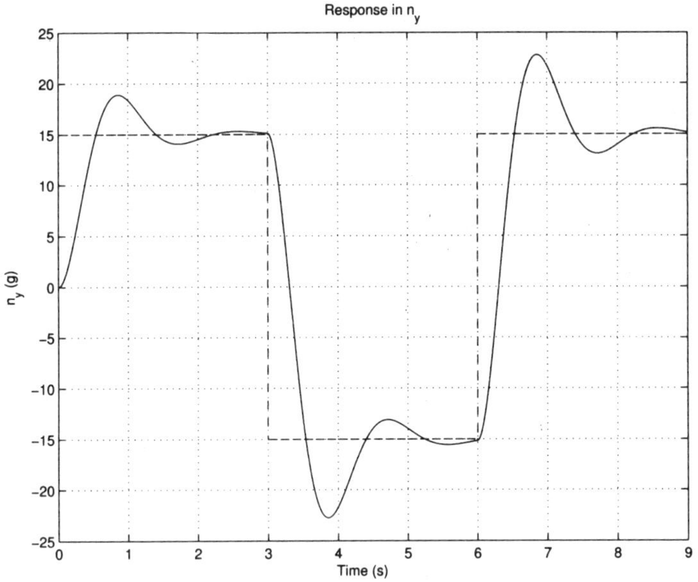

Fig. 12.24. Transient response of the closed-loop system, $t = 15$ s

which corresponds to a change in the normal acceleration with $\pm 15g$. The transient response for this controller is oscillatory. The overshoot is under 30% and the settling time is approximately 2 s. For the same reference signal, the transient response of the pitch rate $\dot{\theta}$ is shown in Figure 12.25. The time response contains high-frequency oscillations.

It is noticed that the $\mathcal{H}_\infty$ controller designed does not satisfy the requirements for the closed-loop system dynamics. Better results are obtained with the μ-controller designed in the next section.

12.5 μ-Synthesis

In this section, we consider the same design problem but using another approach, namely the μ-synthesis method. Because the uncertainties considered in this case are highly structured, better results with respect to the closed-loop performance may be achieved by the μ-synthesis.

The model of the open-loop system has 14 states, 11 inputs and 11 outputs. Uncertainties (*i.e.* parameter variations) are considered with seven parameters $a_{\Theta\Theta}, a_{\Theta\delta_z}, a_{\theta\dot{\theta}}, a_{\theta\theta}, a_{\theta\delta_z}, a_{n_y\alpha}$ and $a_{n_y\delta_z}$.

Denote by $P(s)$ the transfer function matrix of the 11-input, 11-output open-loop system `nominal_dk` and let the block structure of the uncertainty matrix Δ_P be defined as

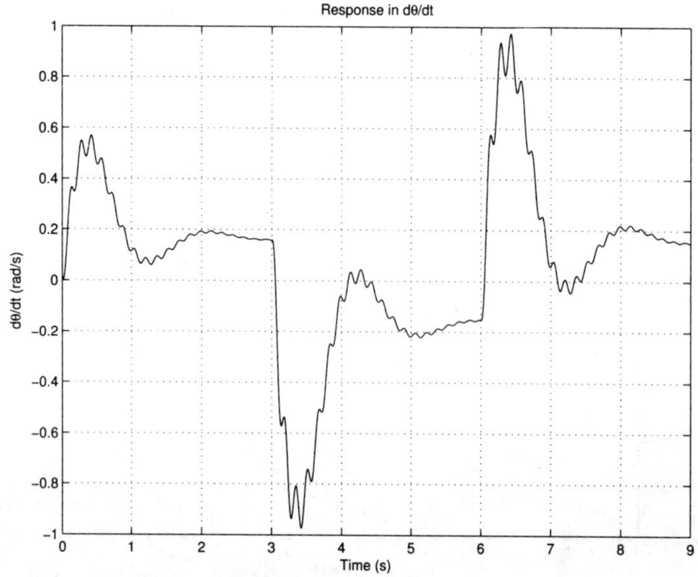

Fig. 12.25. Transient response of $\dot{\theta}$, $t = 15$ s

$$\Delta_P := \left\{ \begin{bmatrix} \Delta & 0 \\ 0 & \Delta_F \end{bmatrix} : \Delta \in \mathcal{R}^{7\times 7},\ \Delta_F \in \mathcal{C}^{2\times 3} \right\}$$

The first block Δ of the matrix Δ_P is diagonal and corresponds to the parametric uncertainties in the vehicle model. The second, diagonal block Δ_F is a fictitious uncertainty block, which is used to introduce the performance requirements in the design framework of the μ-synthesis. To satisfy robust performance requirements it is necessary to find a stabilising controller $K(s)$, such that at each frequency ω over relevant frequency range, the following inequality of the structured singular value holds

$$\mu_{\Delta_P}[F_L(P, K)(j\omega)] < 1$$

The above condition guarantees robust performance of the closed-loop system, *i.e.*,

$$\|\Phi(s)\|_\infty < 1 \tag{12.17}$$

The μ-synthesis of the rocket controller is implemented by the M-file `ms_rock.m` that exploits the function `dkit`. The uncertainty description is defined in the file `dk_rock.m` by the statement

```
BLK_DK = [-1 1;-1 1;-1 1;-1 1;-1 1;-1 1;-1 1;3 2]
```

The results from the iterations after each step are shown in Table 12.4. The controller obtained, after 3 iterations, is of 28th order.

Table 12.4. Results from the D-K iterations

Iteration	Controller order	Maximum value of μ
1	14	1.730
2	18	1.084
3	28	0.810

To check the robust stability and robust performance of the closed-loop system it is necessary to follow μ- analysis using again the file `mu_rock.m`.

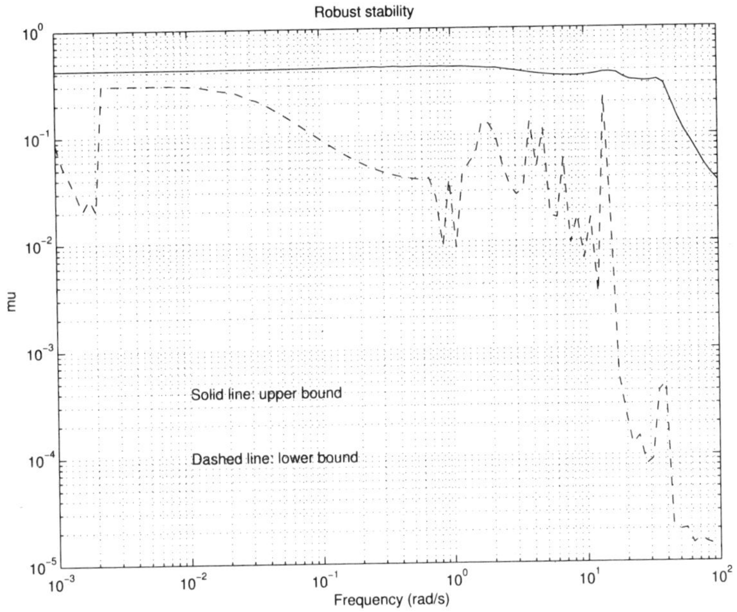

Fig. 12.26. Robust stability of the closed-loop system with K_{mu}

The frequency response of the structured singular value for the case of robust stability is shown in Figure 12.26. The maximum value of μ is 0.441, which shows that under the considered parametric uncertainties the closed-loop system stability is preserved.

The frequency response of the structured singular value for the case of robust performance is shown in Figure 12.27. The maximum value of μ is 0.852, which shows that the closed-loop system achieves robust performance. This value of μ is smaller than the corresponding value of μ, obtained in the $\mathcal{H}_\infty$ design, *i.e.* the μ-controller provides better robustness.

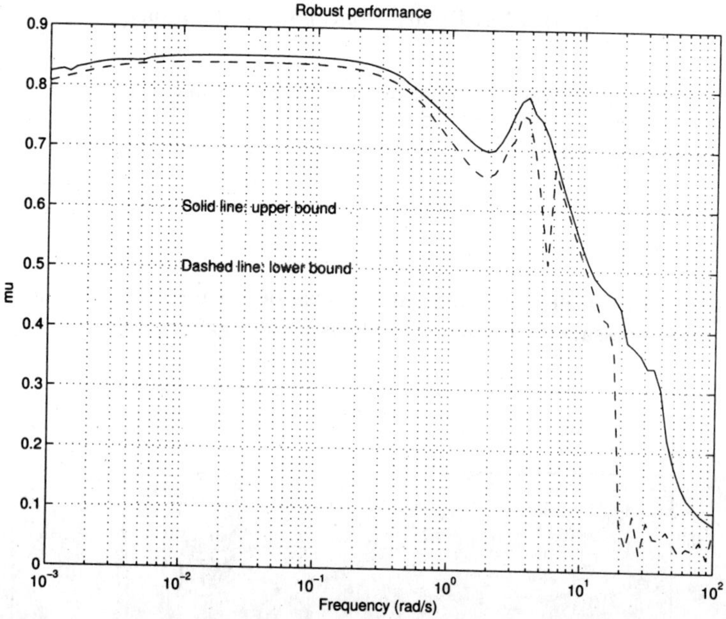

Fig. 12.27. Robust performance of the closed-loop system with K_{mu}

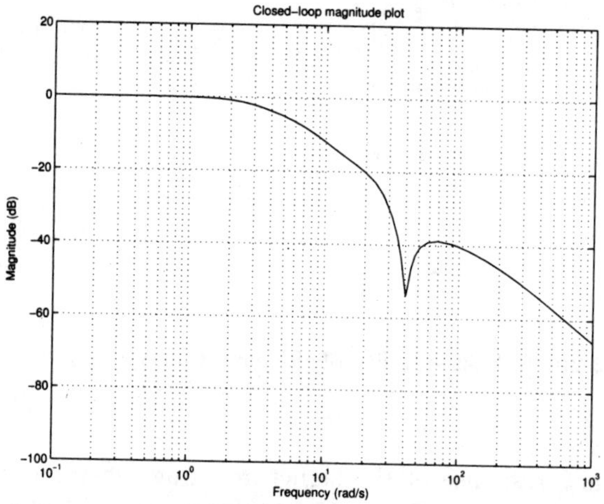

Fig. 12.28. Magnitude response of the closed-loop system

The frequency responses of the closed-loop system with the μ-controller are obtained by using the file `frs_rock.m` and are shown in Figures 12.28 and 12.29. Over the low-frequency region the value of the magnitude response is

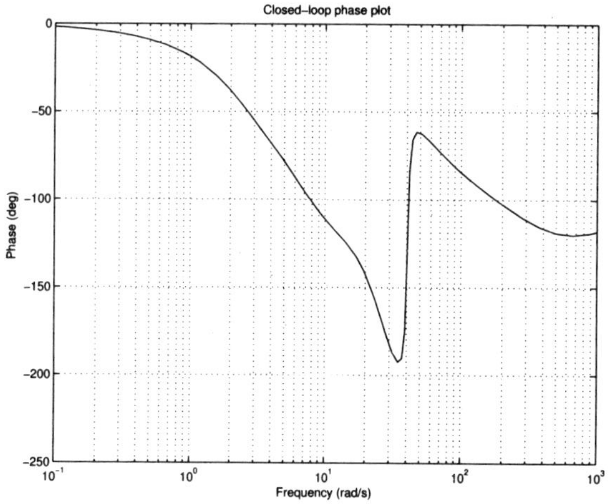

Fig. 12.29. Phase response of the closed-loop system

close to 1, *i.e.* the reference signal can be followed accurately. The closed loop system bandwidth is about 2 rad/s.

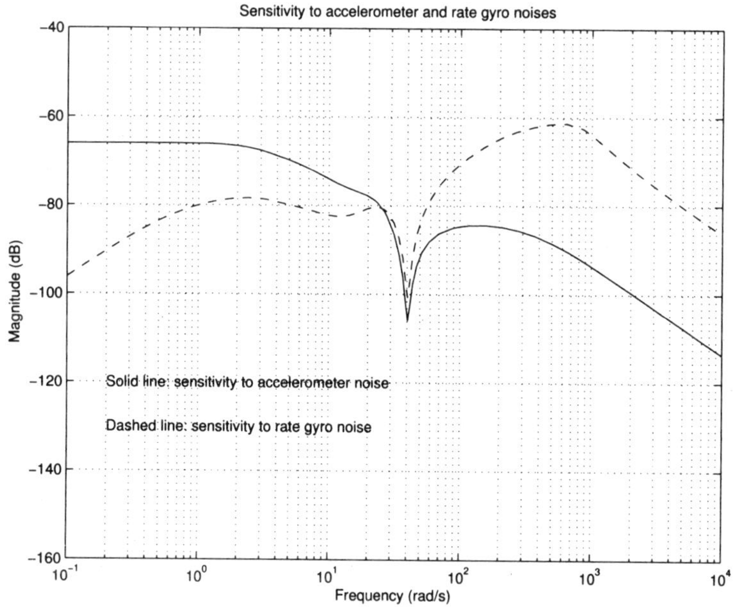

Fig. 12.30. Sensitivity to the noises $\tilde{\eta}_a$ and $\tilde{\eta}_g$

In Figure 12.30 we show the frequency responses of the transfer functions with respect to the noises $\tilde{\eta}_a$ and $\tilde{\eta}_g$. The noises of the accelerometer and of the rate gyro are attenuated 1000 times (60 dB), *i.e.* both noises have a relatively weak influence on the output of the closed-loop system.

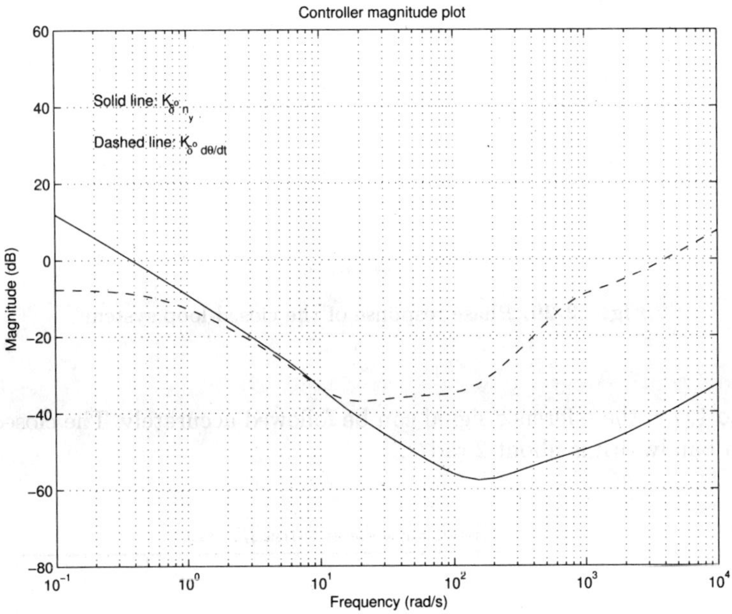

Fig. 12.31. Controller magnitude plots

The magnitude and phase plots of the controller transfer functions are shown in Figures 12.31 and 12.32, respectively.

In Figure 12.33 we show the transient responses of the closed-loop system with the designed μ-controller for a step signal with magnitude

$$r = \pm 15\ g$$

that corresponds to a change in the normal acceleration with $\pm 15g$. The overshoot is under 1% and the settling time is less than 1 s. For the same reference signal, the transient response of the pitch rate $\dot{\theta}$ is shown in Figure 12.34.

The angle of deflation δ_z of the control fins in the closed loop system is shown in Figure 12.35. The magnitude of this angle is less than 30 deg, as required. In Figure 12.36 we show the perturbed transient responses of the closed-loop system for a step reference signal with magnitude $r = \pm 15g$.

The designed μ-controller can be used in cases of a wider range of vehicle coefficient values. In Figure 12.37 we show the transient response with respect to the reference signal for the 15th second of the flight at altitude $H = 1000$ m

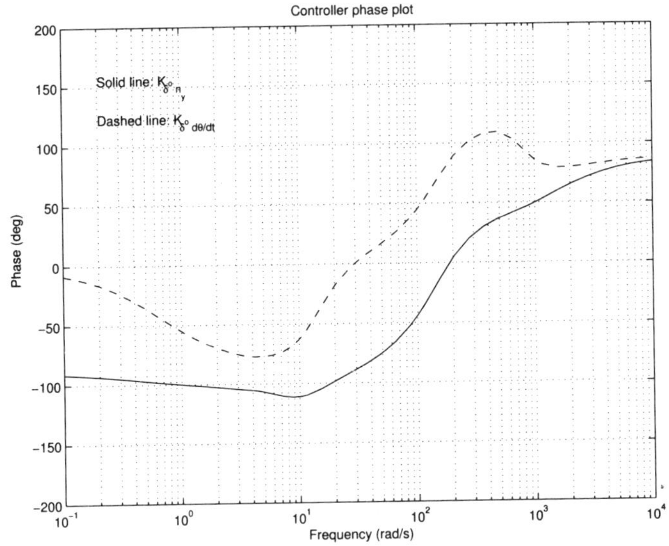

Fig. 12.32. Controller phase plots

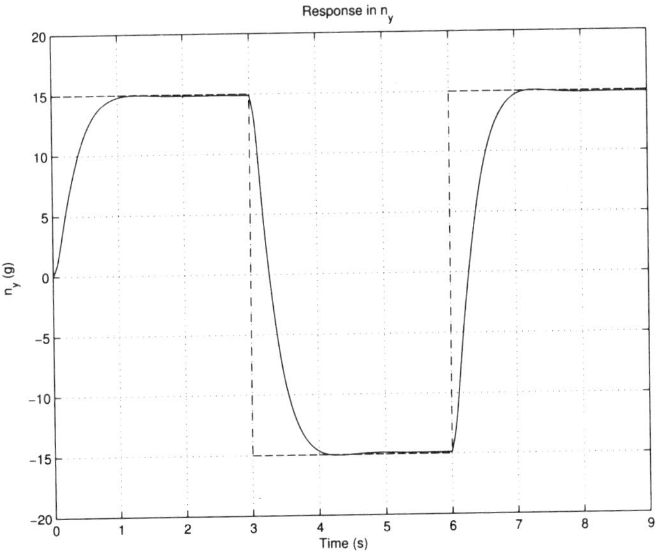

Fig. 12.33. Transient response of the closed loop system, $t = 15$ s

and in Figure 12.38 is the corresponding transient response for the 15th second of the flight at altitude $H = 10000$ m. Due to the achieved robustness of the closed-loop system, both responses are similar to that of the design case,

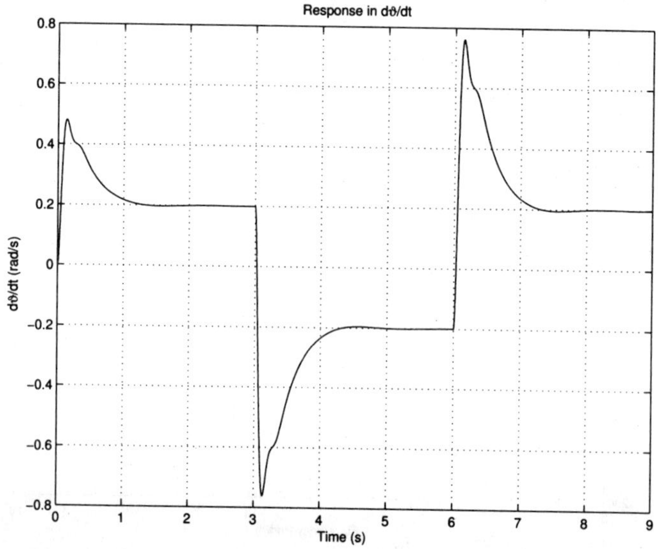

Fig. 12.34. Transient response of $\dot{\theta}$, $t = 15$ s

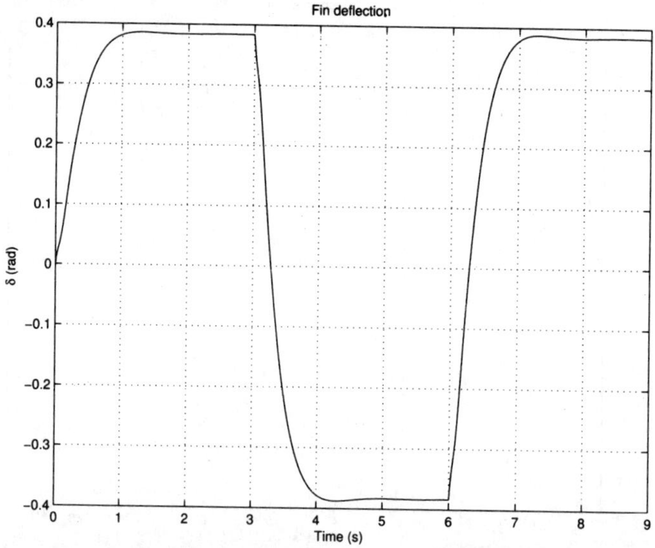

Fig. 12.35. Control action in the closed loop system, $t = 15$ s

though there is significant difference in the rocket dynamics between these two cases and the design case. This shows that the designed controller may be

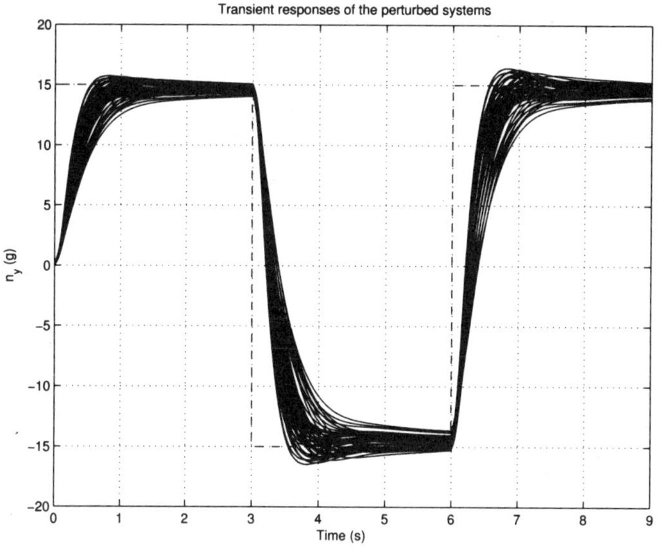

Fig. 12.36. Perturbed transient response of the closed-loop system, $t = 15$ s

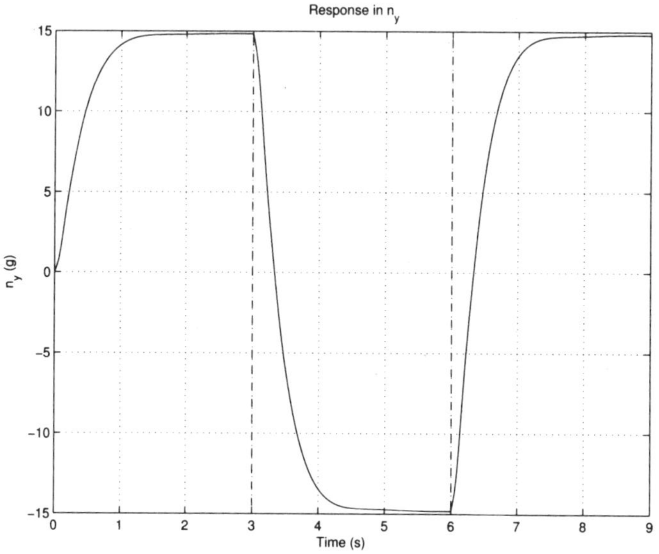

Fig. 12.37. Transient response of the closed-loop system for $H = 1000$ m

employed for system stabilisation for different altitudes and velocities, which would simplify the controller implementation.

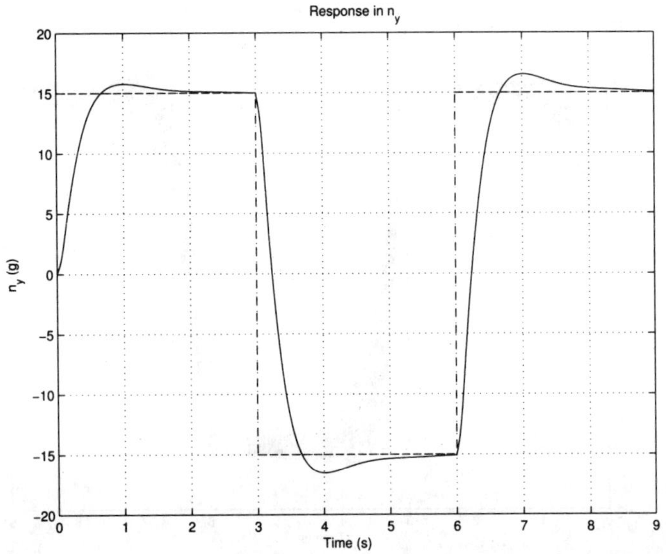

Fig. 12.38. Transient response of the closed-loop system for $H = 10000$ m

As was noted above, the controller obtained by the μ-synthesis is initially of 28th order. In order to reduce the order of the controller, the file `red_rock.m` may be used. This file implements system balancing followed by optimal Hankel approximation, calling functions `sysbal` and `hankmr`. As a result the controller order can be reduced to 8. Further reduction of the controller order leads to deterioration of the closed-loop system dynamics.

The structured singular values of the closed-loop system with the full-order and reduced-order controllers are compared in Figure 12.39. They are close to each other over the whole frequency range of practical interest. The transient responses of the closed loop systems with full-order and reduced-order controllers are indistinguishable, and are thus not included.

12.6 Discrete-time μ-Synthesis

The rocket stabilisation is, in practice, implemented by a digital controller that may be obtained by discretization of a continuous-time (analogue) controller at a given sampling frequency. Another possible approach is to discretise the continuous-time, open loop plant and then synthesize a discrete-time controller directly. In this section we describe the later approach that produces better results in this design exercise.

The discrete-time, open-loop interconnection is obtained by using the file `dlp_rock.m`. Since the frequency bandwidth of the designed closed-loop system in the continuous time case is about 2 Hz, the sampling frequency is

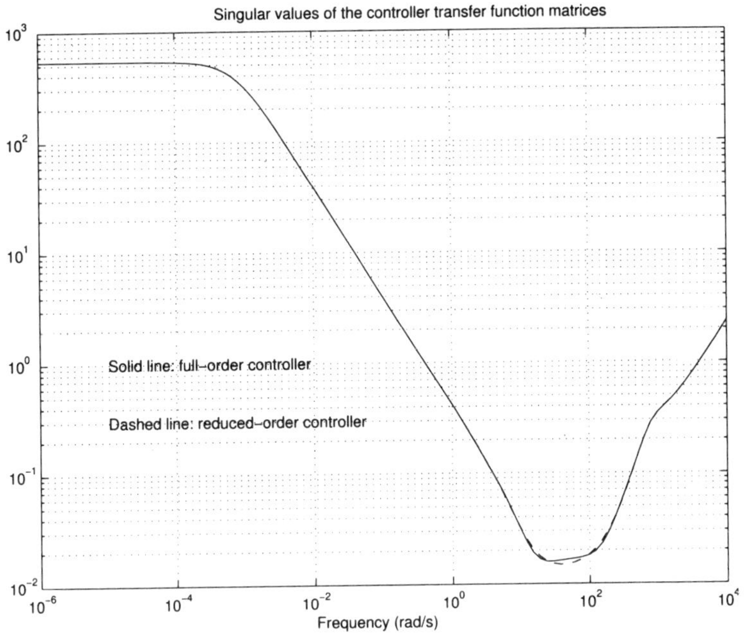

Fig. 12.39. Singular values of the full-order controller and reduced-order controller

chosen equal to 250 Hz that corresponds to a sampling period T_s of 0.004 s. For this frequency we derive a discretised model of the open-loop system by using the function **samhld**. The controller may be synthesised by the aid either of $\mathcal{H}_\infty$ optimisation (by using the function **dhinfsyn**), or μ-synthesis (by using the function **dkit**). In the following, we consider the μ-synthesis of a discrete-time controller that is obtained by implementing the files **dms_rock.m** and **ddk_rock.m**.

As in the μ-synthesis of the continuous-time controller, the synthesis is conducted for the full-order vehicle model. Hence, the uncertainty structure under consideration is of the form

```
BLK_DK = [-1 1;-1 1;-1 1;-1 1;-1 1;-1 1;-1 1;3 2]
```

In the discrete-time case, the frequency range is on the unit circle and chosen as the interval $[0, \pi]$. To set up 100 frequencies, the following line is used

```
OMEGA_DK = [pi/100:pi/100:pi];
```

For the discrete-time design it is necessary to include the command line

```
DISCRETE_DK = 1;
```

The design follows the usual route, by calling the function `dkit`. In Table 12.5 the results of the discrete time D-K iterations are listed.

Table 12.5. Results of the D-K iterations

Iteration	Controller order	Maximum value of μ
1	14	2.101
2	22	0.600
3	28	0.730

The discrete-time controller K_D obtained is of 28th order. Note that better robustness is achieved with the controller obtained at the 2nd step. This controller, however, does not ensure the desired closed-loop dynamics.

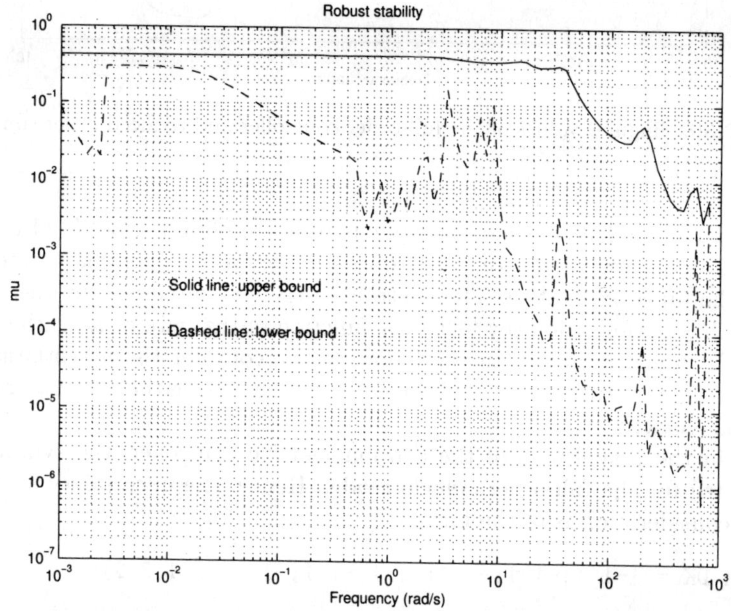

Fig. 12.40. Robust stability of the closed-loop system with K_D

Robust stability and robust performance of the closed-loop system with K_D are shown in Figures 12.40 and 12.41, respectively, in which the μ values over frequency range are calculated by the file `dmu_rock.m`. It is seen that the discrete-time closed-loop system achieves both robust stability and robust performance, just as in the continuous-time case with K_{mu}.

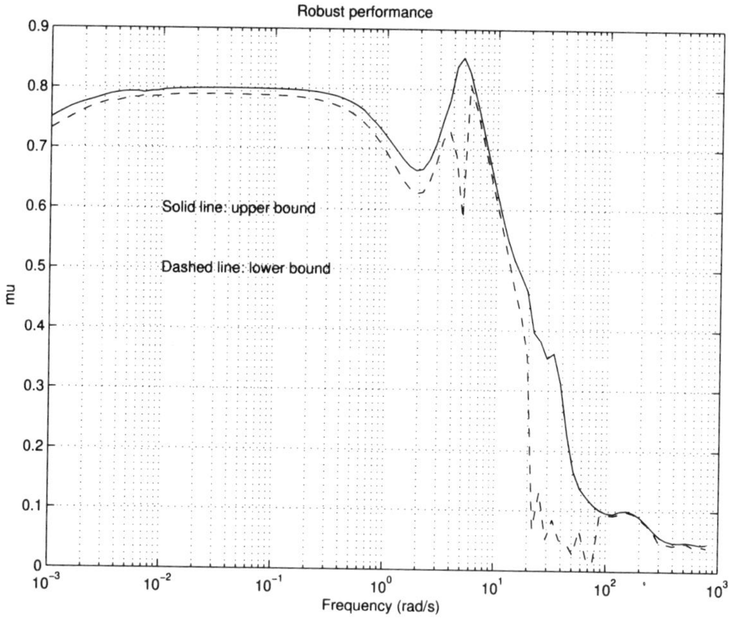

Fig. 12.41. Robust performance of the closed-loop system with K_D

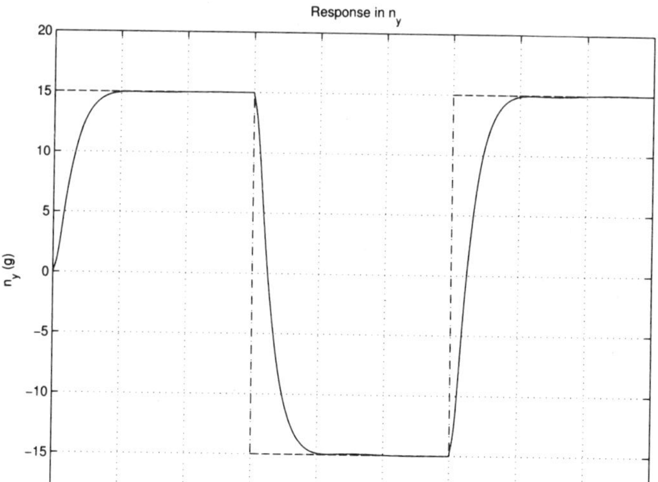

Fig. 12.42. Transient response for K_D, $t = 15$ s

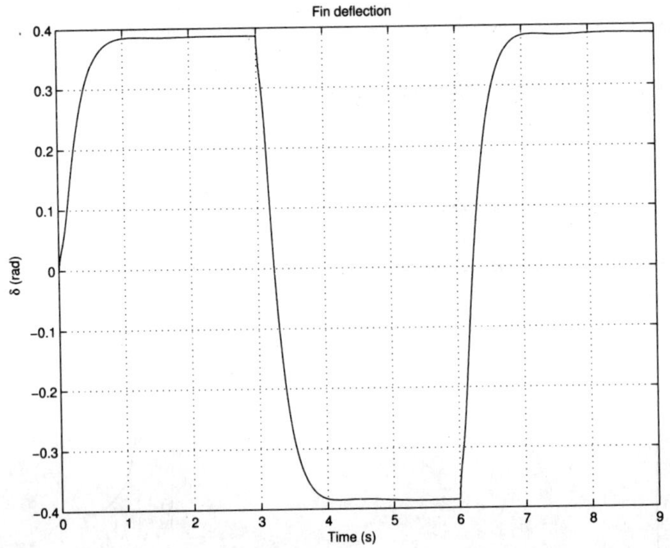

Fig. 12.43. Control action in the closed loop system for K_D, $t = 15$ s

In Figures 12.42 and 12.43 we show the transient response of the closed loop system with respect to the reference signal, and the corresponding control action, respectively. These two figures are generated by the file `dcl_rock.m` that utilises the function `sdtrsp`.

The results displayed above show that the achieved behaviour of the sampled data, closed-loop system are close to those in the continuous-time case, with each corresponding μ-controller.

12.7 Simulation of the Nonlinear System

The dynamics of the nonlinear, time-varying rocket model differs from that of the linear, time-invariant model used in the analysis and design described in the previous sections. Also, for large reference signals there is a strong interconnection between the longitudinal and lateral motions that may affect the stability and performance of the whole system. It is therefore important to study the dynamic behaviour of the closed loop, nonlinear, time-varying system with the designed controller that regulates the six-degree-of-freedom rocket motion.

The closed-loop rocket stabilisation system of the nonlinear, time-varying plant is simulated by using the Simulink® models `c_rock.mdl` (for the continuous-time system with an analogue controller) as well as `d_rock.mdl` (for the sampled data system with a digital controller). Both systems involve two identical controllers for the longitudinal and lateral motion. The sampled-data system also contains 16-bit analogue-to-digital and digital-to-analogue converters. (The outputs of the digital-to-analogue converters are scaled to give the reference input of the servo-actuators.) Both models allow us to simulate the closed-loop system for different reference, disturbance and noise signals. The roll motion is stabilised by a separate gyro with simple lead-lag compensator. (A robust roll controller is also possible to implement.)

In Figure 12.44 we show the Simulink® model `d_rock.mdl` of the nonlinear, sampled-data, closed-loop system.

Before carrying out simulation it is necessary to assign the model parameters by using the M-file `init_c_rock.m` (in the continuous-time case) or the the M-file `init_d_rock.m` (in the discrete-time case).

Before the perturbation motion begins to affect (*i.e.* before the time instance t_0), only the nonlinear equations of the unperturbed (program) motion are solved. (This ensures that the parameters of the linearised model at t_0 are the same as those used in the controller design.) After t_0, the equations of the perturbed motions are solved by using the S-function `s_rock.m`. The initial conditions for the perturbed motion are assigned in the file `inc_rock.m` that is invoked by `s_rock.m`. During the time interval $[0, t_0]$, the equations of the unperturbed motion are solved in the file `inc_rock.m` by using the function `sol_rock.m`. The values of the pitch angle θ for the program motion are assigned in the file `theta_rock.m`.

The simulation of the perturbed motion, which involves the controller action, is based on the nonlinear differential and algebraic equations (12.1)–(12.5).

In Figures 12.45 and 12.46 we show the transient response of the nonlinear sampled-data system with respect to the normal accelerations n_y and n_z, respectively, for a reference step change of $15g$ occurring at $t_0 = 15$ s in each channel. In the simulation we used the discrete-time controller designed for the same moment of the time in the previous section. It is seen that the behaviour of n_z is very close to the behaviour of the corresponding variable shown in Figure 12.42. The small difference in the responses of n_y and n_z is due to the influence of the Earth's gravity on the pitch angle.

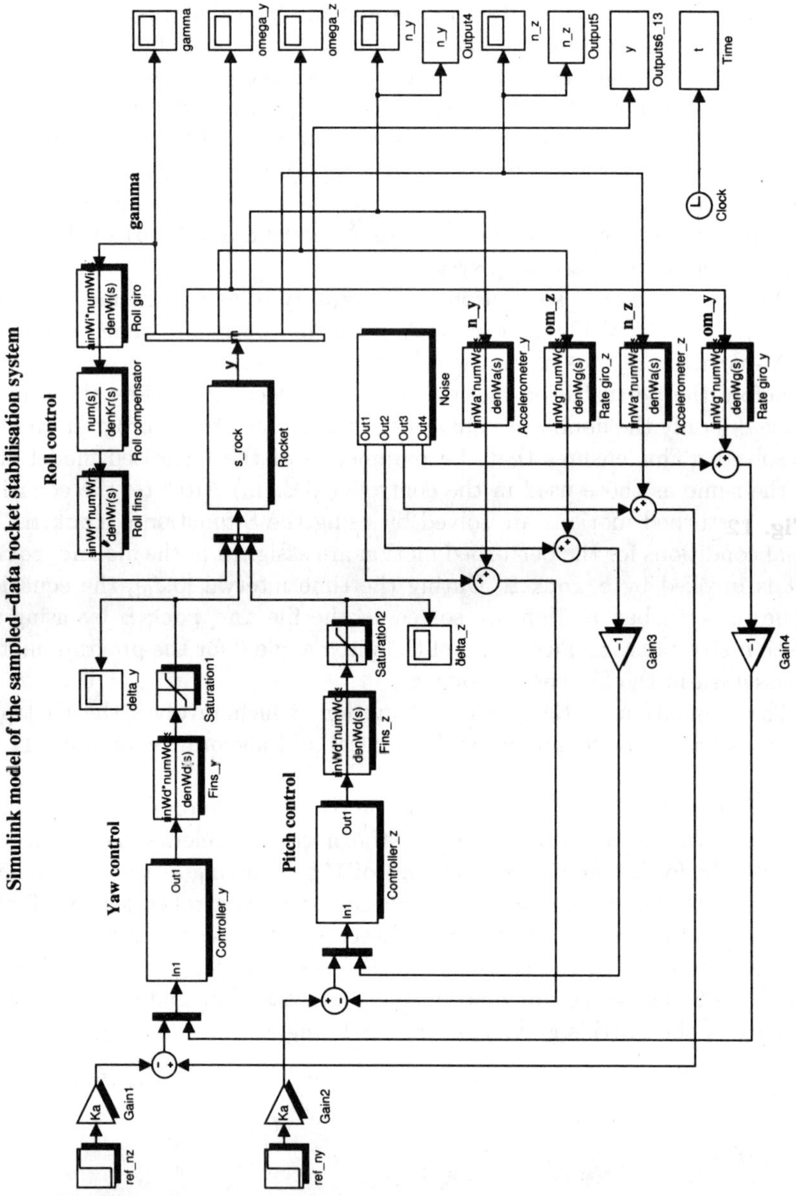

Fig. 12.44. Simulation model of the nonlinear sampled data system

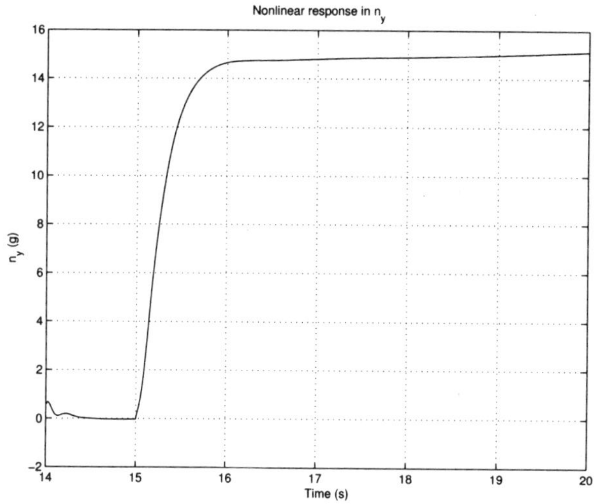

Fig. 12.45. Transient responses in n_y of the nonlinear sampled-data system

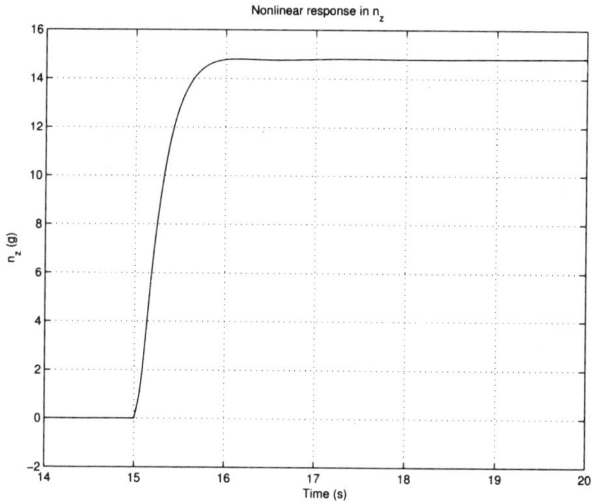

Fig. 12.46. Transient responses in n_z of the nonlinear sampled-data system

12.8 Conclusions

From the results/experiences obtained, the design of a robust stabilisation system for a winged, supersonic rocket may be summarised as the following.

- The linearised equations of the rocket should be arranged in a proper way in order to avoid the appearance of an additional integrator in the uncertainty model. The inclusion of this integrator leads to violation of the conditions for $\mathcal{H}_\infty$ design.
- Both $\mathcal{H}_\infty$ optimisation and μ-synthesis approaches may be used to design controllers that, for a specified moment of flight dynamics, achieve robust stability of the rocket stabilisation system in the presence of disturbances and sensor noises. However, the $\mathcal{H}_\infty$ controller can not ensure robust performance in the given case. The μ-controller achieves both robust stability and robust performance of the closed-loop system.
- The μ-controller obtained may be used successfully for different altitudes and Mach numbers. However, in order to control the rocket efficiently through the whole flight envelope it may be necessary to implement several controllers designed for different flight conditions.
- A digital controller has been successfully designed for a discrete-time model of the open-loop system. The corresponding sampled data, closed-loop system achieves robust stability and robust performance at almost the same as the continuous-time one.
- The μ-synthesis in the discrete-time case shows that achieving a smaller value of μ may lead to the deterioration of the system dynamics, *i.e.* better robustness may be achieved at the price of poorer dynamics. This is why a value of μ slightly less than 1 may be a good trade-off between the requirements for the robustness and dynamic performance.
- The simulation of the nonlinear, time-varying closed-loop system shows that for a sufficiently large interval of time, the dynamics behaviour is close to that of the time-invariant system which has fixed model parameters.

Notes and References

The design of rocket and spacecraft flight-control systems is considered in depth in many books, see for example [11, 12, 54, 172]. The design of robust flight-control systems is presented in [7, 36]. Ensuring good performance of the closed-loop system for the whole range of the flight operating conditions by a fixed controller is rarely possible and, in general, it is necessary to change the controller parameters as the rocket model varies. For this aim, it is possible to use some technique of *gain scheduling*: see the survey papers [84, 131]. The classical approach for gain scheduling is to design several time-invariant controllers for different points in the operational region and then interpolate their parameters for the intermediate values, see, for instance, [130, 118, 13].

This approach has several disadvantages, for instance it is difficult to guarantee robust stability and robust performance in the transition regions, *i.e.* between design points at which the controllers are designed. In this respect robust controllers are more suitable for gain scheduling since they ensure satisfactory performance at least in some neighbourhood around an operating point and thus fewer fixed controllers are needed. Another approach for gain scheduling is to derive a *linear, parameter-varying(LPV)* model of the rocket and then design an LPV controller that hopefully will achieve the desired performance for the whole range of operating conditions. Examples of using this approach may be found in [37, 143].

The elasticity of the rocket body may affect significantly the dynamics of the closed-loop system, see, for instance, [11, 91].

13

Robust Control of a Flexible-Link Manipulator

In this chapter we discuss the robust control system design of a flexible-link manipulator that moves in the horizontal plane.

Lightweight manipulators possess many advantages over the traditional bulky manipulators. The most important benefits include high payload-to-arm weight ratio, faster motion, safer operation, improved mobility, low cost, longer reach and better energy efficiency, *etc.* However, the reduction of weight leads to the increase of the link elasticity that significantly complicates the control of the manipulator. The difficulty in control is caused by the fact that the link model is a distributed parameter plant. In this case, several elastic modes are required to achieve sufficiently high accuracy. Also, the plant has several uncertain parameters (payload mass, hub and structural damping factors, *etc.*) that influence significantly the system performance. The inherent, nonminimum phase behaviour of the flexible manipulator is another obstacle to achieving simultaneously a high-level performance as well as good robustness.

The aim of the present case study is to design a control system for a single-link flexible manipulator. A two-mode dynamic model of the manipulator is first obtained by using the Lagrangian-assumed modes method. This is followed by the modelling of uncertainties involved in the manipulator. The uncertainties include the real parametric uncertainties in the payload mass as well as in the hub and structural damping factors. These parameters are the basic uncertainty source in the dynamic behaviours of the flexible-link manipulators. The μ-synthesis method is then applied to design a robust, noncollocated controller on the feedback signals of joint angle and tip acceleration. In the design, in order to obtain a feasible solution, a simplified uncertainty description is considered in the D-K iterations. Appropriate weighting functions are chosen in the design to ensure robust stability and robust performance. It is shown in this chapter that good robust performance has been achieved in the design. The closed-loop system exhibits excellent tip-motion performance for a wide range of payload mass and the system efficiently suppresses the elastic vibrations during the fast motion of the manipulator tip. For the sake of implementation and reliability in practice, a reduced-order controller is found that maintains the robust stability and robust performance of the closed-loop system. Finally, the advantages of the μ-controller over a conventional, collocated PD controller are demonstrated.

13.1 Dynamic Model of the Flexible Manipulator

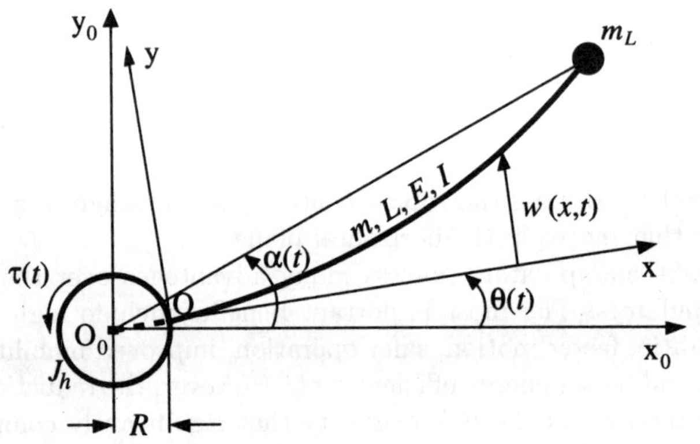

Fig. 13.1. Schematic diagram of the flexible-link manipulator

Figure 13.1 shows the schematic model used to derive the equations of motion for the flexible-link manipulator. The manipulator moves in the horizontal plane. Frame $x_0 - O_0 - y_0$ is the fixed-base frame. Frame $x - O - y$ is the local frame rotating with the hub. The x-axis coincides with the undeformed longitudinal axis of the link. The rotating inertia of the servomotor, the gear box, and the clamping hub are modelled as a single hub inertia J_h. The distance between the hub centre and the root of the link is denoted by R. The flexible link is assumed to be a homogeneous rod with a constant cross-sectional area. L is the length of the link, m is the mass per unit length of the link, I is the link cross-sectional moment of inertia and E is the Young's modulus of elasticity for the material of the link. The payload is modelled as a point mass m_L. The variables $\tau(t)$ and $\theta(t)$ are the driving torque and the joint angle, respectively. The elastic deflection of a point located at a distance x from O along the link is denoted by $w(x,t)$. It is assumed that the elastic deflections of the link lie in the horizontal plane, and are perpendicular to the x-axis and small in magnitude compared to the link length.

The motion equations of the flexible manipulator are to be derived by using the Lagrangian approach combined with the assumed-modes method [94]. The flexible link is modelled as an Euler–Bernoulli beam. The free vibration of the link is described by the partial differential equation [105]

$$EI\frac{\partial^4 w(x,t)}{\partial x^4} + m\frac{\partial^2 w(x,t)}{\partial t^2} = 0$$

with boundary conditions

$$w(0,t) = 0, \quad \frac{\partial w(0,t)}{\partial x} = 0$$

$$\frac{\partial^2 w(L,t)}{\partial x^2} = 0$$

$$\frac{\partial^3 w(L,t)}{\partial x^3} - \frac{m_L}{EI}\frac{\partial^2 w(L,t)}{\partial t^2} = 0$$

According to the assumed-modes method the elastic deflection can be expressed as

$$w(x,t) = \sum_{i=1}^{n} \varphi_i(x)\eta_i(t) \tag{13.1}$$

where $\eta_i(t)$ is the generalised coordinate of the ith mode, $\varphi_i(x)$ is the space eigenfunction of the ith mode, and n is the number of the modes that describe the link deflection. The mode angular frequencies $\omega_i,\ i = 1, ..., n$, of the flexible link are given by

$$\omega_i = \beta_i^2\sqrt{\frac{EI}{m}} \tag{13.2}$$

where $\beta_i,\ i = 1, ..., n$, are the first n positive roots of the transcendental equation

$$1 + \cosh(\beta L)\cos(\beta L) + \frac{m_L}{mL}(\beta L)\left(\sinh(\beta L)\cos(\beta L) - \cosh(\beta L)\sin(\beta L)\right) = 0 \tag{13.3}$$

The shape functions $\varphi_i(x),\ i = 1, ..., n$, satisfy the orthogonality condition

$$m\int_0^L \varphi_i(x)\varphi_j(x)dx + m_L\varphi_i(L)\varphi_j(L) = 0, \quad i \neq j$$

and can be written in the form

$$\varphi_i(x) = \lambda_i\left((\cosh(\beta_i x) - \cos(\beta_i x)) - \frac{\cosh(\beta_i L) + \cos(\beta_i L)}{\sinh(\beta_i L) + \sin(\beta_i L)}(\sinh(\beta_i x) - \sin(\beta_i x))\right) \tag{13.4}$$

A normalisation of the shape functions convenient for the uncertainty modelling is accomplished by determining the coefficients $\lambda_i,\ i = 1, ..., n$, in (13.4) on the basis of the relation

$$m\int_0^L \varphi_i^2(x)dx + m_L\varphi_i^2(L) = 1$$

The joint angle θ and the deflection variables η_i, $i = 1, ..., n$, are used as generalised coordinates in the derivation of the equation of motion. As a result of applying the Lagrangian procedure, the following nonlinear dynamic model of the flexible manipulator is obtained

$$\begin{bmatrix} m_r(\eta) & m_{rf}^T \\ m_{rf} & I_n \end{bmatrix} \begin{bmatrix} \ddot{\theta} \\ \ddot{\eta} \end{bmatrix} + \begin{bmatrix} d_v & 0_n^T \\ 0_n & D_f \end{bmatrix} \begin{bmatrix} \dot{\theta} \\ \dot{\eta} \end{bmatrix}$$
$$+ \begin{bmatrix} 0 & 0_n^T \\ 0_n & C_f \end{bmatrix} \begin{bmatrix} \theta \\ \eta \end{bmatrix} + \begin{bmatrix} h_r(\dot{\theta}, \eta, \dot{\eta}) \\ h_f(\dot{\theta}, \eta) \end{bmatrix} = \begin{bmatrix} 1 \\ 0_n \end{bmatrix} \tau \quad (13.5)$$

where

$$\begin{aligned}
\eta &= [\eta_1 \; ... \; \eta_n]^T \\
m_r(\eta) &= a_0 + \sum_{i=1}^{n} \eta_i^2 \\
a_0 &= J_a + \frac{1}{3} m \left((L+R)^3 - R^3 \right) + m_L (L+R)^2 \\
m_{rf} &= [a_1 \; ... \; a_n]^T \\
a_i &= m \int_0^L (x+R) \varphi_i(x) dx + m_L (L+R) \varphi_i(L) \qquad (13.6) \\
C_f &= \mathrm{diag}(\omega_1^2, \; ..., \; \omega_n^2) \\
D_f &= \mathrm{diag}(d_{f1}, \; ..., \; d_{fn}) \\
h_r(\dot{\theta}, \eta, \dot{\eta}) &= \sum_{i=1}^{n} 2\dot{\theta}\dot{\eta}_i \eta_i \\
h_f(\dot{\theta}, \eta) &= \left[-\dot{\theta}^2 \eta_1 \; ... \; -\dot{\theta}^2 \eta_n \right]^T
\end{aligned}$$

I_n denotes the $n \times n$ identity matrix, 0_n is the n-dimensional null vector, and d_v, d_{f1}, ..., d_{fn} are damping coefficients. The terms $d_v \dot{\theta}$ and $D_f \dot{\eta}$ have been included to account for the viscous friction at the hub and for the structural damping of the flexible link, respectively.

The angle

$$\alpha = \theta + \arctan \frac{w(L,t)}{L+R} \quad (13.7)$$

is chosen as the coordinate that determines the position of the manipulator tip.

The following numerical values of the manipulator parameters are used: $L = 1$ m, $R = 0.4$ m, $J_h = 0.1$ kg m^2, $m = 0.54$ kg/m, flexural rigidity of the flexible link $EI = 18.4$ N m^2. The values of m and EI correspond to an aluminium link with $E = 6.9 \times 10^{10}$ N m^2, density $\rho = 2700$ kg/m^3 and a cross–section 0.004 m$\times$0.05 m.

It is assumed that in performing a given motion the payload mass has a constant but unknown value in the range from 0.125 kg to 0.375 kg. It is also assumed that the coefficients in the damping terms $d_v\dot{\theta}$ and $D_f\dot{\eta}$ are known inaccurately.

The first two natural frequencies of the flexible link, calculated for the average value of the payload according to (13.2) and (13.3), are $\omega_1 = 12.1$ rad/s and $\omega_2 = 99.2$ rad/s. Since the rest natural frequencies are very large ($\omega_3 = 302.5$ rad/s and so on), a two-mode model of the flexible manipulator is used in the controller design.

13.2 A Linear Model of the Uncertain System

In this section we first consider how to model the uncertainties of the flexible-link manipulator and then develop a complete, linear dynamic model of the system in the form of linear fractional transformation (LFT). As mentioned earlier, the uncertainties considered are related to the payload mass, hub damping coefficient and the damping levels of the first two modes. It is important to note that these parameters are the basic source of uncertainty dynamic behaviour of flexible-link manipulators.

In the modelling of uncertain damping levels of the flexible modes, we may set $d_{fi} = d_i\omega_i, i = 1, 2$ where ω_i are the first two natural frequencies and the damping factors d_1, d_2 are considered as uncertain parameters. The uncertain parameters m_L, d_v, d_1, d_2 may be represented as follows

$$m_L = \overline{m}_L(1 + p_m\delta_m),\ d_v = \overline{d}_v(1 + p_{d_v}\delta_{d_v})$$

$$d_1 = \overline{d}_1(1 + p_{d_1}\delta_{d_1}),\ d_2 = \overline{d}_2(1 + p_{d_2}\delta_{d_2})$$

where the uncertain variables $\delta_m, \delta_{d_v}, \delta_{d_1}, \delta_{d_2}$ are real and satisfy the normalised bound

$$-1 \leq \delta_m, \delta_{d_v}, \delta_{d_1}, \delta_{d_2} \leq 1$$

The nominal values and the maximum relative uncertainty bounds of those parameters are set as $\overline{m}_L = 0.25$ kg, $p_m = 0.5$, $\overline{d}_v = 0.15$ kg m^2/s, $p_{d_v} = 0.1$, $\overline{d}_1 = 0.03$ kg m^2, $p_{d_1} = 0.2$, $\overline{d}_2 = 0.1$ kg m^2, $p_{d_2} = 0.2$.

The plant input is the driving torque τ. The controlled variable is the tip position α and the measured variables are the joint angle θ and the tip acceleration $\ddot{\alpha}$.

To obtain a linear model of the manipulator the nonlinear terms $\sum_{i=1}^{n}\eta_i^2$, $h_r(\dot{\theta}, \eta, \dot{\eta})$ and $h_f(\dot{\theta}, \eta)$ in (13.5) are neglected due to the fact that their effects are relatively very small. Also, since $w(L, t) << L$, the function $\arctan(z)$ in (13.7) can be approximated by z. As a result, we may obtain the following equations

$$a_0\ddot{\theta} + d_v\dot{\theta} + a_1\ddot{\eta}_1 + a_2\ddot{\eta}_2 = \tau \qquad (13.8)$$

$$\ddot{\eta}_i + d_i\omega_i\dot{\eta}_i + \omega_i^2\eta_i + a_i\ddot{\theta} = 0, \quad i = 1, 2 \qquad (13.9)$$

$$\alpha = \theta + b_1\eta_1 + b_2\eta_2 \qquad (13.10)$$

$$\ddot{\alpha} = \ddot{\theta} + b_1\ddot{\eta}_1 + b_2\ddot{\eta}_2 \qquad (13.11)$$

where the following notation is used

$$b_i = \frac{\varphi_i(L)}{L+R}, \quad i = 1, 2 \tag{13.12}$$

The difficulty in modelling the uncertainty in this case study comes from the fact that the coefficients a_0, a_1, a_2, ω_1, ω_2, b_1 and b_2 in (13.8)–(13.11) are functions of the payload mass m_L. Among these coefficients, only a_0 explicitly depends on m_L in an affine form. The rest coefficients depend on m_L in a complicated, nonlinear and implicit manner as is seen from (13.2), (13.3), (13.4), (13.6), (13.12). Direct approximation of these coefficients by linear functions of the payload mass leads to a very inaccurate model and the uncertain parameter δ_m would be repeated 13 times. This difficulty is approached here by exploiting the relations between the coefficients as functions of the payload mass. The analysis of these functions shows that there exist sufficiently accurate linear dependencies between appropriately chosen coefficients that can be used to derive a simple and more accurate uncertainty model. The best-suited dependencies are chosen so as to reduce significantly the number of uncertain parameters in the final model achieving in the same time a high accuracy in the description of the uncertainty in the payload mass. This procedure is briefly described in the following.

Equations (13.8) and (13.9) can be rewritten in the form

$$a_0\ddot{\theta} + d_v\dot{\theta} = \tau - a_1\ddot{\eta}_1 - a_2\ddot{\eta}_2 \tag{13.13}$$

$$\frac{1}{\omega_1}\ddot{\eta}_1 + d_1\dot{\eta}_1 + \omega_1\eta_1 = -\frac{a_1}{\omega_1}\ddot{\theta} \tag{13.14}$$

$$\frac{1}{\omega_2}\ddot{\eta}_2 + d_2\dot{\eta}_2 + \omega_2\eta_2 = -\frac{a_2}{\omega_2}\ddot{\theta} \tag{13.15}$$

The coefficients a_1/ω_1, a_2/ω_2 that appear in (13.14) and (13.15) may be expressed as linear functions of a_0, which is in turn a function of m_L, by using the following approximate relationships

$$\frac{a_1}{\omega_1} \approx k_1 a_0 + k_2 \tag{13.16}$$

$$\frac{a_2}{\omega_2} \approx k_3 a_0 + k_4 \tag{13.17}$$

where k_1, k_2, k_3 and k_4 are constants to be determined appropriately. The numerically calculated values of a_1/ω_1 and a_2/ω_2 are shown, as functions of m_L, in Figure 13.2. By using the least square method, linear approximations are obtained (Figure 13.2). The coefficients of the linear approximations are $k_1 = 0.12905$, $k_2 = -0.017706$, $k_3 = 5.5600 \times 10^{-4}$ and $k_4 = 6.8034 \times 10^{-4}$ (all numbers are given to five significant digits).

Accordingly for the variables $(a_i/\omega_i)\ddot{\theta}$, $i = 1, 2$, we have

$$\frac{a_1}{\omega_1}\ddot{\theta} \approx (k_1 a_0 + k_2)\ddot{\theta} \tag{13.18}$$

$$\frac{a_2}{\omega_2}\ddot{\theta} \approx (k_3 a_0 + k_4)\ddot{\theta} \tag{13.19}$$

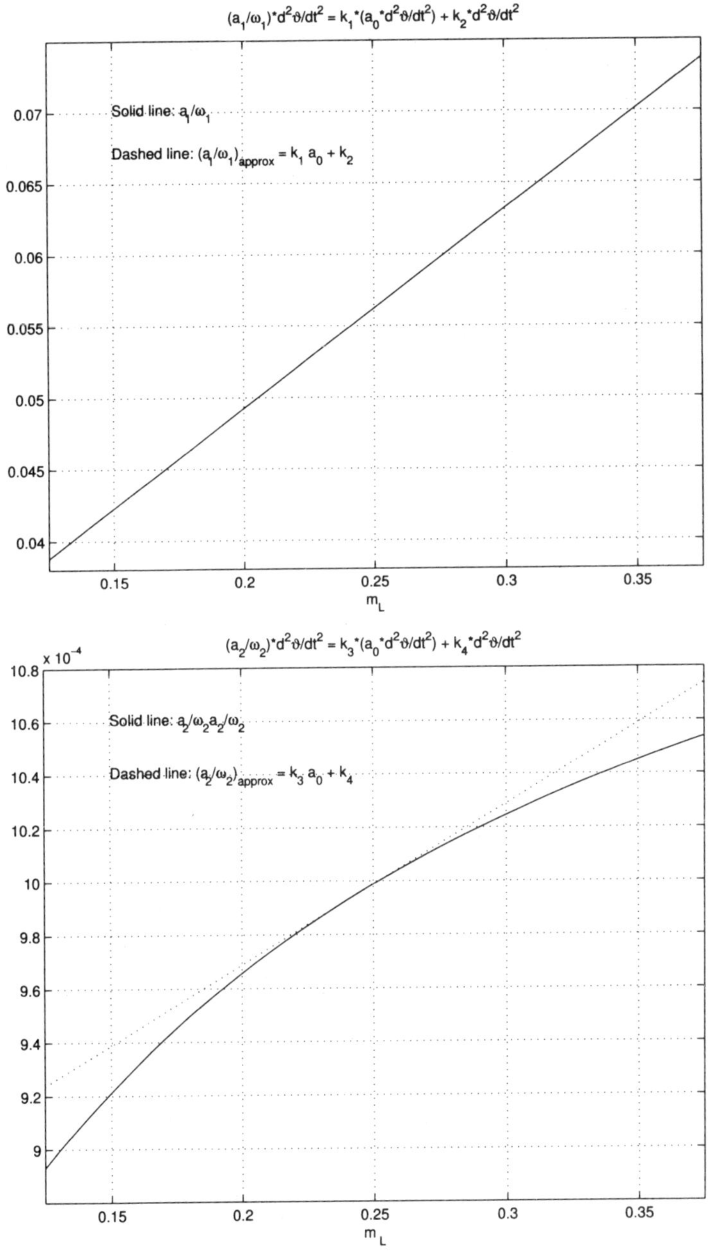

Fig. 13.2. Approximation of a_1/ω_1 (above) and a_2/ω_2 (below) in the expressions for $(a_i/\omega_i)\ddot{\theta}_i,\ i = 1, 2$

Using (13.18) and (13.19), the variables $(a_i/\omega_i)\ddot{\theta}$, $i = 1, 2$, are determined from the variables $a_0\ddot{\theta}$ and $\ddot{\theta}$, and (13.13) can be depicted in a block diagram as in Figure 13.3.

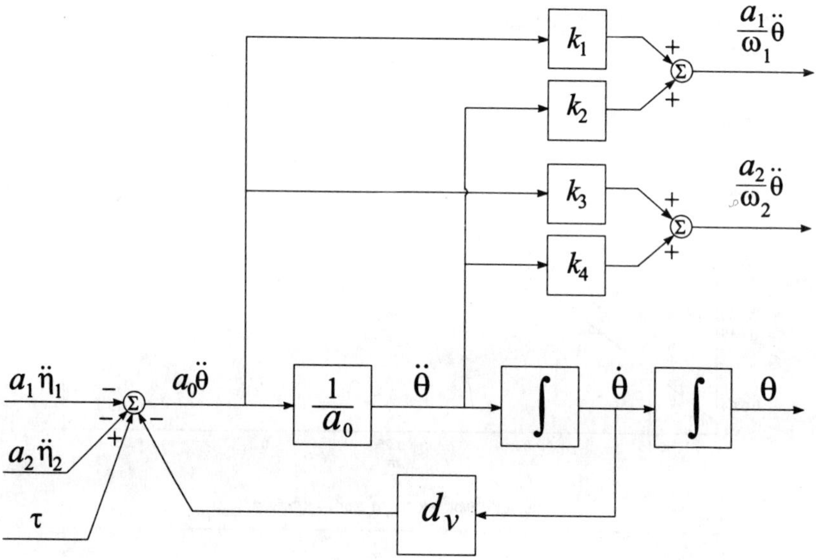

Fig. 13.3. Block diagram of the joint angle equation

In a similar way, the quantities $a_i\ddot{\eta}_i$, appearing in (13.13), can be calculated from $(1/\omega_i)\ddot{\eta}_i$ and $\ddot{\eta}_i$ on the basis of

$$a_1 \approx k_5\frac{1}{\omega_1} + k_6$$

$$a_2 \approx k_{11}\frac{1}{\omega_2} + k_{12}$$

The actual, calculated values and the approximation quantities of a_1 and a_2, as functions of m_L, are shown in Figure 13.4, where $k_5 = 7.2587$, $k_6 = 0.077825$, $k_{11} = 10.165$ and $k_{12} = -0.0033967$.

Again, the block diagrams showing the determination of $a_1\ddot{\eta}_1$ and $a_2\ddot{\eta}_2$ from the variables $\ddot{\eta}_1$ and $\ddot{\eta}_2$ are shown in Figures 13.5 and 13.6, respectively.

The coefficients ω_1 and ω_2 can be expressed in terms of b_1 and b_2, respectively, according to

$$\omega_1 = k_9 b_1 + k_{10}$$

$$\omega_2 = k_{15} b_2 + k_{16}$$

Figure 13.7 plots the actual and approximate quantities of ω_1 and ω_2, as functions of m_L, with $k_9 = 7.7970$, $k_{10} = -0.089136$, $k_{15} = -13.355$ and $k_{16} = 89.500$.

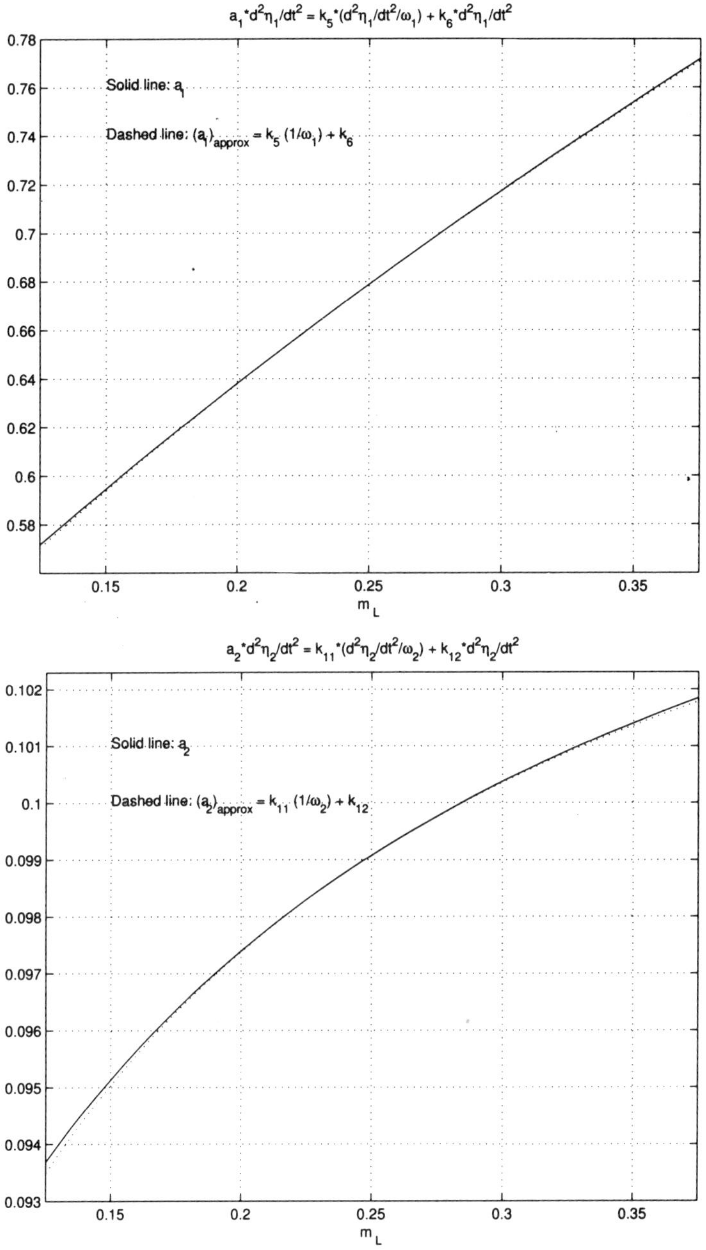

Fig. 13.4. Approximation of a_1 (above) and a_2 (below) in the expressions for $a_i\ddot{\eta}_i$, $i = 1, 2$

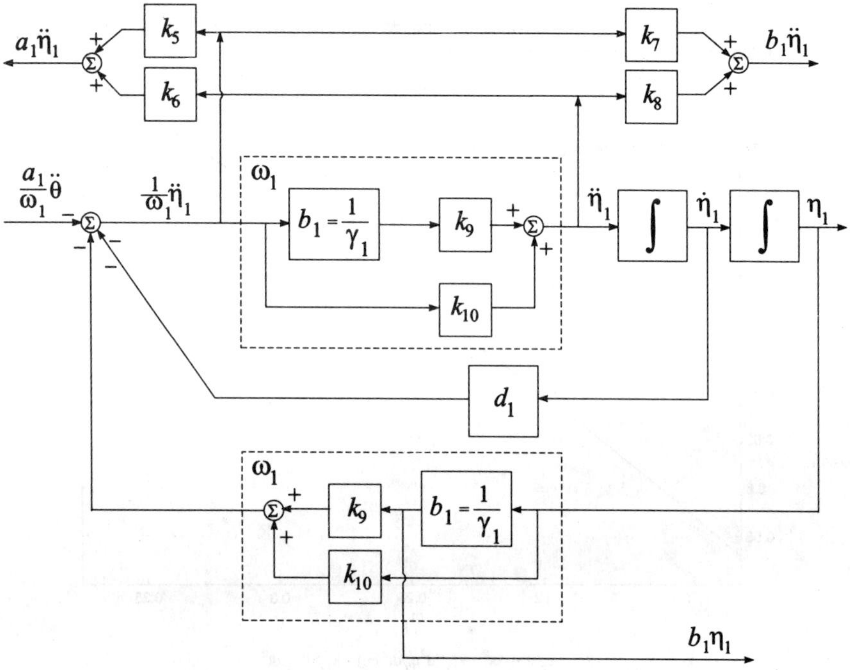

Fig. 13.5. Block diagram of the first elastic mode

In this way we obtain an uncertainty model corresponding to (13.8) and (13.9). In this model only the coefficients a_0^{-1}, ω_1 and ω_2 depend on the payload mass, while the coefficients ω_1 and ω_2 are repeated twice.

The part of the uncertainty model corresponding to (13.10) and (13.11) can be derived without introducing new uncertainty parameters. The terms $b_i\ddot{\eta}_i$, $i = 1, 2$, are expressed by $(1/\omega_i)\ddot{\eta}_i$ and $\ddot{\eta}_i$ using the relationships

$$b_1 \approx k_7 \frac{1}{\omega_1} + k_8$$

$$b_2 \approx k_{13} \frac{1}{\omega_1} + k_{14}$$

Similarly, Figure 13.8 plots the calculated and approximate quantities of b_1 and b_2, as functions of m_L, with $k_7 = -18.980$, $k_8 = 3.1319$, $k_{13} = 7.4054$ and $k_{14} = -8.1915$.

The block diagrams showing the variables $b_1\ddot{\eta}_1$ and $b_2\ddot{\eta}_2$, based on the variables $\ddot{\eta}_1$ and $\ddot{\eta}_2$, are already included in Figures 13.5 and 13.6, respectively.

In the expressions for ω_1, ω_2 and $b_1\eta_1$, $b_2\eta_2$, the coefficients b_1 and b_2 can be represented as LFTs in the real uncertain parameter δ_m by using the following approximate relationships

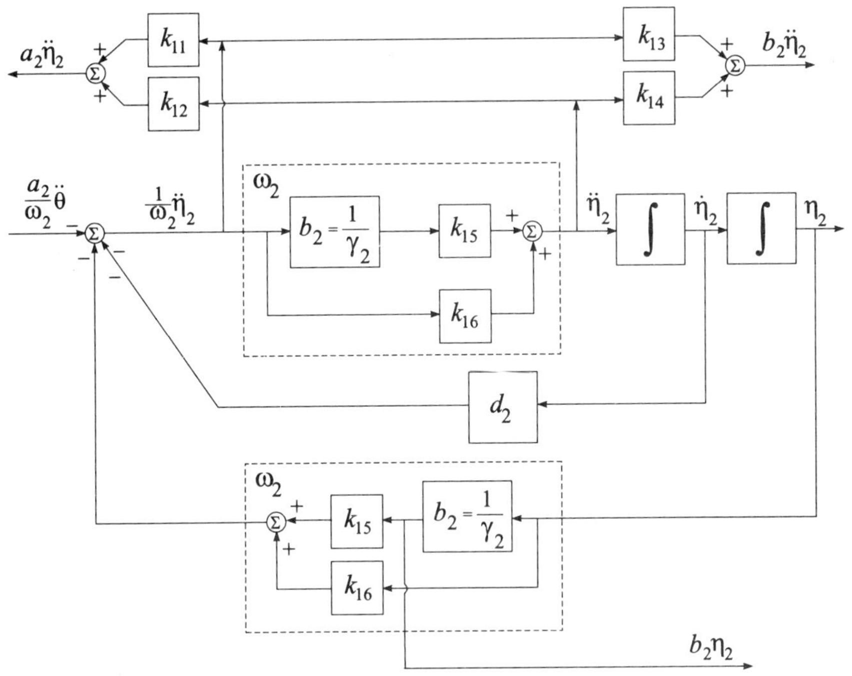

Fig. 13.6. Block diagram of the second elastic mode

$$b_1 \approx \frac{1}{\gamma_1}$$

$$b_2 \approx \frac{1}{\gamma_2}$$

where γ_1 and γ_2 depend affinely on m_L and can be written as

$$\gamma_1 = \overline{\gamma}_1(1 + p_{\gamma_1}\delta_m)$$

$$\gamma_2 = \overline{\gamma}_2(1 + p_{\gamma_2}\delta_m)$$

The calculated and approximate quantities of b_1 and b_2 as functions of m_L are shown in Figure 13.9 for $\overline{\gamma}_1 = 0.64095$, $p_{\gamma_1} = 0.16414$, $\overline{\gamma}_2 = -1.3780$ and $p_{\gamma_2} = 0.36534$.

The constants $k_1, ..., k_{16}, \overline{\gamma}_1, p_{\gamma_1}, \overline{\gamma}_2$ and p_{γ_2} are all determined by least squares approximations in such a way that for the nominal payload mass the corresponding relationships are satisfied perfectly (interpolation conditions). Hence for the nominal payload the manipulator model is accurate. In the approximations, the worst relative error for each relationship is always obtained at the case $m_L = 0.125$ kg, and the largest relative error among all these is less than than 3.4% . These constants are found by using the file `par_flm`.

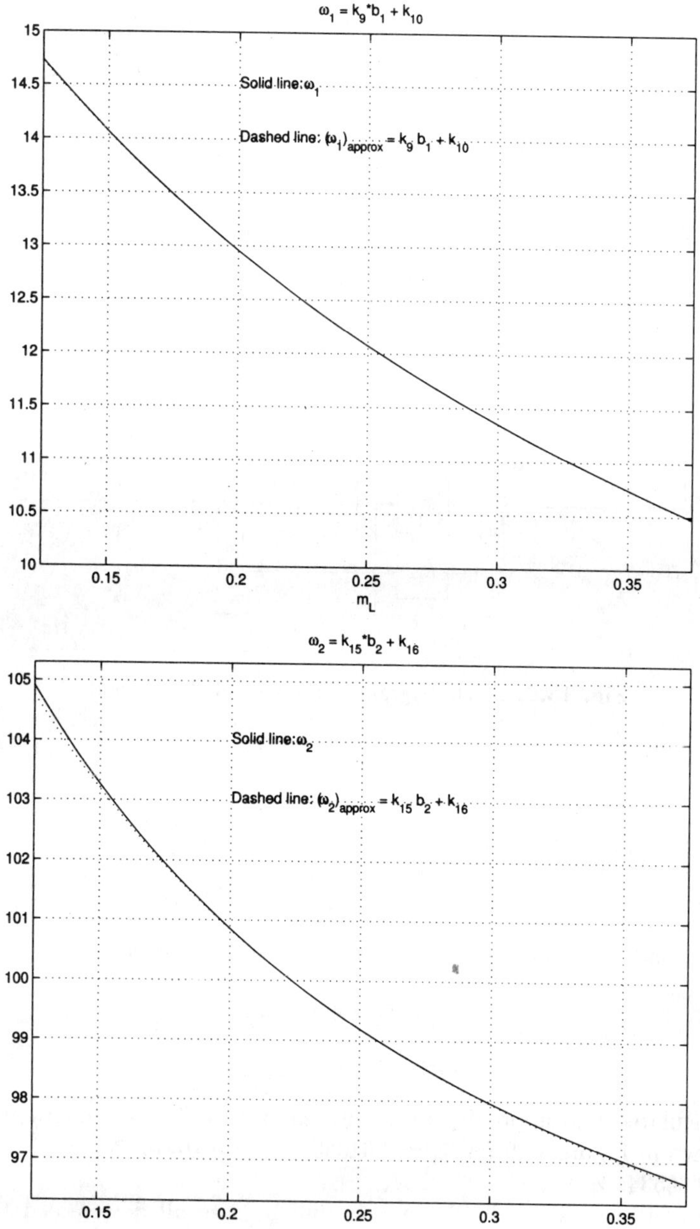

Fig. 13.7. Approximation of ω_1 (above) and ω_2 (below)

Once appropriate approximations of the coefficients have been obtained it is possible to develop the uncertainty models corresponding to (13.13)–(13.15). This is done by using the file `mod_flm`.

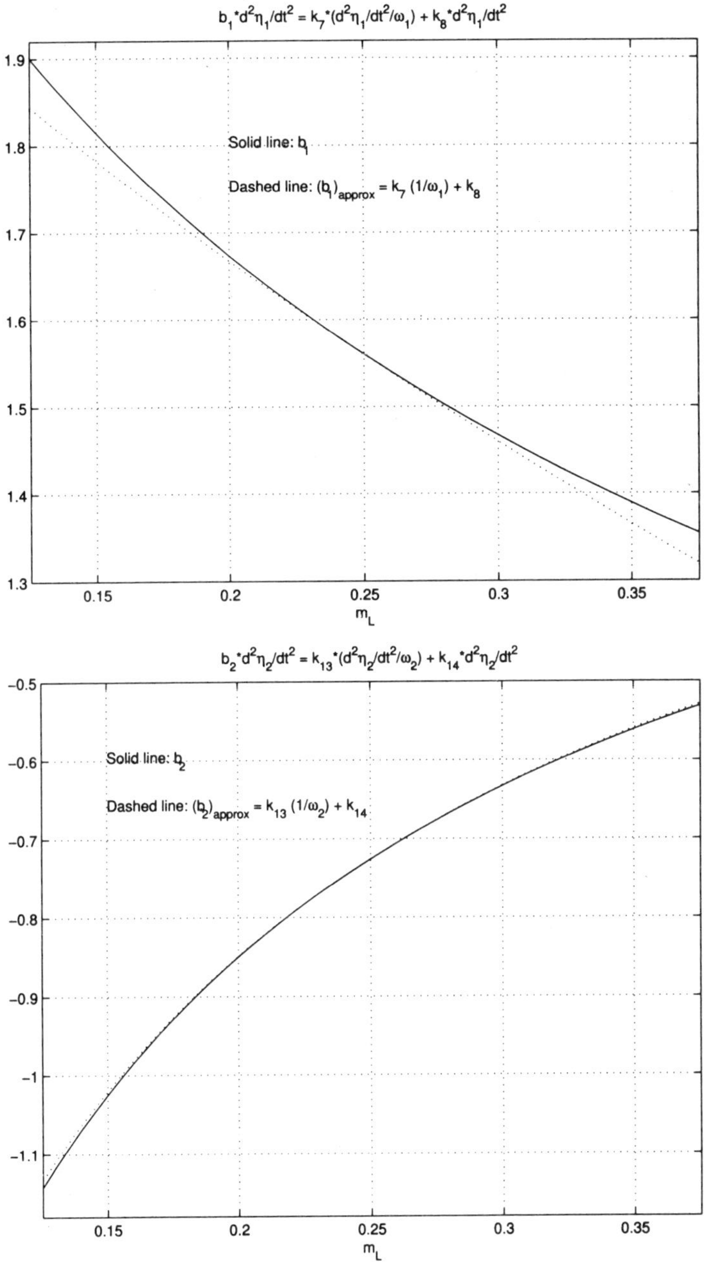

Fig. 13.8. Approximation of b_1 (above) and b_2 (below) in the expressions for $b_i\ddot{\eta}_i,\ i = 1, 2$

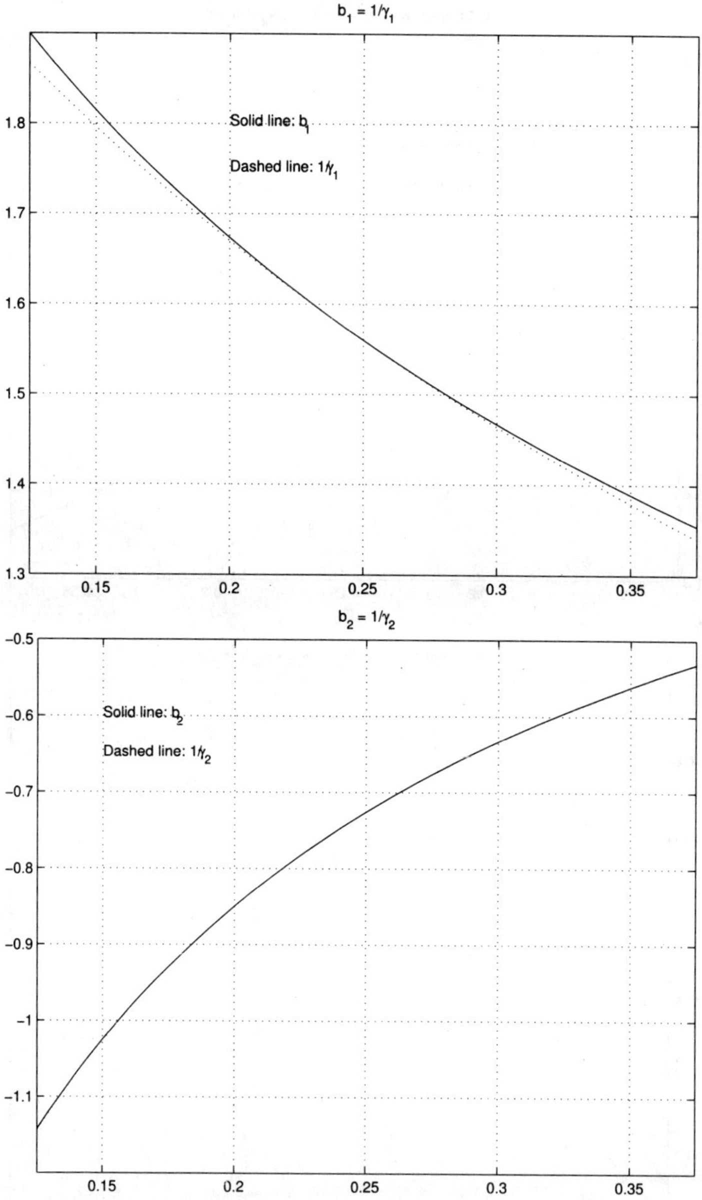

Fig. 13.9. Approximation of b_1 (above) and b_2 (below) in the expressions for ω_i and $b_i\eta_i$, $i = 1, 2$

In Figure 13.10 we show the block diagram of the joint angle with uncertain parameters derived from the block diagram shown in Figure 13.2.

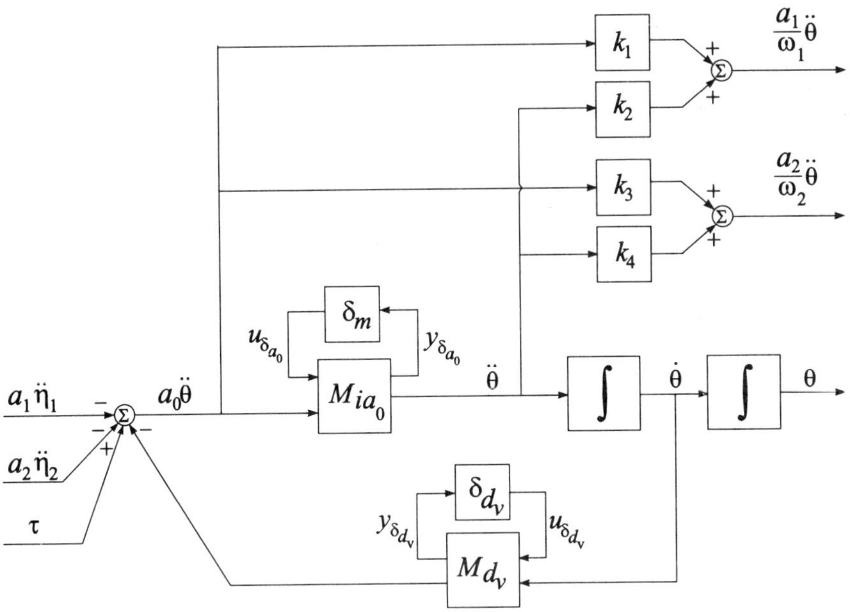

Fig. 13.10. Block diagram of the joint angle with uncertain parameters

The corresponding interconnection is obtained by using the function `sysic`. A schematic diagram of the input/output ordering for this interconnection is shown in Figure 13.11.

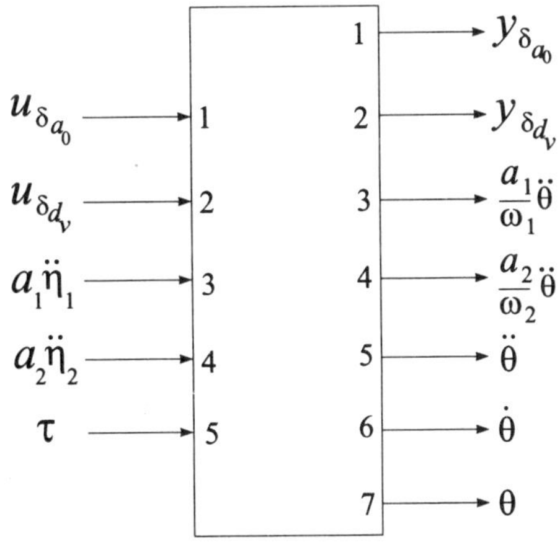

Fig. 13.11. Schematic diagram of the joint angle connection

The uncertainty model of the first elastic mode is given in Figure 13.12 and a schematic diagram of the corresponding interconnection is shown in Figure 13.13.

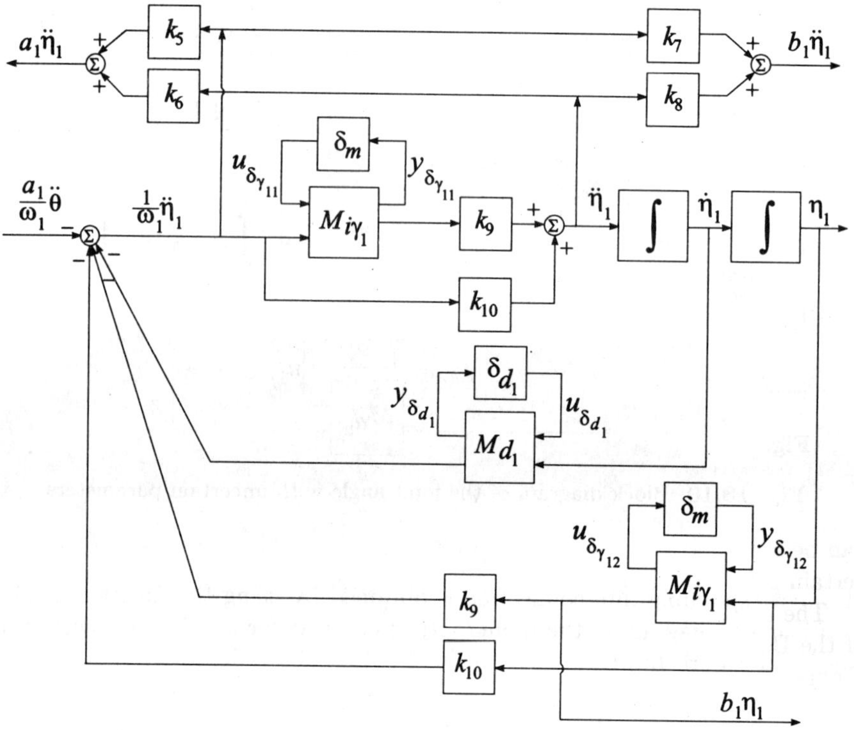

Fig. 13.12. First elastic mode with uncertain parameters

The uncertainty model of the second elastic mode (Figure 13.14) is obtained in a similar way and a schematic diagram of the input/output ordering is shown in Figure 13.15.

A schematic diagram for the output equations interconnection is shown in Figure 13.16.

Finally, the uncertainty model of the flexible link manipulator is obtained by connecting the models shown in Figures 13.10, 13.12 and 13.14. A schematic diagram of the input/output ordering of the corresponding interconnection is shown in Figure 13.17. This model has 9 inputs and 20 outputs.

Based on the above modelling, an LFT model of the flexible-link manipulator with diagonal uncertainty matrix

$$\Delta = \operatorname{diag}(\delta_m I_5, \delta_v, \delta_{d_1}, \delta_{d_2}) \tag{13.20}$$

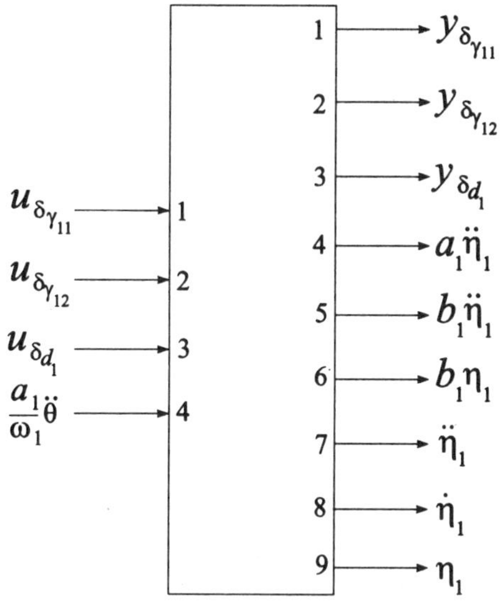

Fig. 13.13. Schematic diagram of the first elastic mode interconnection

can be readily obtained. Note that in this modelling of uncertainties the uncertain parameter δ_m appears repeatedly five times.

The accuracy of such a derived model may be verified by the comparison of the Bode plots of the exact and approximate models of the manipulator. Figure 13.18 shows the Bode plots for the case $m_L = 0.125$ kg. The plots are obtained from the corresponding transfer functions of the input τ and output θ. The Bode plot of the exact model is computed on the basis of (13.8) and (13.9), using the exact values of the coefficients. The Bode plot of the "approximate" model is calculated based on the derived uncertainty model. It can be seen that the match between those two models is very good. Similar closeness of the corresponding plots holds for transfer functions of other plant outputs from the input τ.

It has been found in this case study that if the approximation of the coefficients in (13.8)–(13.11) as functions of the payload mass is conducted directly with linear dependencies, then the model obtained would be very inaccurate. And, in that case, the uncertain parameter δ_m would repeat itself 13 times in the uncertainty model.

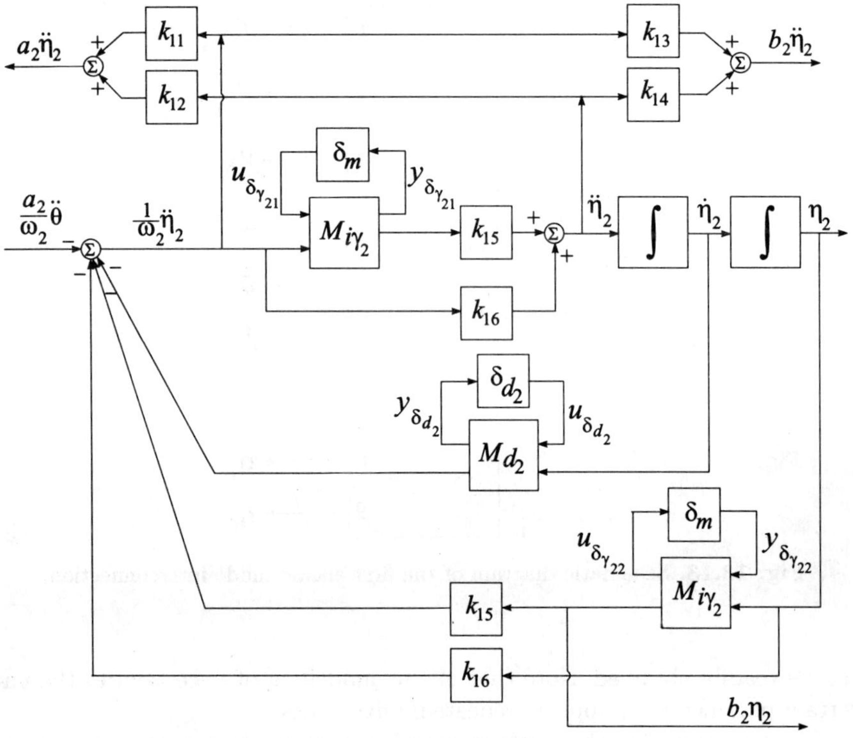

Fig. 13.14. Second elastic mode with uncertain parameters

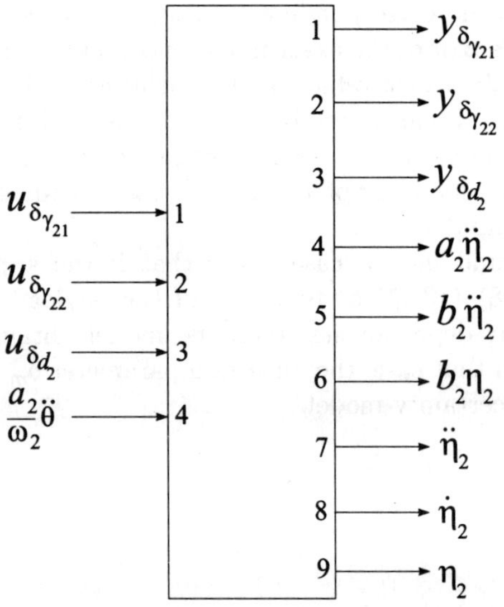

Fig. 13.15. Schematic diagram of the second elastic mode interconnection

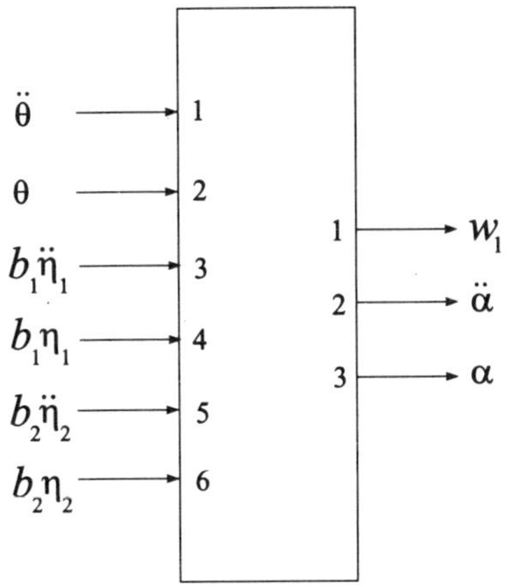

Fig. 13.16. Schematic diagram of the output equations interconnection

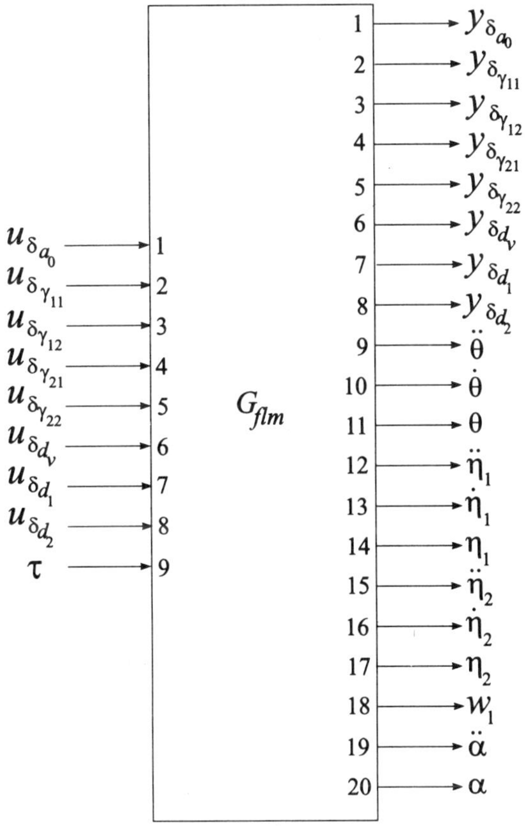

Fig. 13.17. Schematic diagram of the flexible-link manipulator interconnection

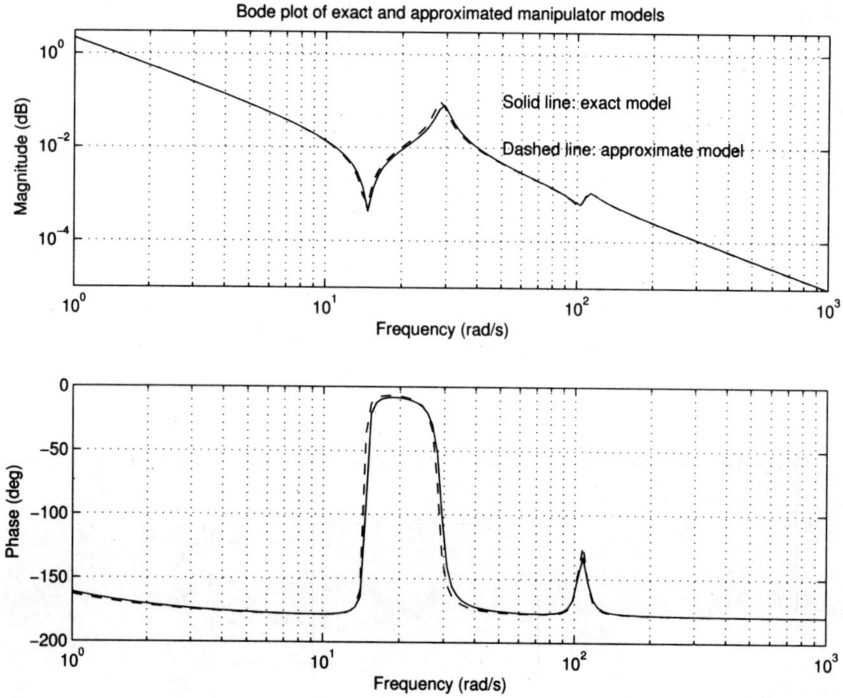

Fig. 13.18. Bode plots of exact and approximated manipulator models for m_L = 0.125 kg

13.3 System-performance Specifications

In this flexible-link manipulator control-system design exercise, the purpose is to find a controller that suppresses efficiently the elastic vibrations of the flexible link in fast motions and moves the tip to a desired position in the presence of uncertainties in the payload mass, hub and structural damping factors. Since the uncertainties considered are real and structured, the most appropriate robust control design method to be applied in the present case is the μ-synthesis.

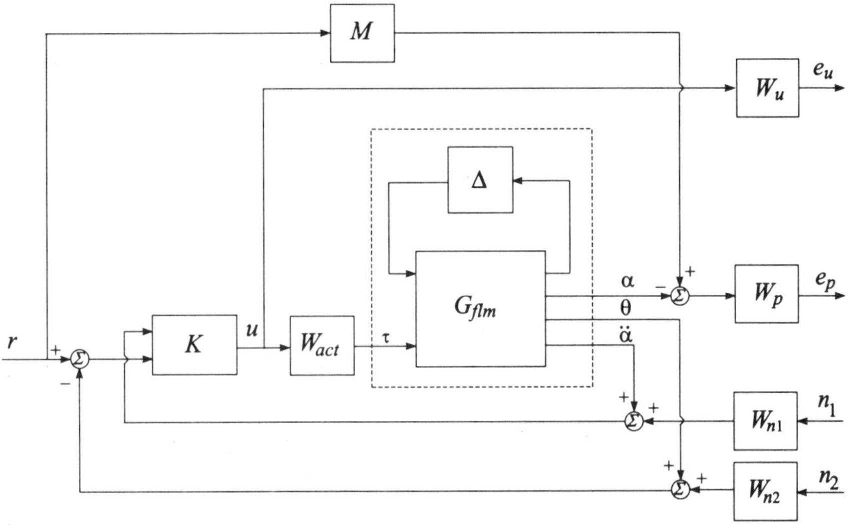

Fig. 13.19. Closed-loop interconnection structure of the flexible-link manipulator system

The block diagram of the closed-loop system incorporating the design requirements is shown in Figure 13.19. The controller K works on the feedback signals of the tip acceleration $\ddot{\alpha}$ and the joint angle θ. The inclusion of the tip acceleration in the control scheme aims to achieve better tip-motion performance and leads to a noncollocated controller structure. Furthermore, in the given design case, we select a suitable dynamic model and target the dynamics of the designed closed-loop system to be close to that model. The use of such a model to represent the desired dynamics allows us to take into account the requirements on system performance more easily and directly. In other words, such a model (named M in Figure 13.19) prescribes the desired dynamic behaviour of the closed-loop system from the reference signal to the tip position. In Figure 13.19 the plant G, enclosed by the dashed rectangle, is in the form of upper LFT, $G = F_U(G_{flm}, \Delta)$, with the nominal model G_{flm}

and the parametric uncertainty matrix Δ (in (13.20)). The internal, current control loop of the servo drive W_{act} is modelled as a first-order lag with the time constant 0.003 s.

Let the 3×1 transfer matrix G be partitioned as

$$G(s) = \begin{bmatrix} G_{\alpha\tau}(s) \\ G_{\theta\tau}(s) \\ G_{\ddot{\alpha}\tau}(s) \end{bmatrix}$$

where $G_{\alpha\tau}$, $G_{\theta\tau}$, $G_{\ddot{\alpha}\tau}$ are the transfer functions from the control torque τ to the outputs α, θ and $\ddot{\alpha}$, respectively, and let the controller

$$K(s) = [K_1(s)\ K_2(s)]$$

It can be shown by direct manipulations that

$$\begin{bmatrix} e_p \\ e_u \end{bmatrix} = \Phi \begin{bmatrix} r \\ n_1 \\ n_2 \end{bmatrix}$$

where

$$\Phi =$$

$$\begin{bmatrix} W_p(SG_{\alpha\tau}W_{act}K_2 - M) & W_pSG_{\alpha\tau}W_{act}K_1W_{n_1} & -W_pSG_{\alpha\tau}W_{act}K_2W_{n_2} \\ W_uSK_2 & W_uSK_1W_{n_1} & -W_uSK_2W_{n_2} \end{bmatrix}$$

and

$$S = \frac{1}{1 - G_{\ddot{\alpha}\tau}W_{act}K_1 + G_{\theta\tau}W_{act}K_2}$$

The design objective for the controller K is thus to be set as

$$\|\Phi\|_\infty < 1 \tag{13.21}$$

for all perturbed $\begin{bmatrix} G_{\alpha\tau}(s) \\ G_{\theta\tau}(s) \\ G_{\ddot{\alpha}\tau}(s) \end{bmatrix} = F_U(G_{flm}, \Delta)$.

It is clear that, with appropriately chosen weighting functions, a controller K satisfying the above (13.21) makes the closed-loop system robustly stable, robustly achieving good matching to the dynamic model M (in terms of e_p), and with restricted control effort (in terms of e_u).

The model transfer function to be matched is taken in this design as

$$M = \frac{625}{s^2 + 50s + 625}$$

The coefficients of this transfer function are chosen to ensure overdamped response with a settling time of about 0.19 s. The performance weighting functions are chosen as

$$W_p = \frac{s^2 + 25s + 150}{s^2 + 22s + 0.15}, \quad W_u = 10^{-5}\frac{5s+1}{0.05s+1}$$

The criterion for the performance weighting function W_p aims to ensure the closeness of the system dynamics to that of the model M over the low-frequency range. The use of the control weighting function W_u allows us to bound the magnitude of the control action in the frequency range containing the natural frequencies of the flexible link. The magnitude plots of the inverses of the performance and control weighting functions are shown in Figure 13.20.

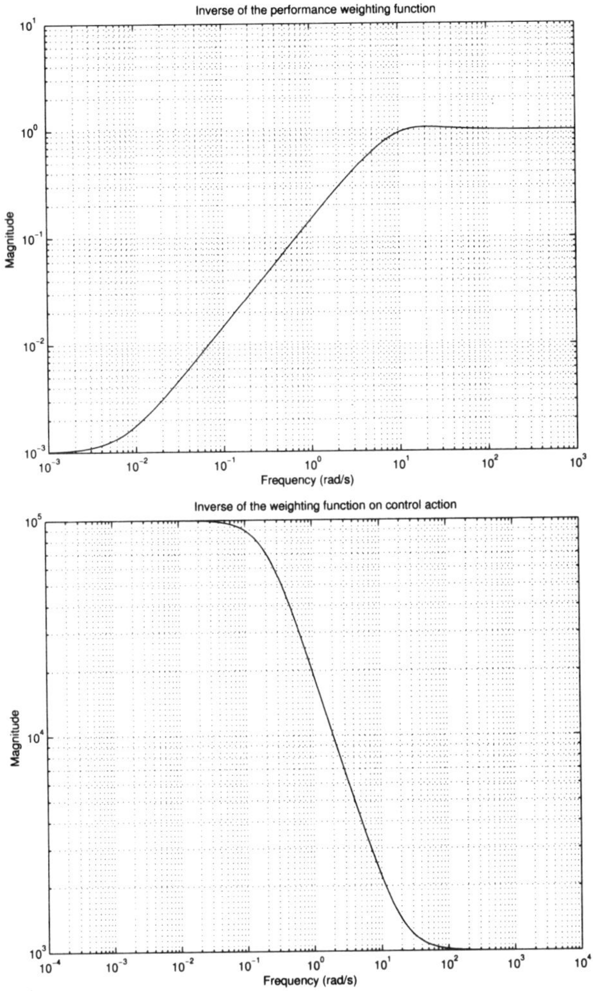

Fig. 13.20. Inverses of performance weighting function (above) and control weighting function (below)

The noise shaping filters

$$W_{n_1} = 2 \times 10^{-5} \frac{s+1}{0.01s+1}, \quad W_{n_2} = 10^{-7} \frac{0.5s+1}{0.005s+1}$$

are determined according to the spectral contents of the sensor noises n_1 and n_2 at the measurements of the tip acceleration and joint angle signals, respectively. The magnitude plots of the noise shaping filters are shown in Figure 13.21.

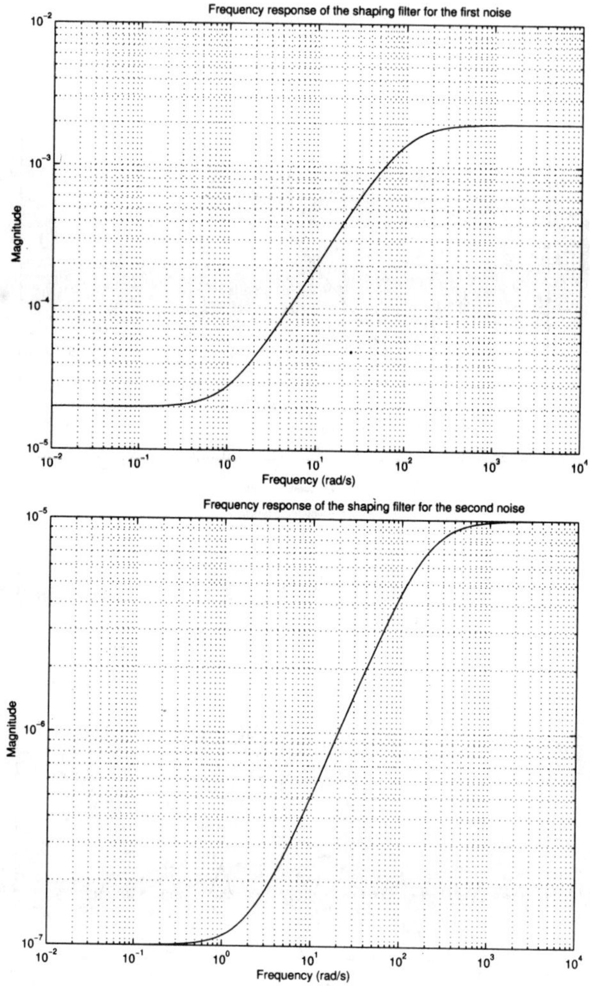

Fig. 13.21. Noise weighting functions

The model transfer function, the performance and control weighting functions as well as the noise shaping filters are assigned in the file `wts_flm.m`.

13.4 System Interconnections

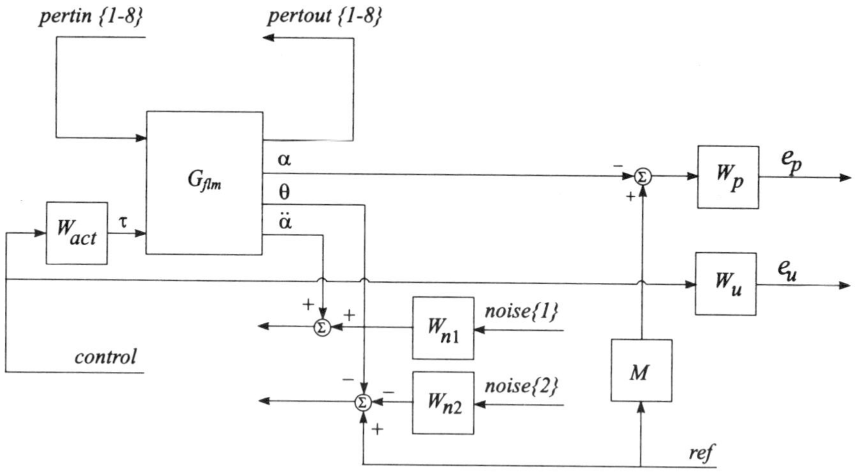

Fig. 13.22. Open-loop interconnection structure of the flexible-link manipulator system

The open-loop interconnection is obtained by the M-file `olp_flm`. The internal structure of the 12-input/12-output open-loop system, which is saved in the variable `sys_ic`, is shown in Figure 13.22. The inputs and outputs of the uncertainties are saved in the variables `pertin` and `pertout`, the reference and the noises saved in the variables `ref`, `noise1` and `noise2`, and the control signal in the variable `control`.

Both variables `pertin` and `pertout` have eight elements, while the rest variables are scalars.

A schematic diagram of the specific input/output ordering for the variable `sys_ic` is shown in Figure 13.23.

The block-diagram used in the simulation of the closed-loop system is shown in Figure 13.24. The corresponding closed-loop interconnection, which is saved in the variable `sim_ic`, is obtained by the M-file `sim_flm`.

A schematic diagram of the specific input/output ordering for the variable `sim_ic` is shown in Figure 13.25.

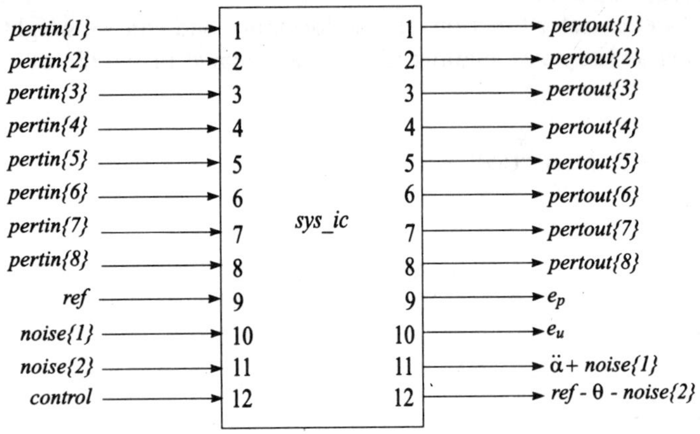

Fig. 13.23. Schematic diagram of the open-loop interconnection

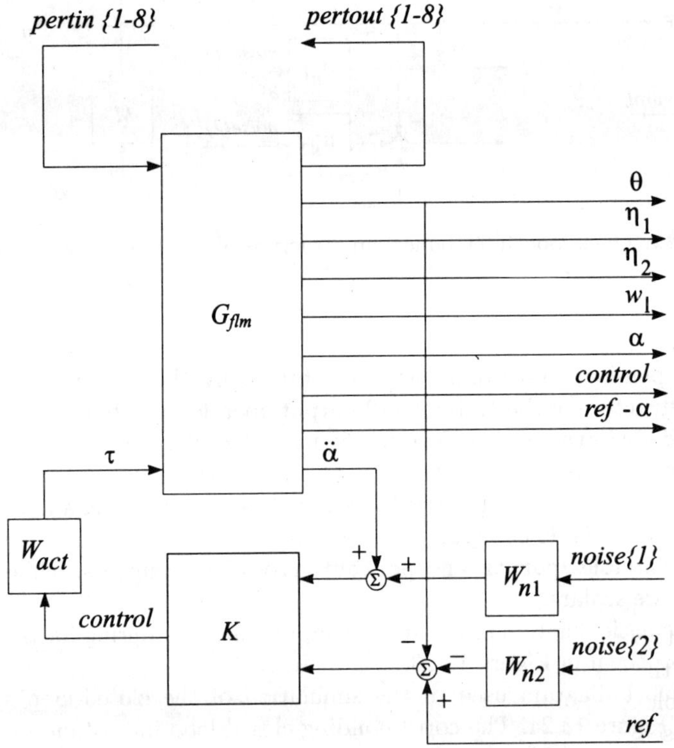

Fig. 13.24. Closed-loop interconnection structure of the flexible-link manipulator system

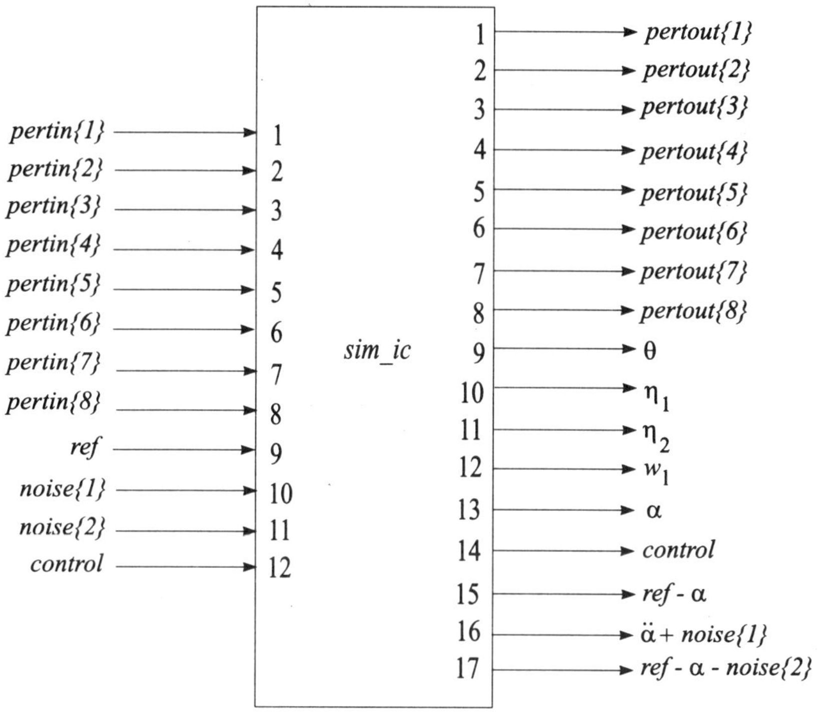

Fig. 13.25. Schematic diagram of the closed-loop interconnection

13.5 Controller Design and Analysis

Let us denote by $P(s)$ the transfer function matrix of the twelve-input, twelve-output open-loop system consisting of the flexible-link manipulator model and the actuator and weighting functions (Figure 13.22). Define a block structure of uncertainty Δ_P as

$$\Delta_P := \left\{ \begin{bmatrix} \Delta & 0 \\ 0 & \Delta_F \end{bmatrix} : \Delta \in \mathcal{R}^{8\times 8},\ \Delta_F \in \mathcal{C}^{3\times 2} \right\}$$

The first part of this matrix corresponds to the uncertain block Δ that is used in the modelling of the uncertainties in the flexible manipulator. The second block Δ_F is a fictitious uncertainty 3×2 block and is introduced to represent the robust performance objective in the framework of the μ-approach. The inputs to the block Δ_F are the weighted error signals e_p and e_u and the outputs from Δ_F are the exogenous signals r, n_1 and n_2 (inputs to the manipulator closed-loop system).

As discussed in previous sections, in order to meet the design objectives a stabilising controller $K = [K_1(s)\ K_2(s)]$ is to be found such that, at each frequency $\omega \in [0, \infty]$, the structured singular value satisfies the condition

$$\mu_{\Delta_P}[F_L(P,K)(j\omega)] < 1$$

The fulfillment of the above condition guarantees the robust performance of the closed-loop system, *i.e.*

$$\|\Phi\|_\infty < 1 \tag{13.22}$$

In the computation of a μ-controller, there is, however, a numerical problem. That is, with the inclusion of the multiple 5×5 real uncertainty block (corresponding to δ_m) the D-K iteration algorithm does not converge. In particular, it is difficult to obtain the approximation of a 5×5 scaling function matrix in the D-step. Hence, in our computation that multiple 5×5 real uncertainty block was removed in the uncertainty matrix during the D-K iteration. It should be stressed that the robust stability and robust performance analysis of the closed-loop system of the designed controller, which will be presented next, is tested with regard to the whole uncertainty structure, *i.e.* with the inclusion of that multiple 5×5 real uncertainty block.

The μ-synthesis is carried out by using the M-file `ms_flm.m`. The uncertainty structure and other parameters used in the D-K iteration are set in the auxiliary file `dk_flm.m`.

The progress of the D-K iteration is shown in Table 13.1.

Table 13.1. Results of.the μ-synthesis

Iteration	Controller order	Maximum value of μ
1	14	1.564
2	18	1.078
3	22	0.618
4	28	0.484

In the design exercise, an appropriate controller is obtained after the fourth D-K iteration. The controller is stable and has an order of 28.

The magnitude and phase plots of the μ-controller are shown in Figure 13.26.

It can be seen from Table 13.1 that after the fourth iteration the maximum value of μ is equal to 0.484. Note that this, however, does not necessarily mean that the robust performance has been achieved since we neglected the multiple 5×5 real uncertainty block in the computation. Hence, additional robust performance analysis is needed as below.

The μ-analysis of the closed-loop system is conducted by the file `mu_flm` that takes into account all uncertainty blocks discussed in Section 13.2.

The frequency response plot of the structured singular value for verification of robust stability is shown in Figure 13.27. The maximum value of μ is 0.444 that means that the stability of the system is preserved under perturbations that satisfy $\|\Delta\|_\infty < \frac{1}{0.444}$.

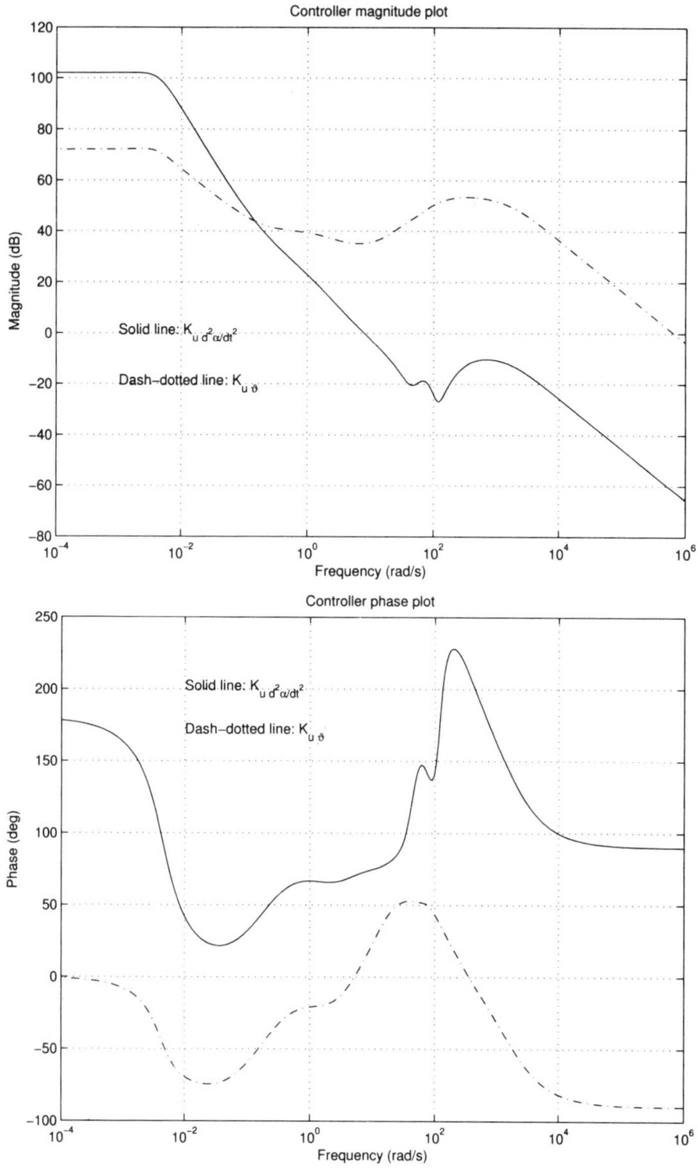

Fig. 13.26. Controller magnitude (above) and phase (below) plots

The frequency response of μ for the robust performance analysis is shown in Figure 13.28. The closed-loop system achieves robust performance since the maximum value of μ is equal to 0.806.

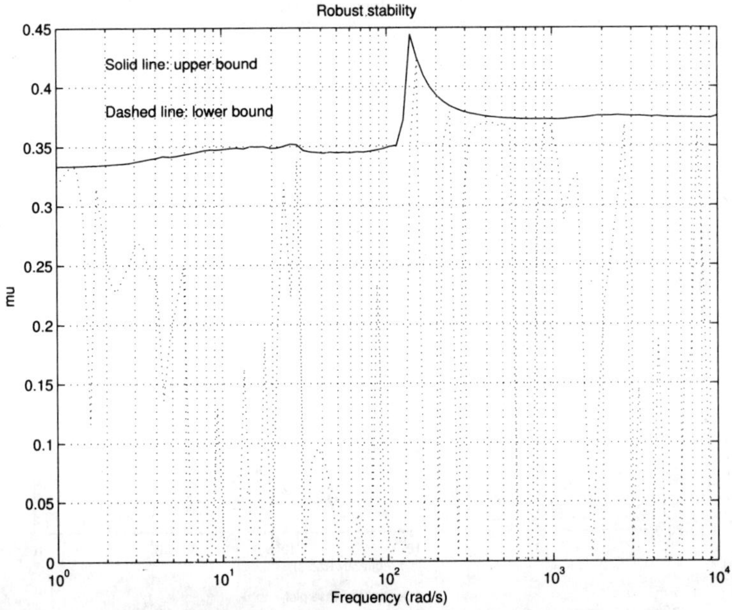

Fig. 13.27. Robust stability for μ-controller

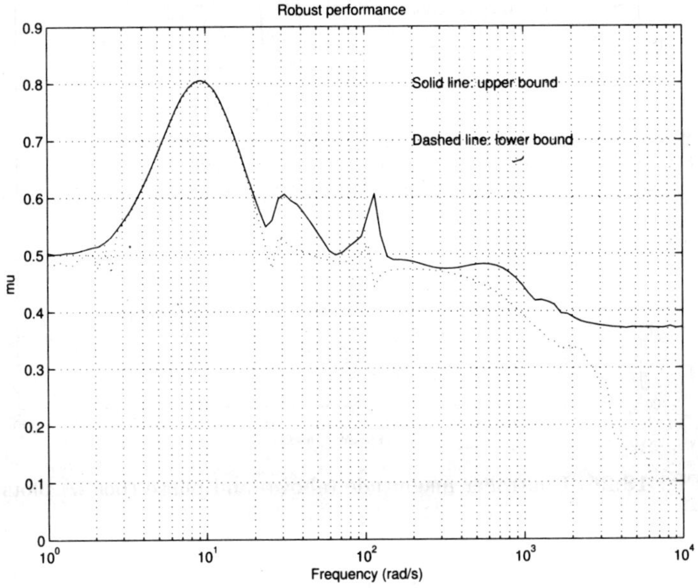

Fig. 13.28. Robust performance for μ-controller

Consider now the closed-loop transient responses that are computed by using the M-file `clp_pend`.

The reference trajectory for the manipulator tip movement in the simulation is chosen in the form

$$r = \begin{cases} at - (a/\psi)\sin(\psi t) + r_0, & 0 \leq t \leq t_m \\ r(t_m), & t_m < t \leq t_f \end{cases}$$

This trajectory allows the tip to be moved smoothly from an arbitrary initial position r_0 to a desired final position $r(t_m) = at_m$, with an appropriate ψ.

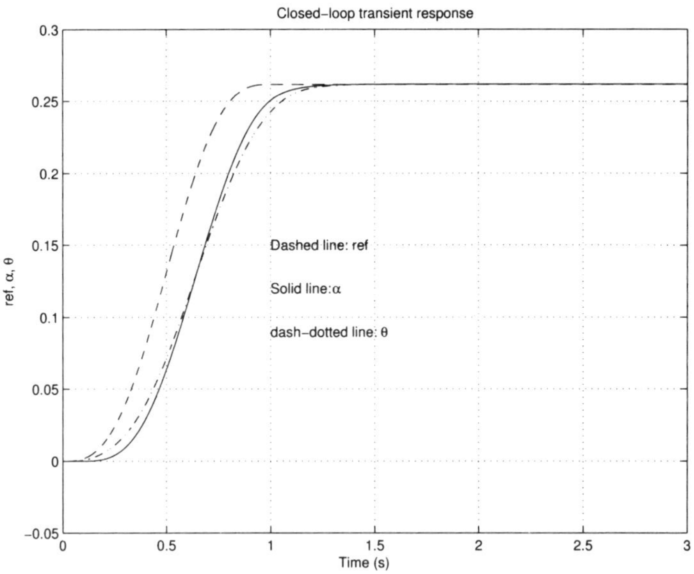

Fig. 13.29. Closed-loop transient response for μ-controller

In Figure 13.29 we show the transient response of the tip position α along with the joint angle θ and the reference r.

The transient response of the tip deflection $w(L,t)$ is shown in Figure 13.30.

The control action generated by the designed μ-controller is shown in Figure 13.31.

The closed-loop frequency responses are obtained by the M-file `frs_flm.m`.

The Bode plot of the closed-loop system is shown in Figure 13.32. The closed-loop bandwidth is about 10 rad/s. Note that a good match in magnitude between the closed-loop system and the dynamic model M is achieved for frequencies up to 70 rad/s.

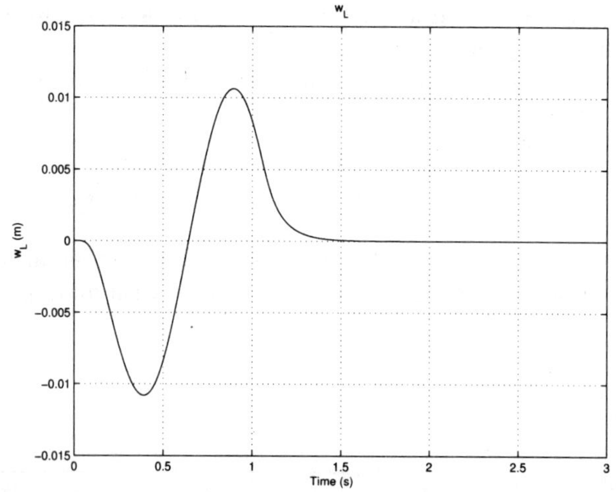

Fig. 13.30. Tip-deflection transient response for μ-controller

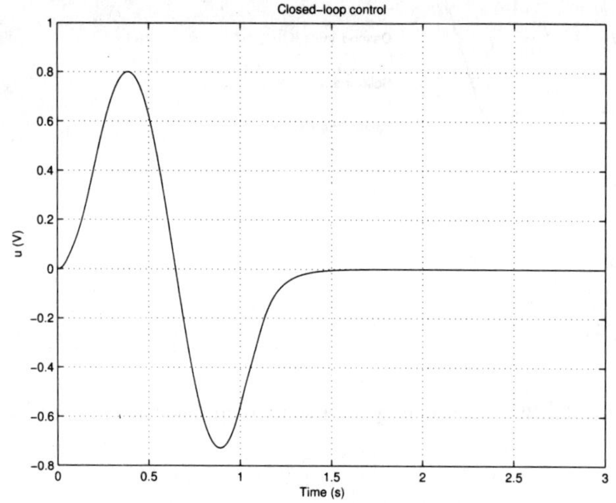

Fig. 13.31. Control action for μ-controller

The Bode plots of the tip-deflection transfer function are shown in Figure 13.33. The maximum amplitude of the tip deflection is observed for the input signal with frequency 40 rad/s.

Finally, in Figure 13.34 we show the magnitude plots with respect to the first and second noise. It is seen from the figure that the noise in measuring the joint angle has a negligible effect on the system output.

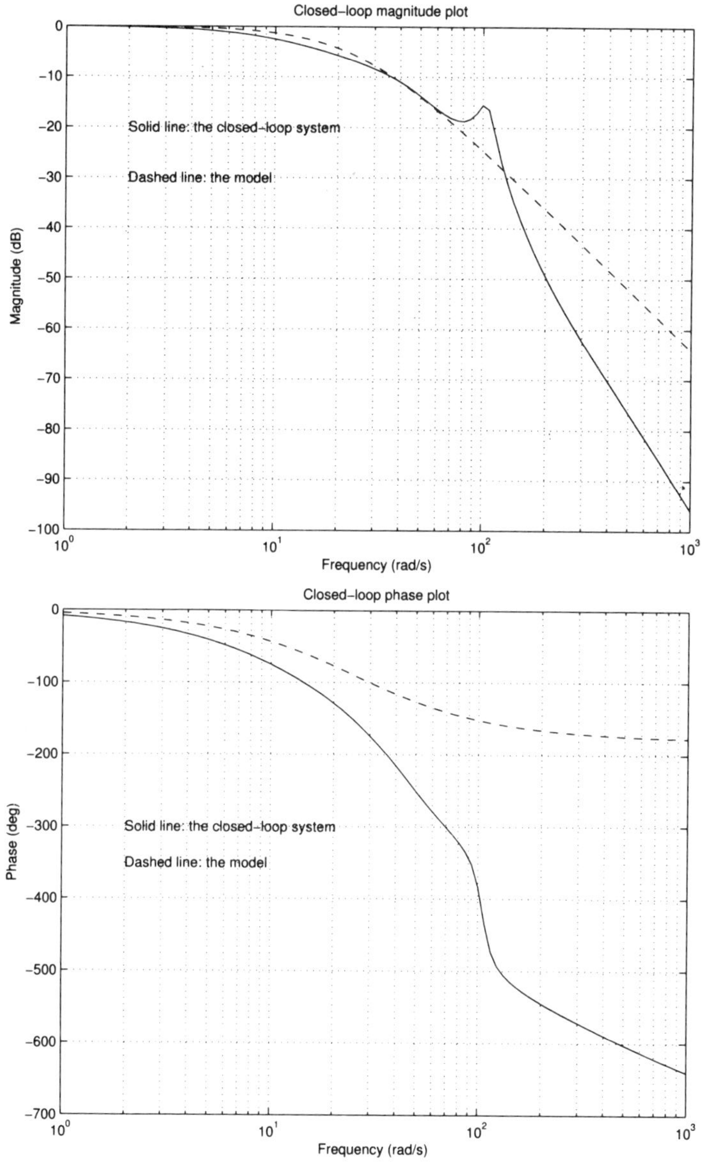

Fig. 13.32. Closed-loop magnitude (above) and phase (below) plots

We consider now the order reduction of the designed controller. As indicated in Table 13.1, the order of the μ-controller is 28. It would be good for implementation if the order could be reduced while essentially keeping the achieved performance. For this aim, we use the M-file **`red_flm.m`**. Af-

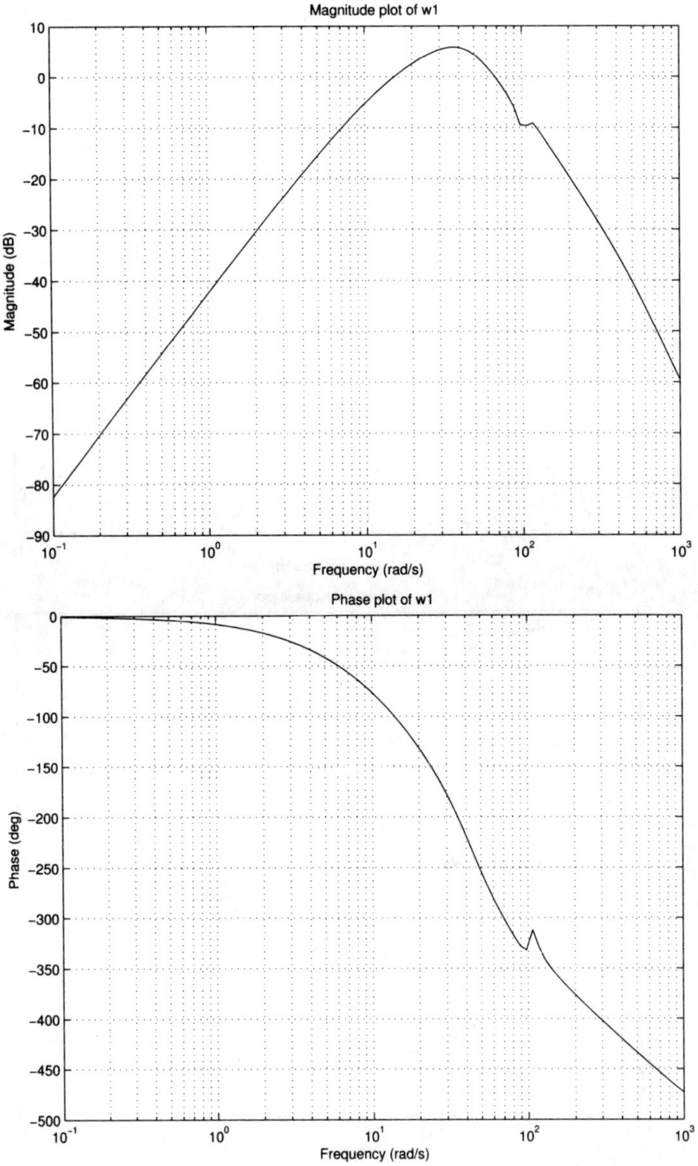

Fig. 13.33. Magnitude (above) and phase (below) plots for the tip deflection

ter the balanced realisation transformation of the controller and by neglecting the small Hankel singular values, the order of the controller can be reduced to 11 without losing too much performance. In Figure 13.35 we compare the frequency responses of the maximum singular values of the full-order

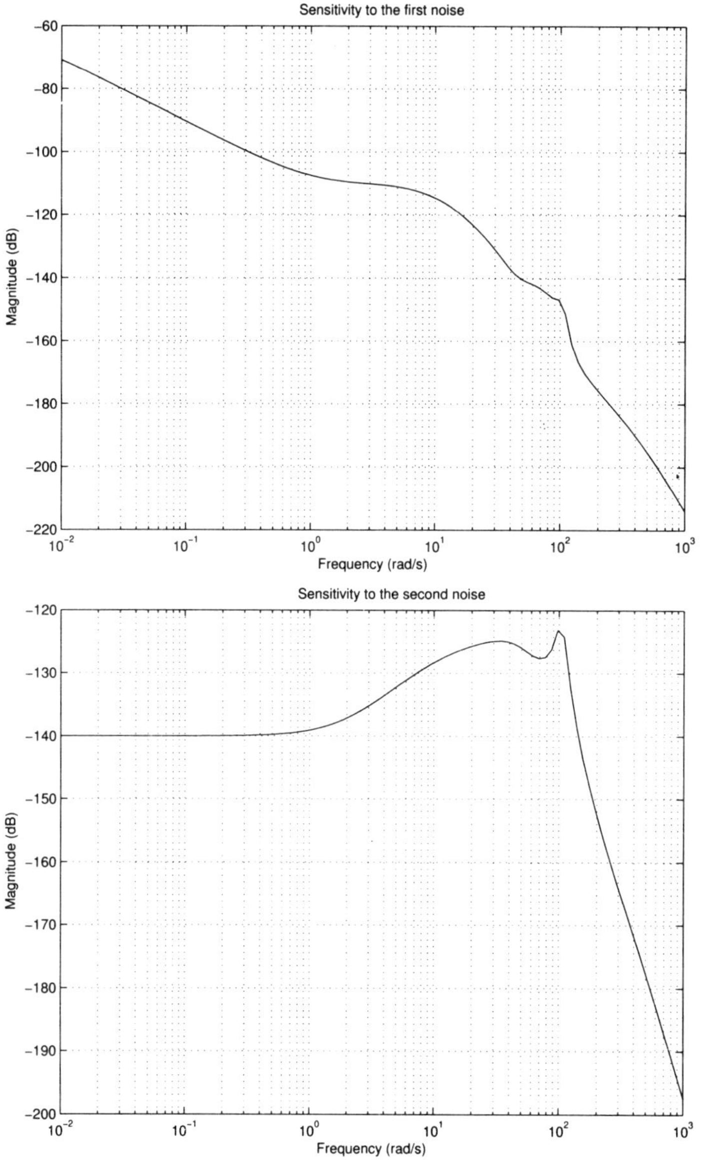

Fig. 13.34. Magnitude plots for the first and second noise

and reduced-order controllers. The frequency responses of both full-order and reduced-order controllers practically coincide. The transient responses of the closed-loop system with full-order and with reduced-order controller are also practically indistinguishable. (Figures are not included here.)

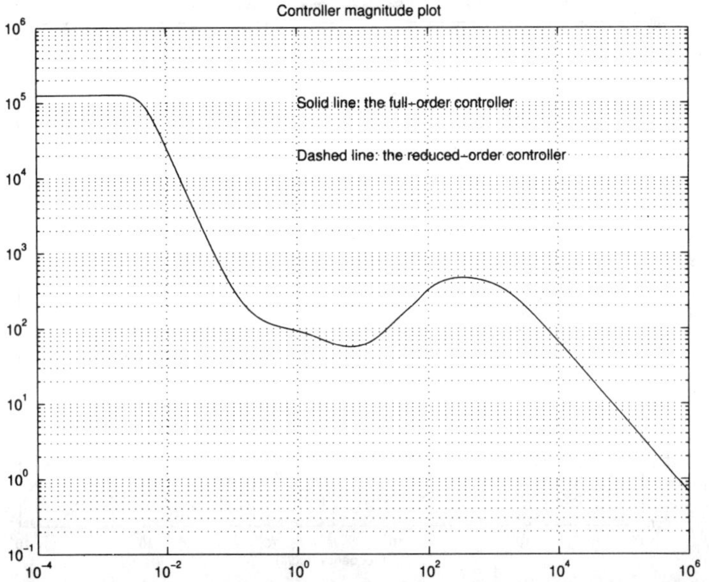

Fig. 13.35. Frequency responses of the full-order and reduced-order controllers

It is interesting to compare the results obtained with the μ-controller with those from the conventional collocated PD controller in the form of

$$u = k_P(r - \theta) - k_D\ddot{\theta}$$

The proportional and derivative coefficients are chosen as $k_P = 358$ N m/rad and $k_D = 28.5$ N m/(rad/s). The values of k_P and k_D are selected such that after neglecting the link flexibility the closed-loop transfer function coincides with the transfer function of the model. The results by using the μ-analysis method in this case (i.e. with the PD controller) are 0.447, 6.82 and 7.41 for the robust stability, nominal performance and robust performance, respectively. Therefore, the PD controller leads to poor nominal performance and poor robust performance, in comparison to the μ-controller designed. This can be seen by comparing Figures 13.36 and 13.37 with Figures 13.29 and 13.30, respectively.

It has to be noticed that good results in the design may also be obtained by using a collocated controller on the feedback from the joint angle θ and the velocity $\dot{\theta}$. The use of the tip acceleration $\ddot{\alpha}$, however, allows better results with respect to the robust performance to be obtained.

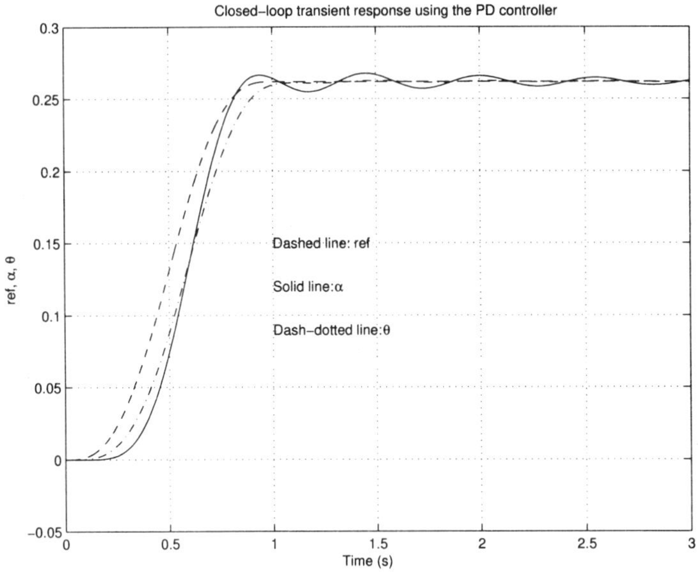

Fig. 13.36. Closed-loop transient response for the PD controller

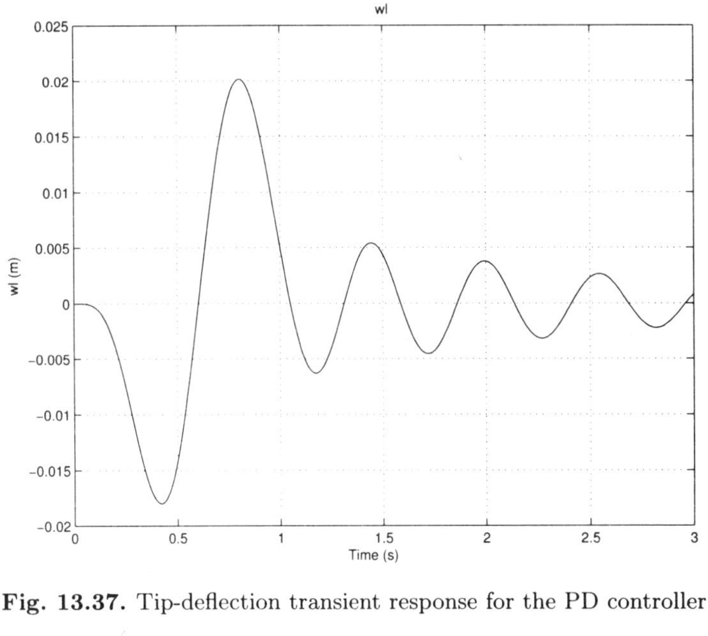

Fig. 13.37. Tip-deflection transient response for the PD controller

13.6 Nonlinear System Simulations

The performance of the μ-controller designed in the previous section is further investigated by simulations of the nonlinear closed-loop system with this controller. The simulation is carried out by the Simulink® model **nls_flm.mdl** using the nonlinear plant model (13.5). A number of simulations may be performed for several values of the payload mass and of the damping coefficients. The Simulink® model **nls_flm.mdl** is shown in Figure 13.38.

Before running the simulation, it is necessary to set the model parameters by using the M-file **init_flm.m**.

The time response of the tip position $\alpha(t)$, along with the joint angle θ and the reference r, for the case of the reduced-order μ-controller and nominal payload mass is given in Figure 13.39.

The tip deflection $w(L,t)$, for the nominal payload mass and the same reference signal, is shown in Figure 13.40. The values of the damping coefficients correspond to the case of light damping of the mechanical structure. In particular, the value of the hub damping coefficient d_r corresponds to a relative uncertainty of –20%. The damping coefficients d_{f_1} and d_{f_2} are taken so that the respective relative perturbations in d_1 and d_2 for the nominal payload are equal to –40%. The parameters of the reference signal used in the simulation ensure fast motion and are set as $a = 0.1\pi$ rad/s, $\psi = 2.5\pi$ s^{-1}, $r_0 = 0$ rad, $t_m = 0.8$ s and $t_f = 3$ s.

In the nonlinear system simulations, it is shown that the μ-controller efficiently suppresses the elastic vibrations during the fast motion of the manipulator tip. It thus justifies that the μ-synthesis is an appropriate robust design method in this exercise. It also confirms the validity of the uncertain model derived.

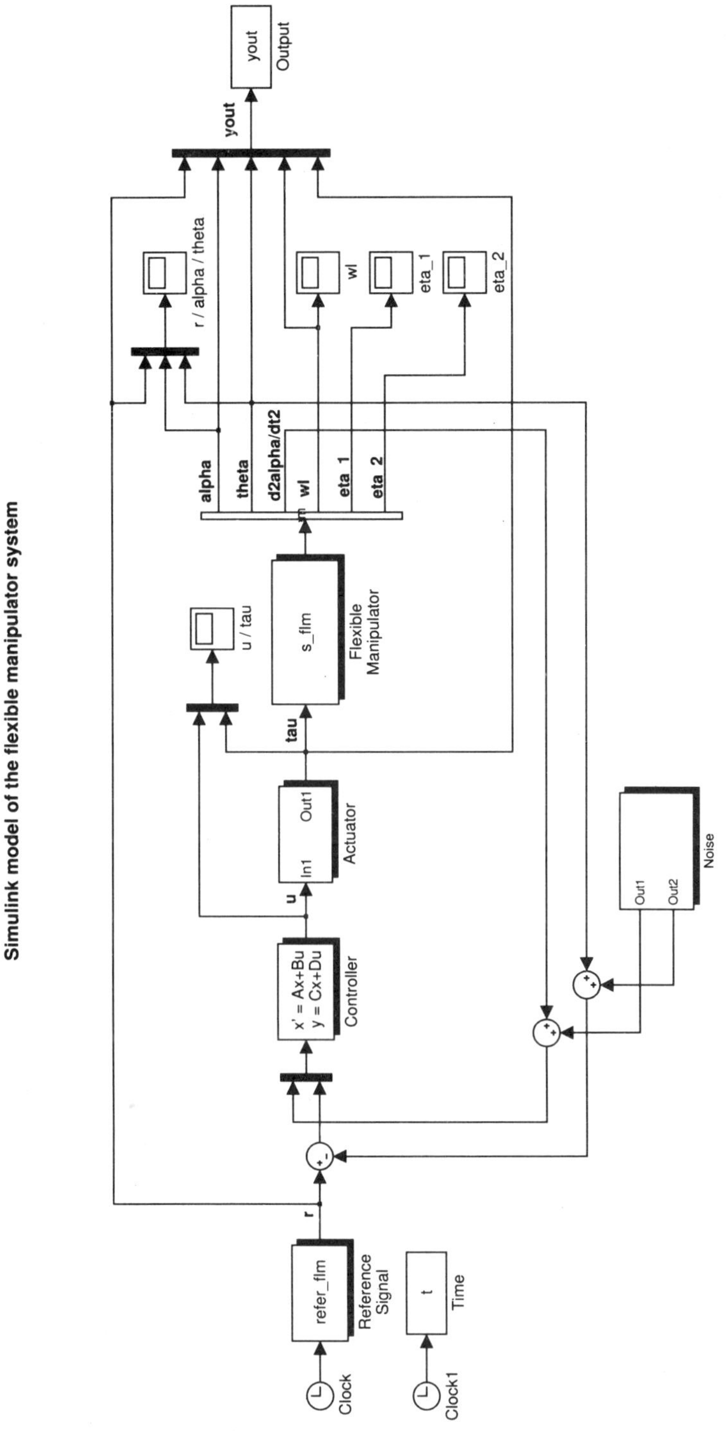

Fig. 13.38. Simulation model of the nonlinear system

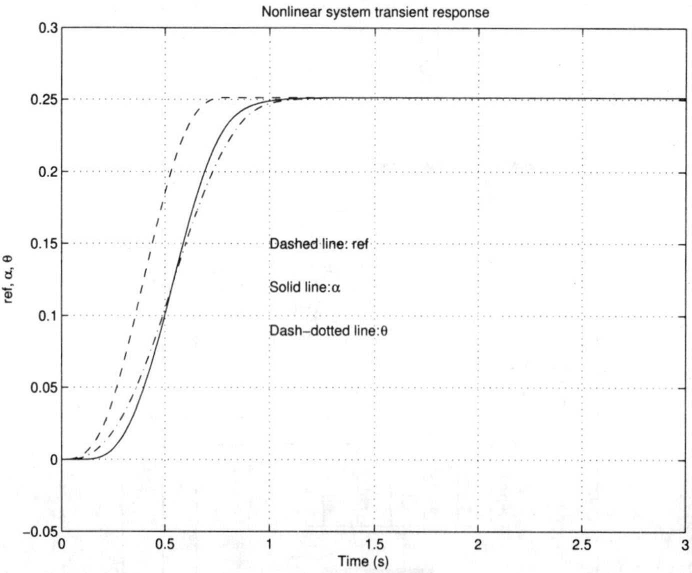

Fig. 13.39. Transient response of the nonlinear system

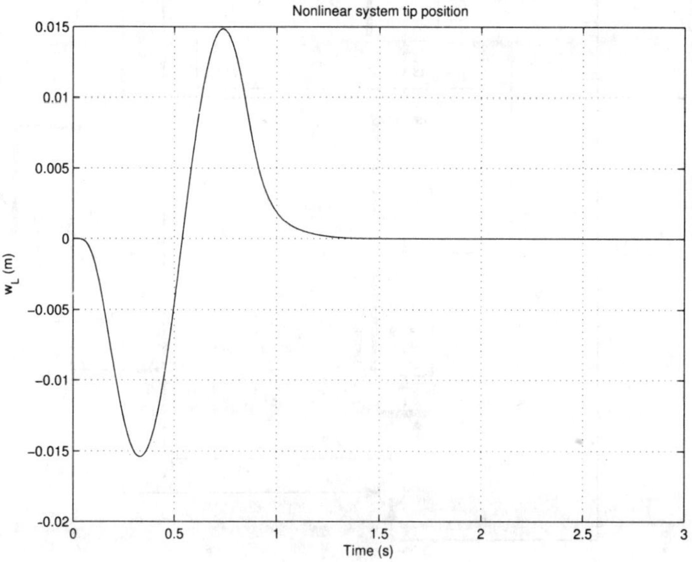

Fig. 13.40. Transient response of the nonlinear system $(w(L, t))$

13.7 Conclusions

A few conclusions may be drawn as the following, based on the analysis and design of the flexible manipulator control system:

- In applying linear robust control system design techniques for a nonlinear plant it is usually unavoidable to derive a complicated uncertainty model, because of the requirement of a sufficiently accurate linear approximation. That would, however, adversely affect the controller design and analysis. It is important, therefore, to simplify the model of uncertainty. Methods such as the numerical approximation used in this study can be considered.
- In contrast to many known models, the uncertainty model derived in this study for the flexible manipulator system contains real parametric uncertainties in a highly structured form. Such a model appeals naturally to the application of μ-synthesis and analysis method that greatly reduces the conservativeness in the controller design.
- A robust noncollocated controller on the feedback signals of joint angle and tip acceleration is designed in this study on the basis of the uncertainty model derived and by using the μ-synthesis. The μ-controller shows very good robust performance on the tip motion for a wide range of payload mass. The controller efficiently suppresses the elastic vibrations during the fast motion of the manipulator tip.
- The nonlinear system simulation results confirm the high performance of the controller designed and also verify the validity of the uncertain model used.
- It is also possible to investigate various noncollocated and collocated controller structures on different output feedback signals, with the uncertainty model and linearised plant derived in this study.

Notes and References

The control of flexible manipulators has been an area of intensive research in recent years. An efficient approach to improve the manipulator performance is to use a feedback from the manipulator tip position [44], tip acceleration [42] or base-strain [43]. The usage of such feedbacks leads to a noncollocated control scheme that may increase the closed-loop system sensitivity to modelling errors or to parameter uncertainties [125].

The necessity to achieve robustness of the manipulator control system in the presence of uncertainties makes it appropriate to apply the robust control design methods. In a few recent papers the authors develop different $\mathcal{H}_\infty$ controllers [45], [85], [146] and μ-synthesis controllers [73] for flexible-link manipulators. A common disadvantage in the previous robust designs for flexible manipulators is the use of unstructured uncertainty model that leads to potentially very conservative results.

References

1. D. Abramovitch and G. Franklin. A brief history of disk drive control. *IEEE Control Systems Magazine*, 22:28–42, 2002.
2. D. Abramovitch and G. Franklin. Disk drive control: The early years. In *Proceedings of the 15th IFAC Congress. Session T-Th-M12*, pages 1–12, Barcelona, Spain, July 2002. CD-ROM.
3. D. Abramovitch, T. Hurst, and D. Henze. An overview of the PES Pareto method for decomposing baseline noise sources in hard disk postion error signals. *IEEE Transactions on Magnetics*, 34:17–23, 1998.
4. D. Abramovitch, F. Wang, T. Hurst, and G. Franklin. Disk drive pivot nonlinearity modelling Part I: Frequency domain. In *Proceedings of the 1994 American Control Conference*, pages 2600–2603, Baltimore, MD, June 1994.
5. D.Y. Abramovitch. Magnetic and optical disk control: Parallels and contrasts. In *Proceedings of the 2001 American Control Conference*, pages 421–428, Arlington, VA, June 2001.
6. V.M. Adamjan, D.Z. Arov, and M.G. Krein. Analytic properties of Schmidt pairs for a Hankel operator and the generalized Schur-Takagi problem. *Mat. USSR Sbornik*, 15:31–73, 1971.
7. R.J. Adams, J.M Buffington, A.G. Sparks, and S.S. Banda. *Robust Multivariable Flight Control.* Springer-Verlag, New York, 1994.
8. U.M. Al-Saggaf and G.F. Franklin. An error bound for a discrete reduced order model of a linear multivariable system. *IEEE Transactions on Automatic Control*, AC-32:815–819, 1987.
9. G.J. Balas, J.C. Doyle, K. Glover, A. Packard, and R. Smith. *μ-Analysis and Synthesis Toolbox: User's Guide.* MUSYN Inc. and The Mathworks, Inc., 1995.
10. D.S. Bernstein and W.M. Haddad. LQG control with an $\mathcal{H}^\infty$ performance bound: A Riccati equation approach. *IEEE Transactions on Automatic Control*, AC-34:293–305, 1989.
11. J.H. Blakelock. *Automatic Control of Aircraft and Missiles.* Wiley, New York, 1991.
12. A.E. Bryson. *Control of Spacecraft and Aircraft.* Princeton University Press, Princeton, NJ, 1994.
13. H. Buschek. Design and flight test of a robust autopilot for the IRIS-T air-to-air missile. *Control Engineering Practice*, 11:551–558, 2003.

14. F.M. Callier and C.A. Desoer. *Linear System Theory.* Springer-Verlag, New York, 1991.
15. B.M. Chen, T.H. Lee, and V. Venkataramanan. *Hard Disk Drive Servo Systems.* Springer-Verlag, Berlin, 2002.
16. K.K. Chew and M. Tomizuka. Digital control of repetetive errors in disk drive systems. *IEEE Control Systems Magazine*, 10:16–20, 1990.
17. B.W. Choi, D.W. Gu, and I. Postlethwaite. Low-order $\mathcal{H}_\infty$ Sub-Optimal Controllers. *IEE Proceedings - Part D*, 141:243–248, 1994.
18. R.R.E. de Gaston and M.G. Safonov. Exact calculation of the multiloop stability margin. *IEEE Transactions on Automatic Control*, AC-33:156–171, 1988.
19. U.B. Desai and D. Pal. A transformation approach to stochastic model reduction. *IEEE Transactions on Automatic Control*, AC-29:1097–1100, 1984.
20. C.A. Desoer and W.S. Chan. The feedback interconnection of lumped linear time-invariant systems. *Journal of the Franklin Institute*, 300:335–351, 1975.
21. C.A. Desoer and M. Vidyasagar. *Feedback Systems: Input-Output Properties.* Academic Press, New York, 1975.
22. J. Ding, M. Tomizuka, and H. Numasato. Design and robustness analysis of dual stage servo system. In *Proceedings of the 2000 American Control Conference*, pages 2605–2609, Chicago, Illinois, June 2000.
23. J.C. Doyle. Analysis of feedback systems with structured uncertainties. *IEE Proceedings - Part D*, 129:242–250, 1982.
24. J.C. Doyle. Structured uncertainty in control system design. In *Proceedings of the 24 IEEE Conference on Decision and Control*, pages 260–265, December 1985.
25. J.C. Doyle. A review of μ: For case studies in robust control. In *Proceedings of 10th IFAC World Congress*, pages 395–402, Munich, Germany, July 1987.
26. J.C. Doyle, B.A. Francis, and A.R. Tannenbaum. *Feedback Control Theory.* Macmillan Publishing Co., New York, 1992.
27. C. Du, J. Zhang, and G. Guo. Vibration analysis and control design comparison of HDDs using fluid bearing and ball bearing spindles. In *Proceedings of the 2002 American Control Conference*, pages 1378–1383, Anchorage, AK, May 2002.
28. K.G. Eltohamy. Nonlinear optimal control of a triple inverted pendulum with single control input. *International Journal of Control*, 69:239–256, 1998.
29. D. Enns. *Model Reduction for Control Systems Design.* PhD thesis, Department of Aeronautics and Astronautics, Stanford University, Stanford, CA, 1984.
30. D. Enns. Model reduction with balanced realizations: An error bound and a frequency weighted generalization. In *Proceedings of the 23rd IEEE Conference on Decision and Control*, pages 127–132, Las Vegas, NV, 1984.
31. R.B. Evans, J.S. Griesbach, and W.C. Messner. Piezoelectric microactuator for dual stage control. *IEEE Transactions on Magnetics*, 35:977–982, 1999.
32. M.K.H. Fan and A.L. Tits. Characterization and efficient computation of the structured singular value. *IEEE Transactions on Automatic Control*, AC-31:734–743, 1986.
33. M.K.H. Fan, A.L. Tits, and J.C. Doyle. Robustness in the presence of mixed parametric uncertainty and unmodeled dynamics. *IEEE Transactions on Automatic Control*, AC-36:25–38, 1991.

34. K.V. Fernando and H. Nicholson. Singular perturbational model reduction of balanced systems. *IEEE Transactions on Automatic Control*, AC-27:466–468, 1982.
35. K.V. Fernando and H. Nicholson. Singular perturbational approximation for discrete-time systems. *IEEE Transactions on Automatic Control*, AC-28:240–242, 1983.
36. G. Ferreres. *A Practical Approach to Robustness Analysis with Aeronautical Applications.* Kluwer Academic/Plenum Publishers, New York, 1999.
37. I. Fialho, G.J. Balas, A.K. Packard, J. Renfrow, and C. Mullaney. Gain-scheduled lateral control of the F-14 aircrfat during powered approach landing. *Journal of Guidance, Control, and Dynamics*, 23:450–458, 2000.
38. B.A. Francis. *A Course in $\mathcal{H}^\infty$ Control Theory*, volume 88 of *Lecture Notes in Control and Information Sciences.* Springer-Verlag, 1987.
39. G.F. Franklin, J.D. Powell, and M.L. Workman. *Digital Control of Dynamic Systems.* Addison-Wesley, Menlo Park, CA, third edition, 1998.
40. K. Furuta, K. Kajiwara, and K. Kosuge. Digital control of a double inverted pendulum on an inclined rail. *International Journal of Control*, 32:907–924, 1980.
41. K. Furuta, T. Ochia, and N. Ono. Attitude control of a triple inverted pendulum. *International Journal of Control*, 39:1351–1365, 1984.
42. E. Garcia-Benitez, J. Walkins, and S. Yurkovich. Nonlinear control with acceleration feedback for a two-link flexible robot. *Control Engineering Practice*, 1:989–997, 1993.
43. S. Ge, T. Lee, and G. Zhu. Improving regulation of a single-link flexible manipulator with strain feedback. *IEEE Transactions on Robotics and Automation*, 14:179–185, 1998.
44. H. Geniele, R. Patel, and K. Khorasani. End-point control of a flexible-link manipulator: Theory and experiment. *IEEE Transactions on Control Systems Technology*, 5:559–570, 1997.
45. M. Ghanekar, D. Wang, and G. Heppler. Scaling laws for linear controllers of flexible link manipulators characterized by nondimensional groups. *IEEE Transactions on Robotics and Automation*, 13:117–127, 1997.
46. K. Glover. All optimal Hankel-norm approximations of linear multivariable systems and their l^∞-error bounds. *International Journal of Control*, 39:1115–1193, 1984.
47. K. Glover. Multiplicative approximation of linear multivariable systems with $\mathcal{L}_\infty$ error bounds. In *Proceedings of the 1986 American Control Conference*, pages 1705–1709, Minneapolis, MN, 1986.
48. K. Glover and J.C. Doyle. State-space formulae for all stabilizing controllers that satisfy an $\mathcal{H}^\infty$ norm bound and relations to risk sensitivity. *Systems & Control Letters*, 11:167–172, 1988.
49. K. Glover and E.A. Jonckheere. A comparison of two Hankel norm methods for approximating spectra. In C.I. Byrnes and A. Lindquist, editors, *Modelling, Identification and Robust Control.* North-Holland, Amsterdam, The Netherlands, 1986.
50. T.B. Goh, Z. Li, B.M. Chen, T.H. Lee, and T. Huang. Design and implementation of a hard disk drive servo system using robust and perfect tracking approach. In *Proceedings of the 38th IEEE Conference on Decision and Control*, pages 5247–5252, Phoenix, Arizona, December 1999.

51. M. Green. Balanced stochastic realizations. *Linear Algebra & its Applications*, 98:211–247, 1988.
52. M. Green. A relative error bound for balanced stochastic truncation. *IEEE Transactions on Automatic Control*, AC-33:961–965, 1988.
53. M. Green and D.J.N. Limebeer. *Linear Robust Control.* Prentice-Hall, Englewood Cliffs, NJ, 1995.
54. A.L. Greensite. *Analysis and Design of Space Vehicle Flight Control Systems.* Spartan, New York, 1970.
55. D.W. Gu, B.W. Choi, and I. Postlethwaite. Low-order stabilizing controllers. *IEEE Transactions on Automatic Control*, AC-38:1713–1717, 1993.
56. D.W. Gu, S.J. Goh, T. Fitzpatrick, I. Postlethwaite, and N. Measor. Application of an expert system for robust controller design. *Control'96*, 2:1004–1009, 1996.
57. D.W. Gu, P.Hr. Petkov, and M.M. Konstantinov. Direct Formulae for the $\mathcal{H}_\infty$ Sub-Optimal Central Controller. Technical Report 1998-7, Niconet Report, 1998. Available at http://www.win.tue.nl/niconet.
58. D.W. Gu, P.Hr. Petkov, and M.M. Konstantinov. Formulae for discrete $\mathcal{H}_\infty$ loop shaping design procedure controllers. In *Proceedings of the 15th IFAC World Congress, paper 276, Session T-Mo-M15*, Barcelona, Spain, July 2002. CD-ROM.
59. D.W. Gu, I. Postlethwaite, and M.C. Tsai. $\mathcal{H}_\infty$ super-optimal solutions. In C.T. Leondes, editor, *Advances in Control and Dynamic Systems*, volume 51, pages 183–246. Academic Press, San Diego, CA, 1992.
60. D.W. Gu, M.C. Tsai, and I. Postlethwaite. State-space formulae for discrete-time optimization. *International Journal of Control*, 49:1683–1723, 1989.
61. L. Guo, H.S. Lee, A. Hudson, and S.-H. Chen. A comprehensive time domain simulation tool for hard disk drive TPI prediction and mechanical/servo enhancement. *IEEE Transactions on Magnetics*, 35:879–884, 1999.
62. S. Hammarling. Numerical solution of the stable non-negative definite Lyapunov equation. *IMA Journal of Numerical Analysis*, 2:303–323, 1982.
63. S. Hara, T. Hara, L. Yi, and M. Tomizuka. Two degree-of-freedom controllers for hard disk drives with novel reference signal generation. In *Proceedings of the 1999 American Control Conference*, pages 4116–4121, San Diego, CA, June 1999.
64. J.W. Helton. Worst case analysis in the frequency domain: The $\mathcal{H}^\infty$ approach to control. *IEEE Transactions on Automatic Control*, AC-30:1154–1170, 1985.
65. J.W. Helton and O. Merino. *Classical Control Using $\mathcal{H}^\infty$ Methods.* Society for Industrial and Applied Mathematics, Philadelphia, PA, 1998.
66. D. Hernandez, S.-S. Park, R. Horowitz, and A.K. Packard. Dual stage track-following servo design for hard disk drives. In *Proceedings of the 1999 American Control Conference*, pages 4116–4121, San Diego, CA, June 1999.
67. D. Hoyle, R. Hyde, and D.J.N. Limebeer. An $\mathcal{H}^\infty$ approach to two-degree-of-freedom design. In *Proceedings of the 30th IEEE Conference on Decision and Control*, pages 1581–1585, Brighton, UK, December 1991.
68. C.S. Hsu, X. Yu, H.H. Yeh, and S.S. Banda. $\mathcal{H}_\infty$ compensator design with minimal order observer. In *Proceedings of the 1993 American Control Conference*, San Francisco, CA, June 1993.
69. Y. Huang, P. Mathur, and W.C. Messner. Robustness analysis on a high bandwidth disk drive servo system with an instrumented suspension. In *Proceedings*

of the 1999 American Control Conference, pages 3620–3624, San Diego, CA, June 1999.

70. J. Ishikawa, Y. Yanagita, and T. Hattori ans M. Hashimoto. Head positioning control for low sampling rate systems based on two degree-of-freedom control. *IEEE Transactions on Magnetics*, 32:1787–1792, 1996.
71. T. Iwasaki and R.E. Skelton. All low order $\mathcal{H}_\infty$ controllers with covariance upper bound. In *Proceedings of the 1993 American Control Conference*, San Francisco, CA, June 1993.
72. H. Kajiwara, P. Apkarian, and P. Gahinet. LPV techniques for control of an inverted pendulum. *IEEE Control Systems Magazine*, 19:44–54, 1999.
73. M. Karkoub, G. Balas, K. Tamma, and M. Donath. Robust control of flexible manipulators via μ-synthesis. *Control Engineering Practice*, 8:725–734, 2000.
74. C. Kempf, W. Messner, M. Tomizuka, and R. Horovitz. Comparison of four discrete-time repetetive control algorithms. *IEEE Control Systems Magazine*, 13:48–54, 1993.
75. S.W. Kim, B.D.O. Anderson, and A.G. Madievski. Error bound for transfer function order reduction using frequency weighted balanced truncation. *Systems & Control Letters*, 24:183–192, 1995.
76. M. Kobayashi, T. Yamaguchi, and R. Horowitz. Track-seeking controller design for dual-stage actuator in magnetic disk drives. In *Proceedings of the 2000 American Control Conference*, pages 2610–2614, Chicago, Illinois, June 2000.
77. C.M. Kozierok. *The PC Guide. The Reference Section on Hard Disk Drives.* January 2003. Available at http://www.PCGuide.com.
78. S. Kung. A new low-order approximation algorithm via singular value decomposition. In *Proceedings of the 18th IEEE Conference on Decision and Control*, Ft. Lauderdale, Florida, December 1979.
79. S. Kung and D.W. Lin. Optimal Hankel norm model reduction: Multivariable systems. *IEEE Transactions on Automatic Control*, AC-26:832–852, 1981.
80. P.J. Larcombe. On the generation and solution of the symbolic, open-loop charactersitic equation for a double inverted pendulum. *International Journal of Systems Science*, 24:2379–2390, 1993.
81. A.J. Laub. On computing balancing transformations. In *Proceedings of the Joint 1980 American Control Conference*, pages FA8–E, San Francisco, CA, August 1980.
82. A.J. Laub, M.T. Heath, C.C. Paige, and R.C. Ward. Computation of system balancing transformations and other applications of simultaneous diagonalization algorithms. *IEEE Transactions on Automatic Control*, AC-32:115–121, 1987.
83. H.S. Lee. Controller optimization for minimum position error signals of hard disk drives. In *Proceedings of the 2000 American Control Conference*, pages 3081–3085, Chicago, Illinois, June 2000.
84. D.J. Leith and W.E. Leithead. Survey of gain-scheduling analysis and design. *International Journal of Control Control*, 73:1001–1025, 2000.
85. Y. Li, B. Thang, Z. Shi, and Y. Lu. Experimental study for trajectory tracking of a two-link flexible manipulators. *International Journal of Systems Science*, 31:3–9, 2000.
86. D.J.N. Limebeer. The specification and purpose of a controller design study. In *Proceedings of the 30th IEEE Conference on Decision and Control*, pages 1579–1580, Brighton, UK, December 1991.

87. D.J.N. Limebeer, E.M. Kasenally, and J.D. Perkins. On the design of robust two degree of freedom controllers. *Automatica*, 29:157–168, 1993.
88. C.-A. Lin and T.-Y. Chiu. Model reduction via frequency weighted balanced realization. *CONTROL – Theory and Advanced Technology*, 8:341–351, 1992.
89. J.L. Lin. *Control System Design for Robust Stability and Robust Performance*. PhD thesis, Department of Engineering, University of Leicester, Leicester, UK, May 1992.
90. J.L. Lin, I. Postlethwaite, and D.W. Gu. μ-K iteration: a new algorithm for μ-synthesis. *Automatica*, 29:219–244, 1993.
91. R. Lind and M. Brenner. *Robust Aeroservoelastic Stability Analysis. Flight Test Applications*. Springer-Verlag, London, 1999. ISBN 1-85233-096-1.
92. A. Lindquist and G. Picci. On the stochastic realization problem. *SIAM Journal of Control and Optimization*, 17:365–389, 1979.
93. Yi Liu and B.D.O. Anderson. Singular perturbation of balanced systems. *International Journal of Control*, 50(4):1379–1405, 1989.
94. A. De Luca and B. Siciliano. Trajectory control of a non-linear one-link flexible arm. *International Journal of Control*, 50:1699–1715, 1989.
95. P. Lundström, S. Skogestad, and J.C. Doyle. Two-degree-of-freedom controller design for an ill-conditioned distillation process using μ-synthesis. *IEEE Transactions on Control Systems Technology*, 7:12–21, 1999.
96. J. Lunz. *Robust Multivariable Feedback Control*. Prentice Hall, London, 1989.
97. C.C. MacDuffee. *The Theory of Matrices*. Chelsea Publishing Co., New York, 1950.
98. J.M. Maciejowski. *Multivariable Feedback Design*. Addison-Wesley, Wokingham, England, 1989. ISBN 0-201-18243-2.
99. D.P. Magee. Optimal filtering to improve performance in hard disk drives: Simulation results. In *Proceedings of the 1999 American Control Conference*, pages 71–75, San Diego, CA, June 1999.
100. D. McFarlane and K. Glover. A loop shaping design procedure using $\mathcal{H}_\infty$ synthesis. *IEEE Transactions on Automatic Control*, AC-37:749–769, 1992.
101. D.C. McFarlane and K. Glover. *Robust Controller Design Using Normalized Coprime Factor Plant Descriptions*, volume 138 of *Lecture Notes in Control and Information Sciences*. Springer-Verlag, 1990.
102. G.A. Medrano-Cerda. Robust stabilization of a triple inverted pendulum-cart. *International Journal of Control*, 68:849–865, 1997.
103. G.A. Medrano-Cerda. Robust computer control of an inverted pendulum. *IEEE Control Systems Magazine*, 19:58–67, 1999.
104. Z. Meier, H. Farwig, and H. Unbehauen. Discrete computer control of a triple-inverted pendulum. *Optimal Control Applications & Methods*, 11:157–171, 1990.
105. L. Meirovitch. *Elements of Vibration Analysis*. McGraw-Hill, New York, 1986.
106. W. Messner and R. Ehrlich. A tutorial on controls for disk drives. In *Proceedings of the 2001 American Control Conference*, pages 408–420, Arlington, VA, June 2001.
107. D.G. Meyer. Fractional balanced reduction: Model reduction via fractional representation. *IEEE Transactions on Automatic Control*, AC-35(3):1341–1345, 1990.
108. B. C. Moore. Principle component analysis in linear systems: Controllability, observability, and model reduction. *IEEE Transactions on Automatic Control*, AC-26:17–31, 1981.

109. M. Morari and E. Zafiriou. *Robust Process Control.* Prentice-Hall, Englewood Cliffs, NJ, 1989.
110. S. Mori, H. Nishihara, and K. Furuta. Control of an unstable mechanical system. Control of pendulum. *International Journal of Control*, 23:673–692, 1976.
111. G. Murad. *Robust Multivariable Control of Industrial Production Processes: A Discrete-Time Multi-Objective Approach.* PhD thesis, Department of Engineering, University of Leicester, Leicester, UK, 1995.
112. G. Murad, D-W. Gu, and I. Postlethwaite. A Direct Model Reduction Approach for Discrete-Time Non-Minimal State-Space Systems. Internal Report 94-23, University of Leicester, Leicester, UK, September 1994.
113. G. Murad, D.-W. Gu, and I. Postlethwaite. Robust internal model control of a binary distillation column. In *IEEE International Conference on Industrial Technology*, pages 194–198, Shanghai, China, December 1996.
114. G. Murad, I. Postlethwaite, D.-W. Gu, and R. Samar. On the structure of an $\mathcal{H}^\infty$ two-degree-of-freedom internal model-based controller and its application to a glass tube production process. In *Proceedings of the Third European Control Conference*, pages 595–600, Rome, September 1995.
115. G. Murad, I. Postlethwaite, D.-W. Gu, and J.F. Whidborne. Robust control of a glass tube shaping process. In *Proceedings of the Second European Control Conference*, volume 4, pages 2350–2355, Gröningen, The Netherlands, June-July 1993.
116. Z. Nehari. On bounded bilinear forms. *Annals of Mathematics*, 65:153–162, 1957.
117. W.Y. Ng. *Interactive Multi-Objective Programming as a Framework for Computer-Aided Control System Design.* Lecture Notes in Control and Information Sciences. Springer-Verlag, 1989.
118. R.A. Nichols, R.T. Reichert, and W.J. Rugh. Gain scheduling for H_∞ controllers: A flight control example. *IEEE Trans. Control Systems Technology*, 1:69–53, 2001.
119. International Society Niconet. *SLICOT: Numerical Subroutine Library for Control and Systems Theory.* Available at http://www.win.tue.nl/niconet, 2001.
120. A. Packard and J.C. Doyle. The complex structured singular value. *Automatica, Special Issue on Robust Control*, 29:71–109, 1993.
121. A. Packard, M.K.H. Fan, and J.C. Doyle. A power method for the structured singular value. In *Proceedings of the 27th IEEE Conference on Decision and Control*, pages 2132–2137, Austin, Texas, December 1988.
122. A. Packard and P. Pandey. Continuity properties of the real/complex structured singular value. *IEEE Transactions on Automatic Control*, AC-38:415–428, 1993.
123. L.Y. Pao and G.F. Franklin. Time-optimal control of flexible structures. In *Proceedings of the 29th IEEE Conference on Decision and Control*, pages 2580–2581, Honolulu, HI, December 1990.
124. T. Pappas, A.J. Laub, and N.R. Sandell. On the numerical solution of the discrete-time algebraic Riccati equation. *IEEE Transactions on Automatic Control*, AC-25:631–641, 1980.
125. J. Park and H. Asada. Dynamic analysis of noncollocated flexible arms and design of torque transmission mechanisms. *ASME Journal of Dynamic Systems, Measurements and Control*, 116:201–207, 1994.

126. L. Pernebo and L.M. Silverman. Model reduction via balanced state space representations. *IEEE Transactions on Automatic Control*, AC-27:382–387, 1982.
127. I. Postlethwaite, J.L. Lin, and D.W. Gu. Robust control of a high purity distillation column using μ-K iteration. In *Proceedings of the 30th IEEE Conference on Decision and Control*, Brighton, UK, December 1991.
128. R.M. Redheffer. Remarks on the basis of network theory. *Journal of Mathematics and Physics*, 28:237–258, 1950.
129. R.M. Redheffer. On a certain linear fractional transformation. *Journal of Mathematics and Physics*, 39:269–286, 1960.
130. R.T. Reichert. Dynamic scheduling of modern-robust-control autopilot design for missiles. *IEEE Control Systems Magazine*, 12:35–42, 1992.
131. W.J. Rugh and J.S. Shamma. Research on gain scheduling. *Automatica*, 36:1401–1425, 2000.
132. M.G. Safonov. Stability margins of diagonally perturbed multivariable feedback systems. *IEE Proceedings - Part D*, 129:251–256, 1982.
133. M.G. Safonov and R.Y. Chiang. A Schur method for balanced-truncation model reduction. *IEEE Transactions on Automatic Control*, AC-34:729–733, 1989.
134. M.G. Safonov, R.Y. Chiang, and D.J.N. Limebeer. Hankel model reduction without balancing – A descriptor approach. In *Proceedings of the 26th IEEE Conference on Decision and Control*, Los Angeles, CA, December 1987.
135. M.G. Safonov, D.J.N. Limebeer, and R.Y. Chiang. Simplifying the $\mathcal{H}^\infty$ theory via loop-shifting, matrix-pencil and descriptor concepts. *International Journal of Control*, 50:2467–2488, 1989.
136. R. Samar, I. Postlethwaite, and D.-W. Gu. Model reduction with balanced realizations. *International Journal of Control*, 62:33–64, 1995.
137. R.S. Sánchez-Peña and M. Sznaier. *Robust Systems. Theory and Applications.* John Wiley & Sons, Inc., New York, 1998.
138. J. Sefton and K. Glover. Pole-zero cancellations in the general $\mathcal{H}_\infty$ problem with reference to a two block design. *Systems & Control Letters*, 14:295–306, 1990.
139. R.S. Sezginer and M.L. Overton. The largest singular value of $e^X A_o e^{-X}$ is convex on convex sets of commuting matrices. *IEEE Transactions on Automatic Control*, AC-35:229–230, 1990.
140. S. Skogestad. Dynamics and control of distillation columns. A tutorial introduction. *Trans. IChemE*, 75:1–36, 1997.
141. S. Skogestad, M. Morari, and J.C. Doyle. Robust control of ill-conditioned plants: High purity distillation. *IEEE Transactions on Automatic Control*, 33:1092–1105, 1988.
142. S. Skogestad and I. Postlethwaite. *Multivariable Feedback Control.* John Wiley and Sons Ltd, Chichester, UK, 1996.
143. M.S. Spillman. Robust longitudinal flight control design using linear parameter-varying feedback. *Journal of Guidance, Control, and Dynamics*, 23:101–108, 2000.
144. V. Sreeram, B.D.O. Anderson, and A.G. Madievski. New results on frequency weighted balanced reduction technique. In *Proceedings of the 1995 American Control Conference*, pages 4004–4009, Seattle, WA, June 1995.
145. A.A. Stoorvogel. *The $\mathcal{H}_\infty$ Control Problem: A State Space Approach.* Prentice Hall, Englewood Cliffs, NJ, 1992.

146. R. Sutton, G. Halikias, A. Plummer, and D. Wilson. Modelling and $\mathcal{H}_\infty$ control of a single-link flexible manipulator. *Proceedings of the Institution of Mechanical Engineers, Part I, Journal of Systems and Control Engineering*, 213:85–104, 1999.
147. K.S. Tang, K.F. Man, and D.-W. Gu. Structured Genetic Algorithm for Robust $\mathcal{H}_\infty$ Control Systems Design. *IEEE Transactions on Industrial Electronics*, 43(5):575–582, 1996.
148. M.S. Tombs and I. Postlethwaite. Truncated balanced realization of a stable non-minimal state space system. *International Journal of Control*, 46(4):1319–1330, 1987.
149. V.A. Tsacouridis and G.A. Medrano-Cerda. Discrete-time H_∞ control of a triple inverted pendulum with single control input. *IEE Proceedings-Control Theory Appl.*, 146:567–577, 1999.
150. G.-W. van der Linden and P.F. Lambrechts. H_∞ control of an experimental inverted pendulum with dry friction. *IEEE Control Systems Magazine*, 19:44–50, 1993.
151. A. Varga. Balancing free square-root algorithm for computing singular perturbation approximations. In *Proceedings of the 30th IEEE Conference on Decision and Control*, pages 1062–1065, Brighton, UK, December 1991.
152. A. Varga. On stochastic balancing related model reduction. In *Proceedings of the 39th IEEE Conference on Decision and Control*, pages 2385–2390, Sydney, Australia, December 2000.
153. M. Vidyasagar and H. Kimura. Robust controllers for uncertain linear multivariable systems. *Automatica*, 22:85–94, 1986.
154. M. Vidyasagar, H. Schneider, and B. Francis. Algebraic and topological aspects of feedback stabilization. *IEEE Transactions on Automatic Control*, AC-27:880–894, 1982.
155. D.J. Walker. Relationship between three discrete-time $\mathcal{H}_\infty$ algebraic Riccati equation solutions. *International Journal of Control*, 52:801–809, 1990.
156. F. Wang, T. Hurst, D. Abramovitch, and G. Franklin. Disk drive pivot nonlinearity modelling Part II: Time domain. In *Proceedings of the 1994 American Control Conference*, pages 2604–2607, Baltimore, MD, June 1994.
157. G. Wang, V. Sreeram, and W.Q. Liu. A new frequency-weighted balanced truncation method and an error bound. *IEEE Transactions on Automatic Control*, 44:1734–1737, 1999.
158. J.F. Whidborne, D.-W. Gu, and I. Postlethwaite. Algorithms for solving the method of inequalities – A comparative study. In *Proceedings of the 1995 American Control Conference*, June 1995.
159. J.F. Whidborne and G.P. Liu. *Critical Control Systems.* Research Studies Press Ltd, Taunton, Somerset, UK, 1993.
160. J.F. Whidborne, G. Murad, D.-W. Gu, and I. Postlethwaite. Robust control of an unknown plant – the IFAC 93 benchmark. *International Journal of Control*, 61:589–640, 1994.
161. J.F. Whidborne, I. Postlethwaite, and D.-W. Gu. Robust controller design using $\mathcal{H}^\infty$ loop-shaping and the method of inequalities. *IEEE Transactions on Control Systems Technology*, 2:455–461, 1994.
162. J.F. Whidborne, I. Postlethwaite, and D.-W. Gu. Multiobjective control system design – a mixed optimization approach. In R.P. Agarwal, editor, *Recent Trends in Optimization Theory and Applications*, volume 5 of *World Scientific Series in Applicable Analysis*, pages 467–482. World Scientific, 1995.

163. J.F. Whidborne, I. Postlethwaite, and D.-W. Gu. Robust control of a paper machine. *Control Engineering Practice*, 3:1475–1478, 1995.
164. M.T. White, M. Tomizuka, and C. Smith. Rejection of disk drive vibration and shock disturbances with a disturbance observer. In *Proceedings of the 1999 American Control Conference*, pages 4127–4131, San Diego, CA, June 1999.
165. W.N. White and R.C. Fales. Control of a double inverted pendulum with hydraulic actuation: A case study. In *Proceedings of the 1999 American Control Conference*, pages 495–499, San Diego, CA, June 1999.
166. M. Yamakita, T. Hoshino, and K. Furuta. Control practice using pendulum. In *Proceedings of the 1999 American Control Conference*, pages 490–494, San Diego, CA, June 1999.
167. D.C. Youla, H.A. Jabr, and J.J. Bongiorno. Modern Weiner-Hopf design of optimal controllers, Part ii: The multivariable case. *IEEE Transactions on Automatic Control*, AC-21:319–338, 1976.
168. D.C. Youla, H.A. Jabr, and C.N. Lu. Single-loop feedback stabilization of linear multivariable dynamical plants. *Automatica*, 10:159–173, 1974.
169. V. Zakian and U. Al-Naib. Design of dynamical and control systems by the method of inequalities. *IEE Proceedings - Control & Science*, 120:1421–1427, 1973.
170. G. Zames. Feedback and optimal sensitivity: Model reference transformations, multiplicative seminorms and approximate inverses. *IEEE Transactions on Automatic Control*, AC-26:301–320, 1981.
171. G. Zames and B.A. Francis. Feedback, minimax sensitivity, and optimal robustness. *IEEE Transactions on Automatic Control*, AC-28:585–600, 1983.
172. P. Zarchan. *Tactical and Strategic Missile Guidance*, volume 176 of *Progress in Astronautics and Aeronautics*. AIAA, Washington, DC, 3rd edition, 1998.
173. H.P. Zeiger and A.J. McEwen. Approximate linear realization of given dimension via Ho's algorithm. *IEEE Transactions on Automatic Control*, AC-19:153, 1974.
174. K. Zhou and J.C. Doyle. *Essentials of Robust Control*. Prentice Hall, Upper Saddle River, NJ, 1998.
175. K. Zhou, J.C. Doyle, and K. Glover. *Robust and Optimal Control*. Prentice Hall, Upper Saddle River, NJ, 1995.

Index